AF389213

Water Flow, Solute and Heat Transfer in Groundwater

Water Flow, Solute and Heat Transfer in Groundwater

Editor

Alexander Yakirevich

MDPI • Basel • Beijing • Wuhan • Barcelona • Belgrade • Manchester • Tokyo • Cluj • Tianjin

Editor
Alexander Yakirevich
Ben-Gurion University of the
Negev
Israel

Editorial Office
MDPI
St. Alban-Anlage 66
4052 Basel, Switzerland

This is a reprint of articles from the Special Issue published online in the open access journal *Water* (ISSN 2073-4441) (available at: https://www.mdpi.com/journal/water/special_issues/Water_Flow_Solute_and_Heat_Transfer_Groundwater).

For citation purposes, cite each article independently as indicated on the article page online and as indicated below:

LastName, A.A.; LastName, B.B.; LastName, C.C. Article Title. *Journal Name* **Year**, *Article Number*, Page Range.

ISBN 978-3-03936-896-9 (Hbk)
ISBN 978-3-03936-897-6 (PDF)

Contents

About the Editor

Alexander Yakirevich has a M.S. in Fluid Mechanics (1973) from the Voronezh State University and a Ph.D. in Soil Physics (1981) from the All-Russian Institute of Hydraulic Engineering and Land Reclamation. He is currently a professor emeritus in the J. Blaustein Institutes of Desert Research, Ben-Gurion University of the Negev (Israel). His research focuses on the experimental and theoretical investigation of water flow, solute, and heat transfer, in porous and fractured media; the development of mathematical models for multiphase and multicomponent flow and transport in the vadose zone and groundwater system; the simulation of overland flow; and solutes transport in runoff and streams.

Editorial

Water Flow, Solute and Heat Transfer in Groundwater

Alexander Yakirevich

Zuckerberg Institute for Water Research, The Jacob Blaustein Institutes for Desert Research, Ben-Gurion University of the Negev, Sede Boqer 8499000, Israel; alexy@bgu.ac.il

Received: 11 June 2020; Accepted: 24 June 2020; Published: 28 June 2020

Abstract: Groundwater is an essential and vital water resource for drinking water production, agricultural irrigation, and industrial processes. The better understanding of physical and chemical processes in aquifers enables more reliable decisions and reduces the investments concerning water management. This Special Issue on "Water Flow, Solute and Heat Transfer in Groundwater" of *Water* focuses on the recent advances in groundwater dynamics. In this editorial, we introduce 12 high-quality papers that cover a wide range of issues on different aspects related to groundwater: protection from contamination, recharge, heat transfer, hydraulic parameters estimation, well hydraulics, microbial community, colloid transport, and mathematical models. By presenting this integrative volume, we aim to transfer knowledge to hydrologists, hydraulic engineers, and water resources planners who are engaged in the sustainable development of groundwater resources.

Keywords: groundwater; porous and fractured media; contaminant transport; heat transfer; parameters; colloids; microbial community; field and laboratory studies; mathematical modeling

---

1. Introduction

Worldwide, groundwater provides essential and valuable water resources for drinking water production, agricultural irrigation [1], and industrial processes [2]. Particularly in arid and semi-arid areas where ample surface water sources do not exist, groundwater is often the major, or even the only, drinking water resource. However, human activities, due to increasing livelihood and development demands, have contributed to the deterioration of groundwater quality. Aquifer systems are thus driven out of their natural equilibrium condition both from a quantitative and qualitative point of view. Careless operations of groundwater resources result in inadequate management and protection measures during the planning and implementation of the development projects in the recharge areas. An overexploitation of the aquifers leads to the deterioration of the groundwater quality and the almost irreversible contamination of the aquifer system. Experience in recent decades has shown that once groundwater is contaminated by chemical, biological, or radiological agents, it is always difficult to clean up these contaminants, and the remediation involves high costs [3]. Therefore, protecting groundwater from depletion and pollution is one of the major tasks for sustainable water development.

The increasing concern regarding problems related to environmental protection from pollution and the high costs of remediation policies stimulate the development and application of groundwater management models, which combine optimization procedures with groundwater flow and transport models. In decision-making that requires knowledge of the hydro-geological environment, one of the most important features is uncertainty, which mainly stems from inadequate concepts or description of the processes and inherent lack of information about the aquifer's properties. Understanding the physical and chemical processes occurring in the aquifer system enables more reliable decisions and reduces the investments concerning water management.

The papers included in this Special Issue cover a wide range of subjects that are relevant to groundwater dynamics: well hydraulics, heat transfer, protection from contamination, recharge, parameter estimation, microbial community and colloids transport, models, and numerical solution methods.

2. Special Issue Overview

This Special Issue on "Water Flow, Solute and Heat Transfer in Groundwater" contains 12 papers. In the following, we shortly describe each study.

Wang et al. (2020) [4] investigated the seepage characteristics of single ascending partially and fully penetrating relief wells using a series of laboratory sand-tank experiments and numerical simulations. They found that the seepage characteristics of ascending wells dewatering an overlying aquifer are different from those of the conventional pumping wells descending from the ground surface into the underlying aquifer because of the pronounced influence of the seepage face boundary condition along the seepage boundary of the ascending dewatering well. Modified versions of the Dupuit and Dupuit–Thiem formulae for a single ascending relief well for a degree of penetration less than the critical one for unconfined, confined, and confined-unconfined aquifers were developed.

Two papers address the topics of groundwater contamination and protection. Li et al. (2019) [5] used the numerical model of density-dependent groundwater flow and solute transport to analyze the influence of tidal fluctuation and water curtain systems on the temporal-spatial variations in seawater intrusion in an island oil storage cavern in China. The results show that the operation of an underground water-sealed oil storage cavern in an island environment has a risk of inducing seawater intrusion. The water curtain system can decrease seawater intrusion and reduce the influence of tidal fluctuation on the seepage field inside the island. Liu et al. (2019) [6] compared different wellhead protection area (WHPA) delineation methods, addressing the delineation uncertainty due to the uncertainty from the input parameters. Comparisons were performed at two pumping sites–a single well and a wellfield consisting of eight wells in an unconfined coastal aquifer in Israel. The results from single well and wellfield indicated that interferences between wells are important for WHPA delineation, and thus that only semi-analytical and numerical modelling are recommended for WHPA delineation at wellfields.

Princ et al. (2020) [7] investigated experimentally the relationship between the entrapped air content and the corresponding hydraulic conductivity for two coarse sands. The amount and distribution of air bubbles were quantified by micro-computed X-ray tomography (CT) for the selected runs. The relationship between the initial and residual gas saturation was successfully fitted with a linear model. The combination of X-ray-computed tomography and infiltration experiments has a large potential in exploring the effects of entrapped air on water flow.

Two papers consider problems related to the technical applications of heat transfer in groundwater. Zang et al. (2019) [8] presented a new approach aiming at using geothermal water resources for residential heating. Simulations based on mathematical models of groundwater flow and heat transfer show that well spacing, reinjection temperature, and production rate are the most significant factors influencing thermal breakthrough in geothermal reservoirs. It was shown that in the case of Xinji, China, an indirect geothermal district heating system is much better than a direct geothermal district heating system, both technically and economically. The developed approach can be applied to other regions with geothermal energy utilization. Hu et al. (2018) [9] considered the effect of groundwater flow on the artificial ground freezing (AGF) projects in highly permeable formations. The numerical model to simulate groundwater flow and temperature changes was developed and used to assess the efficiency of AGF for strengthening a metro tunnel in South China. The simulation results show that the freezing wall appears in an asymmetrical shape with horizontal groundwater flow normal to the axial of the tunnel, while along the groundwater flow direction, a freezing wall forms slowly and on the upstream side the thickness of the frozen wall is thinner than that on the downstream side. The pipes' spacing influences the temperature field and closure time of the frozen wall.

Balaban et al. (2019) [10] present the results of field research related to microbial community in a fractured chalk aquitard below the industrial zone Neot Hovav in Israel, which is polluted by a wide variety of hazardous organic compounds. The spatial variations in indigenous bacterial population and their structure were assessed as a function of distance from the polluting source. The de-halogenating potential of the microbial population was tested through a series of lab microcosm experiments, thus exemplifying the potential and limitations for the bioremediation of the site. The dehalogenation of halogenated ethylene was demonstrated in contrast with the persistence of brominated alcohols. The persistence is likely due to the chemical characteristics of the brominated alcohols and not because of the absence of active de-halogenating bacteria.

Bagalkot and Kumar (2018) [11] developed a numerical model to investigate the influence of gravitational force on the transport of colloids in a single horizontal fracture–matrix system. Results suggest that the gravitational force significantly alters and controls the velocity of colloids in the fracture. The mass flux across the fracture–matrix interface is predominantly dependent on the colloidal size. An as large as 80% reduction in the penetration of colloids in the rock matrix was observed when the size of the colloid was increased from 50 to 600 nm.

Groundwater recharge (*GR*) is an important parameter affecting the use of groundwater resources. Two papers draw attention to the significance of factors influencing *GR*. Wang et al. (2019) [12] use a one-dimensional process-based vadose zone model with generated soil hydraulic parameters to simulate the soil moisture, actual evapotranspiration (Et_a), and *GR*. The simulations showed that the dependence of ET_a and *GR* on the soil hydraulic properties varied considerably with the climatic conditions and greatly weakened at the site with an arid climate. In contrast, the distribution of the mean relative difference in soil moisture was still significantly correlated with the soil hydraulic properties (most notably the residual soil moisture content) under arid climatic conditions. Adane et al. (2019) [13] studied the impact of climate forecasts on *GR* within a probabilistic framework in a site-specific study in the Nebraska Sand Hills (NSH), the largest stabilized sand dune region in the USA, containing the greatest recharge rates within the High Plains Aquifer. A total of 19 downscaled climate projections were used to evaluate the impact of precipitation and reference evapotranspiration on the *GR* rates simulated by using a HYDRUS 1-D model. To present the results at a sub-annual time resolution, three representative climate projections (dry, mean, and wet scenarios) were selected from the statistical distribution of the cumulative *GR*. In the dry scenario, the excessive evapotranspiration demand in the spring and precipitation deficit in the summer can cause plant withering due to excessive root-water stress. This may pose a significant threat to the survival of the native grassland ecology in the NSH and potentially lead to desertification processes if climate change is not properly addressed.

The simulation of two- and three-dimensional water flow in the vadose zone-groundwater system using the Richards' equation is computationally intensive and time consuming. Pinzinger and Blankenburg (2020) [14] used the free software library AMGCL (algebraic multigrid C++ library) and developed a numerical algorithm that allows a significant reduction in the computational running time without losing accuracy. This was mostly pronounced for large-scale models.

Contaminant hydrogeology models of non-aqueous phase liquid (NAPL) transport in saturated porous media account for the dissolution of residual NAPL by a Sherwood–Gilland empirical model. The standard methods of volume averaging to derive upscaled transport equations describing the same dissolution and transport phenomena typically yield forms of equations that are seemingly incompatible with Sherwood–Gilland source models. Hansen (2019) [15] developed new simplification approaches (including a physics-preserving transformation of the domain and a new geometric lemma) that allow one to avoid solving traditional closure boundary value problems and to obtain a general, volume-averaged governing equation that does not reduce to the advection-dispersion-reaction equation with a Sherwood–Gilland source.

3. Conclusions

The objective of this Special Issue was to focus on the recent advances in groundwater studies related to fundamental investigations of water flow, solute transport, and heat transfer using various experimental techniques, mathematical models of physical mechanisms, management strategies, and experience gained from case studies. Twelve high-quality papers were included in this Special Issue relating to well hydraulics, freshwater–saltwater interactions, groundwater contamination and protection, the impact of climate change on groundwater recharge, microbial community structure, hydraulic parameter estimation, colloid transport, and numerical modeling techniques.

The overview of papers published in this SI shows that fundamental research on groundwater is very important for understanding the behavior of aquifers, most of which, nowadays, are under stress conditions. The results of these studies can be directly used for developing an improved picture of groundwater dynamics and aquifer use. Future works should continue towards integrative investigations of groundwater interactions with other components of the hydro-geochemical cycle, such as atmosphere, surface waters, soils, and lithosphere.

Funding: This research received no external funding.

Acknowledgments: We thank all the authors who contributed their research works to this Special Issue.

Conflicts of Interest: The author declares no conflict of interest.

References

1. Siebert, S.; Burke, J.; Faures, J.M.; Frenken, K.; Hoogeveen, J.; Döll, P.; Portmann, F.T. Groundwater use for irrigation—A global inventory. *Hydrol. Earth Syst. Sci.* **2010**, *14*, 1863–1880. [CrossRef]
2. Shah, T.; Burke, J.; Villholth, K.G.; Angelica, M.; Custodio, E.; Daibes, F.; Hoogesteger, J.; Giordano, M.; Girman, J.; Van Der Gun, J.; et al. *Groundwater: A Global Assessment of Scale and Significance*; Earthscan: London, UK; International Water Management Institute (IWMI): Colombo, Sri Lanka, 2007; pp. 395–423.
3. World Water Assessment Programme (United Nations). *Water: A Shared Responsibility*; The United Nations World Water Development Report 2; UN-HABITAT: Nairobi, Kenya, 2006.
4. Wang, W.; Faybishenko, B.; Jiang, T.; Dong, J.; Li, Y. Seepage characteristics of a single ascending relief well dewatering an overlying aquifer. *Water* **2020**, *12*, 919. [CrossRef]
5. Li, Y.; Zhang, B.; Shi, L.; Ye, Y. Dynamic variation characteristics of seawater intrusion in underground water-sealed oil storage cavern under island tidal environment. *Water* **2019**, *11*, 130. [CrossRef]
6. Liu, Y.; Weisbrod, N.; Yakirevich, A. Comparative study of methods for delineating the wellhead protection area in an unconfined coastal aquifer. *Water* **2019**, *11*, 1168. [CrossRef]
7. Princ, T.; Fideles, H.M.R.; Koestel, J.; Snehota, M. The impact of capillary trapping of air on satiated hydraulic conductivity of sands interpreted by X-ray microtomography. *Water* **2020**, *12*, 445. [CrossRef]
8. Zhang, L.; Wang, R.; Song, H.; Xie, H.; Fan, H.; Sun, P.; Du, D. Numerical investigation of techno-economic multiobjective optimization of geothermal water reservoir development: A case study of China. *Water* **2019**, *11*, 2323. [CrossRef]
9. Hu, R.; Liu, Q.; Xing, Y. Case study of heat transfer during artificial ground freezing with groundwater flow. *Water* **2018**, *10*, 1322. [CrossRef]
10. Balaban, N.; Yankelzon, I.; Adar, E.; Gelman, F.; Ronen, Z.; Bernstein, A. The spatial distribution of the microbial community in a contaminated aquitard below an industrial zone. *Water* **2019**, *11*, 2128. [CrossRef]
11. Bagalkot, N.; Kumar, G.S. Colloid transport in a single fracture–matrix system: Gravity effects, influence of colloid size and density. *Water* **2018**, *10*, 1531. [CrossRef]
12. Wang, Z.; Wang, T.; Zhang, Y. Interplays between state and flux hydrological variables across vadose zones: A numerical investigation. *Water* **2019**, *11*, 1295. [CrossRef]
13. Adane, A.; Zlotnik, V.A.; Rossman, N.R.; Wang, T.; Nasta, P. Sensitivity of potential groundwater recharge to projected climate change scenarios: A site-specific study in the Nebraska sand hills, USA. *Water* **2019**, *11*, 950. [CrossRef]
14. Pinzinger, R.; Blankenburg, R. Speeding up the computation of the transient Richards' equation with AMGCL. *Water* **2020**, *12*, 286. [CrossRef]
15. Hansen, S.K. Exploring compatibility of Sherwood-Gilland NAPL dissolution models with micro-scale physics using an alternative volume averaging approach. *Water* **2019**, *11*, 1525. [CrossRef]

Article

Seepage Characteristics of a Single Ascending Relief Well Dewatering an Overlying Aquifer

Wenxue Wang [1,*], Boris Faybishenko [2,*], Tong Jiang [1], Jinyu Dong [1] and Yang Li [3]

[1] Henan Province Key Laboratory of Rock and Soil Mechanics and Structural Engineering, North China University of Water Resources and Electric Power, 136 Jinshui East Rd, Zhengzhou 450045, China; jiangtong@ncwu.edu.cn (T.J.); dongjinyu@ncwu.edu.cn (J.D.)
[2] Energy Geosciences Division, Earth and Environmental Sciences Area, Lawrence Berkeley National Laboratory, University of California, Berkeley, CA 94720, USA
[3] Hydrology and Water Resources Bureau of Henan Province, Zhengzhou 450003, China; laymanlee13@163.com
*   Correspondence: wangwenxue@ncwu.edu.cn (W.W.); bafaybishenko@lbl.gov (B.F.); Tel.: +86-15905211945 (W.W.)

Received: 10 February 2020; Accepted: 17 March 2020; Published: 24 March 2020

Abstract: The application of groundwater relief, i.e., dewatering, ascending wells, drilled upward from the mining tunnel into the overlying aquifer, is common in underground mining engineering. In this study, the seepage characteristics of single ascending partially and fully penetrating relief wells are investigated using a series of laboratory sand-tank experiments and numerical simulations. The seepage characteristics of ascending wells dewatering an overlying aquifer are different from those of conventional pumping wells descending from the ground surface into the underlying aquifer, because of the pronounced influence of the seepage face boundary condition along the seepage boundary of the ascending dewatering well. The seepage face of the ascending well is formed as the well casing remains open and water is discharged under the action of gravity through the well casing. The results of laboratory sand-tank experiments and modeling show that when the degree of penetration of an ascending relief well does not exceed a critical value, the effect of the seepage face cannot be ignored. In particular, the seepage flux increases as the degree of penetration increases following an exponential function, and the relationship between the seepage flux and the well radius can be described using a power law function. The results of numerical simulations are used to develop a series of type curves to evaluate the effects of the critical degree of penetration for different well radii and different aquifer water levels. Modified versions of the Dupuit and Dupuit–Thiem formulae for a single ascending partially well for the degree of penetration less than the critical one for the unconfined, confined, and confined-unconfined aquifers are developed.

Keywords: ascending relief well; groundwater; seepage; sand-tank; modeling; Dupuit formula; Dupuit-Thiem formula

---

1. Introduction

Construction of many underground facilities and tunnels, deep foundation pits, and underground coal mines, which may be affected by underground water intrusion, require dewatering of an overlying aquifers [1–11]. A common method to reduce the groundwater level to prevent inundation of underground excavations is the use of water pumping wells, which are drilled from the land surface into the aquifer [5,7,8]. Another method applied in the mining industry for dewatering and depressurization of overlying aquifers is to drill dewatering wells from the underground mines' tunnels upward into the overlying aquifer [1,11–19].

In this paper, the dewatering wells drilled upward into the overlying aquifer are called "ascending relief wells" (ARWs), and they are used to partially or completely dewater an aquifer above an active or abandoned underground mine. The ARWs can also be used in combination with pumping wells. For example, a comprehensive drainage from pumping wells and the panels was used in the 8th mining area of Taiping Coal Mine, China [17]. The ascending wells were drilled below the mining face to test hydrological parameters in many underground coal mines [11,15]. The drainage water discharged under gravity from the overlying aquifer through ARWs is collected in the underground tunnel and then pumped out through the drainage system. Schematic diagrams of the partially penetrating ARWs in an unconfined aquifer, a confined aquifer, and a confined-unconfined aquifer are shown in Figure 1.

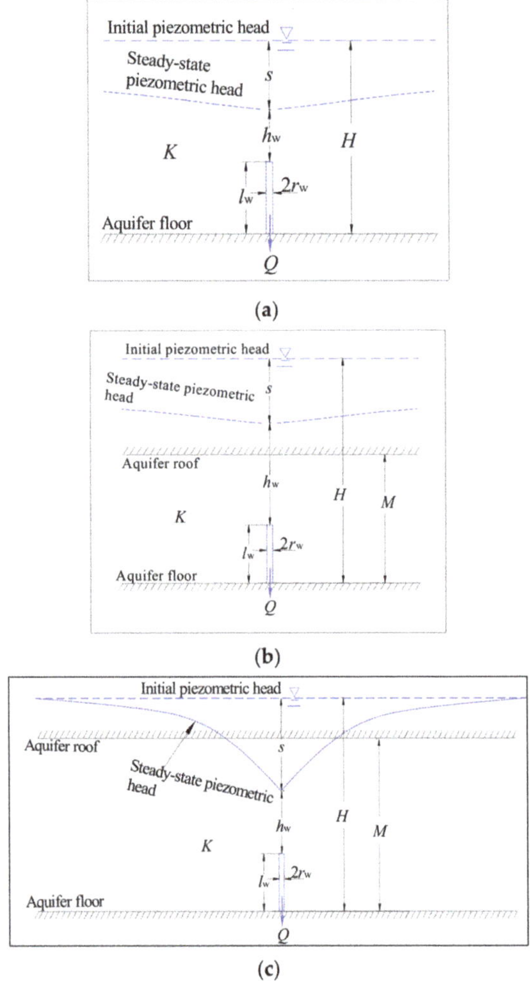

Figure 1. Schematic diagrams of a partially penetrating ascending relief well Ascending Relief Wells (ARW) in (**a**) an unconfined aquifer, (**b**) a confined aquifer, and (**c**) a confined-unconfined aquifer. On the figures, s is the drawdown of water table, h_w is the height of the water table above the well, l_w is well screen length, r_w is well radius, H is the initial piezometric head, M is the aquifer thickness, and K is hydraulic conductivity.

Contrary to the conventional pumping wells, the casing of the ARW remains open to the ambient air with no standing water. The seepage characteristics of the ARWs are different from those of conventional pumping wells, because of downward water flow by the action of gravity into the underground tunnel. The physics of seepage through the walls of the ARWs is generally similar to that of seepage through the walls of the open tunnels. In particular, because of the formation of the seepage face at the tunnel walls [20–24], boundary conditions for numerical simulations of seepage into the tunnels are based on assigning the atmospheric pressure at the tunnel walls and a constant hydraulic head boundary condition along the tunnel perimeter [25–28]. The literature review shows that there is no generalized theory to assess the performance of ARWs [11,29,30], and determination of seepage characteristics of ARWs is based on using methods developed for conventional pumping wells [4,31–39].

The overall objectives of this paper are (1) to study the evolution of the seepage characteristics of single ascending partially and fully penetrating relief wells, and (2) to determine the critical degree of penetration of a partially penetrating ARW needed to obtain a maximum outflow under different far-field water level boundary conditions, (3) to assess the applicability and to modify the Dupuit and Dupuit–Thiem formulae for a single ascending partially well, in an unconfined, homogeneous and isotropic aquifer. The Dupuit and Dupuit–Thiem formulae were considered as a basis to calculate seepage fluxes of the ARW, and laboratory sand-tank experiments and numerical simulations were also conducted.

The following simplifying assumptions were taken into consideration in this study: (1) the aquifer is homogeneous, isotropic, and with no regional flow; (2) there is no water leakage through the top and bottom of the aquifer, i.e., through the underlying and overlying aquicludes of the confined aquifer; (3) the ARW is vertical with a screen along its entire length of the ARW penetration into the overlying aquifer, and no water column inside the ARW, allowing for free drainage (under the gravity), through the open well bottom, into the underlying tunnel, and there is no enforced pumping; (4) the atmospheric pressure boundary condition at the well-aquifer interface; and (5) groundwater flow in the aquifer toward the well is axisymmetric, and can be described by the Darcy law.

2. Methods of Analytical, Experimental, and Modeling Investigations

2.1. Dupuit and Dupuit–Thiem Formulae

Analysis of pumping wells drilled from the surface, in general, relies on the Dupuit assumption of a constant hydraulic head along the vertical profile of the aquifer. This assumption makes possible to use a depth-integrated equation for the flux evaluation into the well. Based on the original analytical solution for unconfined flow by Dupuit [40], Thiem [41] was likely the first to estimate aquifer parameter from pumping tests in a confined aquifer. Boulton [42] developed the first transient well test solution to analyze the unconfined aquifer. Streltsova [43] proposed the solution to consider unsteady radial flow in an unconfined aquifer. Neuman [44,45] developed solutions considering both confined storage and delayed yield from the unconfined aquifer. A variety of new analytical approaches were applied to deal with well hydraulics models involving mixed and complicated boundaries, which gained more accurate solutions for particular conditions [7,36,39,46–51]. Mishra and Kuhlman [52] have recently summarized the application of the Dupuit theory for an unconfined aquifer from Dupuit to the present. Charnyi [53] showed that the Dupuit formula for an unconfined aquifer is valid not only for a hydraulic approximation, i.e., assuming that the groundwater velocity is independent of the height above the aquitard, but also for rigorous hydrodynamic calculations. Although, the Dupuit formula is not accurate for calculations of the phreatic line for $r < H$, it provides accurate calculations of the seepage flux [53]. Shercliff [54] conducted one—and two-dimensional modeling of steady seepage flow in unconfined aquifers, and provided a proof of Charnyi's result that one—and two-dimensional theory yield the same value for the flow rate in a horizontal aquifer or porous bed between vertical ends, and showed the extent to which it can be generalized to non-uniform or anisotropic media.

The forms of the Dupuit and Dupuit–Thiem formulae are summarized in Table 1, and were used in this paper as a basis to estimate seepage fluxes into ARWs, and were then modified based on the results of laboratory sand-tank experiments and modeling (see Section 3).

Table 1. Formulae for steady-state seepage flux in unconfined, confined, and confined-unconfined aquifers.

Types of Aquifers	Seepage Flux (Q)
Unconfined (Dupuit)	$Q = 1.366 \dfrac{K(2H - s)s}{\lg \dfrac{R}{r_w}}$
Confined (Dupuit–Thiem)	$Q = 2.73 \dfrac{KMs}{\lg \dfrac{R}{r_w}}$
Confined-Unconfined (Dupuit–Thiem)	$Q = 1.366 \dfrac{K(2HM - M^2 - h_w^2)}{\lg \dfrac{R}{r_w}}$

Notes: Q is the well seepage flux, K is hydraulic conductivity, H is the initial piezometric head, s is the drawdown (i.e., dewatering depth), r_w is the well radius, M is the thickness of the aquifer, R is the radius of influence of the pumping well, h_w is the water level in the pumping well.

2.2. Laboratory Experiments

2.2.1. Seepage Sand-Tank and Boundary Conditions

Laboratory sand-tank axisymmetric radial water flow experiments were conducted to investigate the seepage to the ARW in an unconfined aquifer. The outer boundary head was at the elevation of 60 cm and kept constant for the entire test. The water entering the ARW was discharged freely under gravity, and the piezometric head inside the ARW was zero. The boundary conditions are shown Figure 2. The sand-tank model is generally a downscaled model of the aquifer [55].

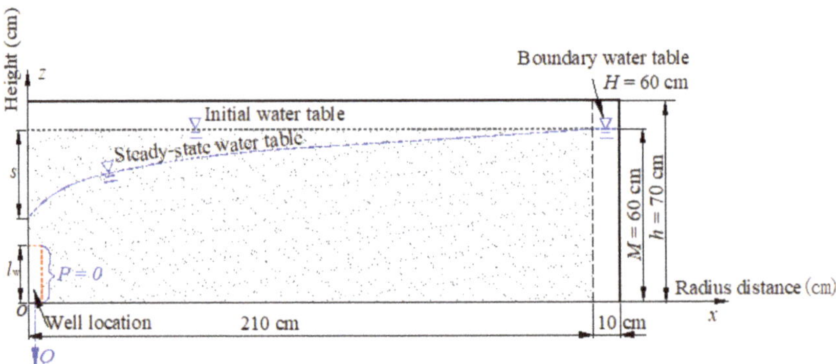

Figure 2. Schematic depicting initial and boundary conditions of the radial flow to a partially penetrating well in the sand-tank tests.

The tank was fabricated from steel plates and acrylic sheets, representing a 60° sector, i.e., a 1/6 portion of a circular flow system for saving experimental materials, with a radius of 210 cm and height of 70 cm. A design schematic and a photograph of the sand-tank assembly are shown in Figure 3. The outer boundary water head was maintained at the elevation of 60 cm above the bottom of the sand-tank, and water was supplied into the sand-tank from a plastic cistern. The outflow from the sand-tank was monitored by an electronic scale. The overflowing water and the well discharge were collected in separate water collection vessels for cyclic utilization. To measure the hydraulic pressure,

96 piezometers with water-level monitoring tubes were installed in six layers at different elevations and distances from the modeled ARW well. The inside diameter of the piezometers was 4 mm, shown in Figure A1 (see Appendix A). The piezometer storage effect is irrelevant because steady state water levels were measured. Water levels in the piezometer tubes were recorded using a camera, which is shown in Figure 3a.

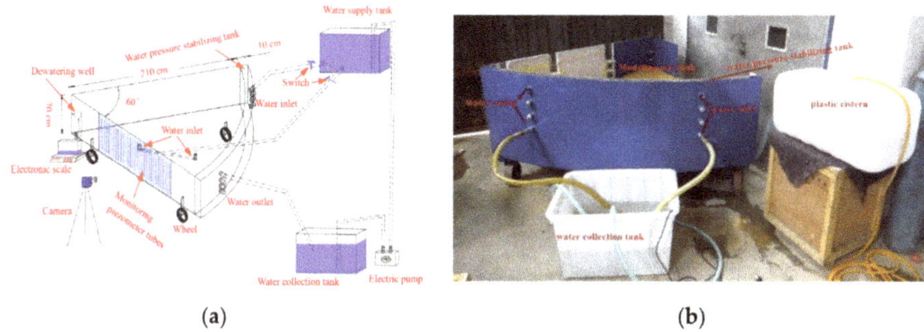

(a)   (b)

Figure 3. Schematic design diagram (**a**) and a photograph (**b**) of the seepage sand-tank used for laboratory experiments.

2.2.2. Ascending Relief Well (ARW) Design

Laboratory tests were conducted to investigate the performance of 24 types of ARWs with different radii—4.5 cm, 9.5 cm, 14.5 cm and 19.5 cm, and lengths—2 cm, 5 cm, 10 cm, 20 cm, 30 cm and 60 cm, corresponding to the degrees of penetration 1/30, 1/12, 1/6, 1/3, 1/2 and 1, as shown in Figure A2. To prevent migration of sand particles into the testing wells, 1-mm radius holes with 4 mm pitches were made in the well screens and tops, and the well screens were wrapped up by stainless steel filters with 0.25 mm diameter openings and 0.18 mm diameter steel wire.

2.2.3. Laboratory Testing Protocol

Loading the Sand-Tank

The tank was filled up with preliminary washed river coarse sand, the particle size distribution curve of a sand sample used for the sand-tank experiment is shown in Figure A3. To preclude possible seepage along the sand-tank wall, a thin layer of vaseline was first besmeared along the inner surface of the tank wall. Six 10 cm thick sand layers were loaded into the tank layer by layer (in total 1.38 m^3), and 16 piezometers were installed between each two layers–in all 96 piezometers.

Sand Saturation

To saturate the sand, water was injected through two bottom inlet ports, while the water supply level was set at the 60 cm elevation above the tank bottom. During the tank saturation, the rising water level in the tank allowed for air escaping upward from the sand, which reduced the volume of entrapped air [56]. When the water table reached the elevation of 60 cm, water injection was continued for about two hours to allow for expelling the remaining entrapped air from the sand. Then, the bottom water inlets were shut off and kept closed during the dewatering process, and the water was diverted from the plastic cistern to the water pressure-stabilizing tank to maintain the stable upper boundary water level. The estimated saturated hydraulic conductivity of sand was about 0.6 cm/s. During dewatering, to assess the flux, water samples were collected and weighed, simultaneously with camera recording and measurements of piezometric heads in monitoring tubes.

2.3. Numerical Simulations

Numerical simulations of flow in the aquifer and the seepage from the ARW were carried out using a software package, MIDAS GTS NX. This software package is based on finite-element simulations for geotechnical design applications, such as deep foundations, excavations, complex tunnel systems, seepage, consolidation analysis, embankment design, dynamic and slope stability analysis—see http://en.midasuser.com/product/gtsnx_overview.asp. Two types of 3-D numerical simulations were conducted: one was to replicate the sand-tank laboratory experiment (Figure A4a), and the other one was to simulate the extended flow field domain (Figure A4b).

The 3-D extended modeling domain was 1200 m by 1200 m in the plan and 120 m in the vertical direction, with a 60 m homogeneous aquifer and a 60 m aquiclude overlying the aquifer. The ARW was located at the center of the domain, shown in Figure A4b. Simulations were performed to assess water flow to a single partially ARW and a fully penetrating ARW with different radii (0.1 m, 0.2 m, 0.5 m, 1 m, 2 m, 4 m, 6 m and 10 m), the well length (from 1 to 60 m with the 1 interval), and the outer boundary heads (60 m, 70 m, 90 m, 120 m, 150 m, 180 m, 210 m, and 240 m). To simulate the discharge from the ARW under gravity, the piezometric head inside the ARW was set to zero. The aquifer saturated hydraulic conductivity was 0.4 m/d. The fixed water pressure head (Dirichlet-type) boundary conditions were assigned at the outer boundary of the flow domain and the ARW well screen.

3. Results of Laboratory Sand-Tank Tests

To present the results of investigations, all distances and hydraulic heads are given as dimensionless values being normalized by the aquifer thickness M for the confined aquifer or the original water level, H, above the unconfined aquifer bottom. For example, for the confined aquifer: $R_{nor} = \frac{R_x}{M}$, $H_{nor} = \frac{H}{M}$, $p_{nor} = \frac{p}{M}$, $l_{nor} = \frac{l_w}{M}$, $s_{nor} = \frac{s}{M}$ and $h_{nor} = \frac{h}{M}$, where R_x is the radial distance, H is the outer boundary water level, p is the piezometric head, l_w is the ARW penetrating length into the aquifer, s is the drawdown, and h is the height above the aquifer bottom. Here, l_{nor} is also the ARW degree of penetration into the aquifer. The well radius r_w was normalized as $r_{nor} = \frac{r_w}{r_1}$, using $r_1 = 1$ cm for the lab experiment, and 1 m for numerical modeling. The seepage fluxes Q_{r1} for the fully penetrating well of radius $r_1 = 1$ cm (for lab experiment) and 1 m (numerical modeling), which were calculated for the case of complete drawdown, were used to determine the dimensionless seepage fluxes given by $Q_{nor} = \frac{Q}{Q_{r1}}$. The results are given below using dimensionless parameters.

3.1. Piezometric Head Distribution

Figure 4 demonstrates the piezometric head distribution vs. the distance from the well in the sand tank experiments for the case of the large diameter fully penetrating well, with $r_{nor} = 19.5$ and $l_{nor} = 1$. For large values of the well radius and length, the piezometric heads were zero at different heights above the ARW center with a maximum seepage flux. The seepage characteristics in the vicinity and at the ARW are practically the same as those in the case of a fully penetrating pumping well. Figure 5 shows the piezometric head distribution vs. distance from a partially penetrating ARW with a small degree of penetration of $l_{nor} = 1/30$ and a radius of $r_{nor} = 4.5$. The figure shows the zero piezometric head inside and at the top of the ARW, and the piezometric heads above the well top being greater than zero, indicating the downward flow. Figure 6 depicts the piezometric head distribution along the height of the sand-tank at the well center $R_{nor} = 0$ and at the radial distance $R_{nor} = 1/6$. In particular, the piezometric heads at the heights of $h_{nor} = 1/6$ and $h_{nor} = 1/3$ above the ARW center were nearly the same.

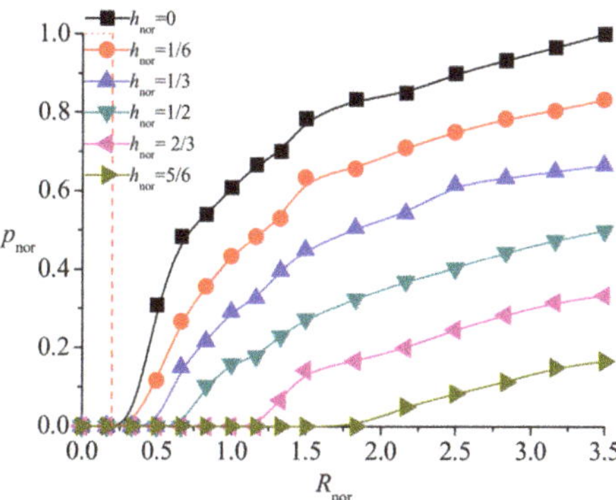

Figure 4. Piezometric head p_{nor} distribution for the case of $r_{\text{nor}} = 19.5$ and $l_{\text{nor}} = 1$, i.e., a fully penetrating well in the laboratory sand-tank experiments.

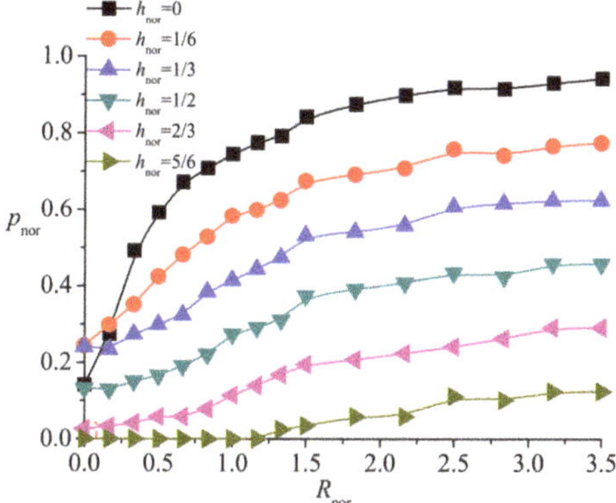

Figure 5. Piezometric head p_{nor} distributions for the case of $r_{\text{nor}} = 4.5$ and $l_{\text{nor}} = 1/30$ in the laboratory sand-tank experiments.

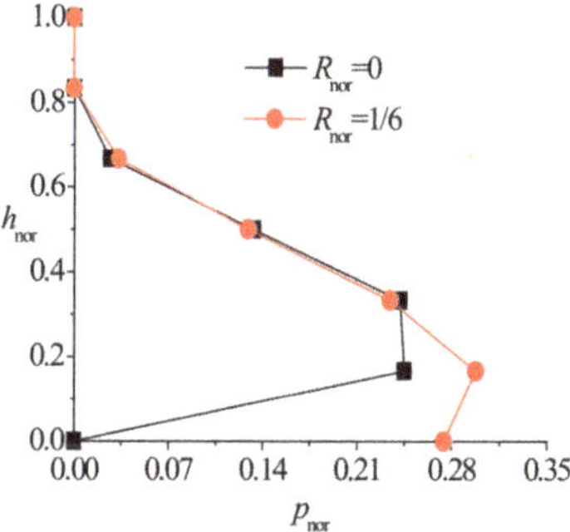

Figure 6. Piezometric head p_{nor} distribution at radial distances R_{nor} of 0 and 1/6 for the case of $r_{\text{nor}} = 4.5$ and $l_{\text{nor}} = 1/30$ in the laboratory experiment.

Figure 7 shows the 2-D plots of the piezometric head distribution (for the case of $r_{\text{nor}} = 4.5$) for different values of the ARW degree of penetration. The experimental results show that the piezometric head inside the ARW was zero, and a piezometric head dropped in the vicinity of the partially penetrating ARW. The water table is determined at the elevation where the piezometric heads is zero.

Figures 5 and 6 show that the piezometric heads in the vicinity and above the partially penetrating ARW were greater in the middle and smaller at the top and bottom of the well. With the increase of the radial distance from the ARW, this effect gradually disappeared, and the piezometric heads at different depths increased with depth (see Figures 4 and 5).

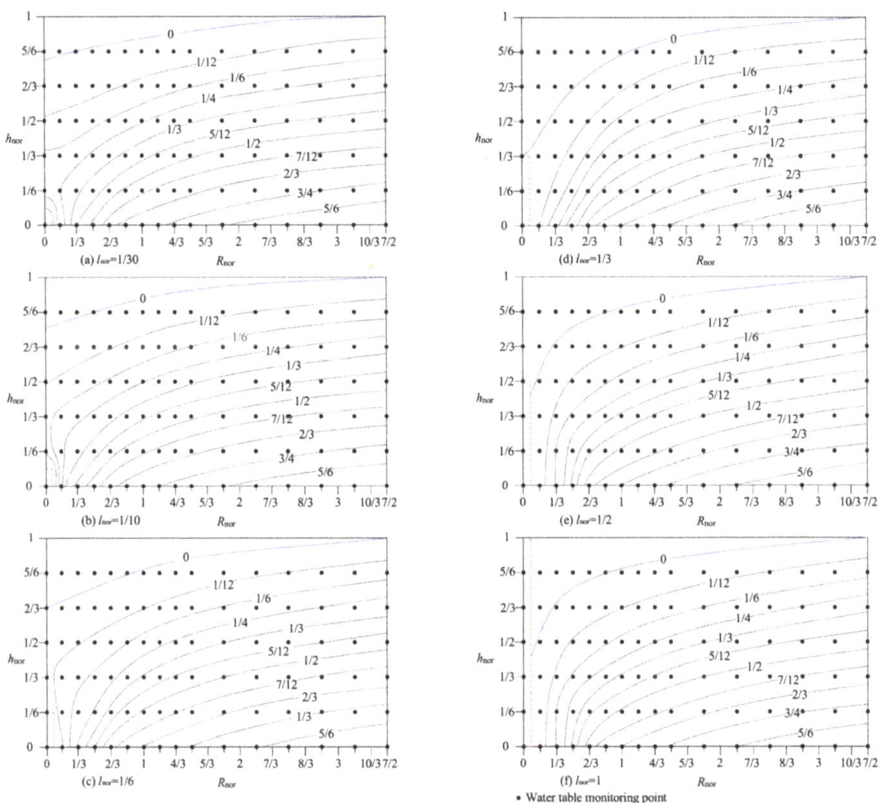

Figure 7. Piezometric head p_{nor} distribution for the case of r_{nor} = 4.5 for different penetrating lengths l_{nor}. For partially penetrating wells: (**a**) l_{nor} = 1/30, (**b**) l_{nor} = 1/12, (**c**) l_{nor} = 1/6, (**d**) l_{nor} = 1/3, (**e**) l_{nor} = 1/2, and for a fully penetrating well: (**f**) l_{nor} = 1. The dashed red lines indicate the length of the well screened interval.

The effect of the well length on the water head distribution can be described as follows. For $l_{nor} \geq$ 1/2, the piezometric head above the well is zero. For $l_{nor} \leq$ 1/3, the piezometric head above the well is greater than zero. Because experiments were conducted for only six values of ARW length for r_{nor} = 4.5, it can be deduced that the critical length of the ARW penetration l_{nor} ranges from 1/3 to 1/2.

The critical degree of ARW penetration of a confined aquifer $\zeta_c = l_c / M$ is defined as the ratio of the well critical length to the aquifer thickness M. For an unconfined aquifer, $\zeta_c = l_c / H$, where H is the original water level above the unconfined aquifer bottom.

The influence of the well radius, length and boundary aquifer head on the critical degree of penetration of an ARW will be evaluated using the results of numerical modeling summarized below in Section "Numerical simulation results."

3.2. Seepage Flux

Figure 8 demonstrates that the seepage flux (Q_{nor}) increases asymptotically with the increase of the degree of penetration l_{nor}, which can be described by an exponential relationship (with $R^2 > 0.96$) given by:

$$\begin{cases} Q_{nor} = Q_{nor}^0 + A_0 \times \exp(B_0 \times l_{nor}), \text{ for } l_{nor} < \zeta_c \\ Q_{nor} = Q_{nor}^1 + A_1 \times \exp(B_1 \times \zeta_c), \text{ for } l_{nor} \geq \zeta_c \end{cases} \tag{1}$$

where Q_{nor}^0, A_0 and B_0, Q_{nor}^1, A_1 and B_1 are fitting parameters, and ζ_c is the critical degree of penetration.

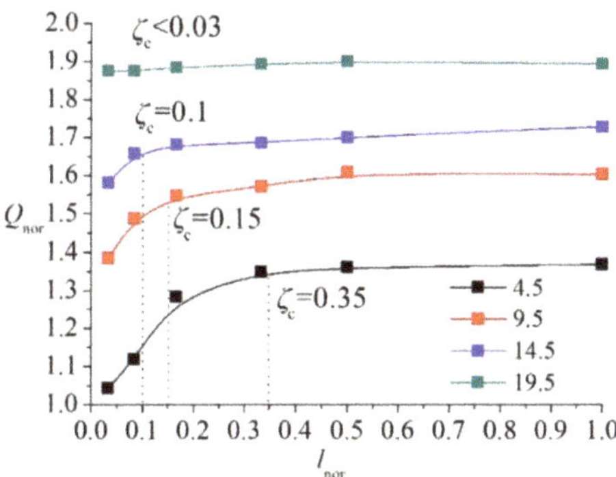

Figure 8. Relationship between the seepage flux Q_{nor} and the degree of penetration l_{nor} for different well radii.

Figure 8 demonstrates that for the small well radii, the seepage flux increased initially rapidly, and then practically stabilized for $\zeta_c > 0.35$. When the well radius is large, the seepage flux reaches the maximum value at a relatively small degree of penetration. The critical degree of penetration ζ_c decreases with the increase in the well radius, such as $\zeta_c = 0.1$ for $r_{nor} = 14.5$, $\zeta_c = 0.15$ for $r_{nor} = 9.5$, and $\zeta_c = 0.35$ for $r_{nor} = 4.5$. (The dependence of the critical degree of penetration ζ_c on the well radius and the boundary head, based on the results of numerical simulations, is described below in Section "Numerical simulation results"). Figure 9 demonstrates a relationship between the seepage flux Q_{nor} and the well radius r_{nor} given by

$$Q_{nor} = Q_{nor}^2 + A_2 \times r_{nor}^{B_2} \tag{2}$$

with $R^2 > 0.96$ and $B_2 < 1$, where Q_{nor}^2, A_2 and B_2 are fitting parameters, r_{nor} is the well radius.

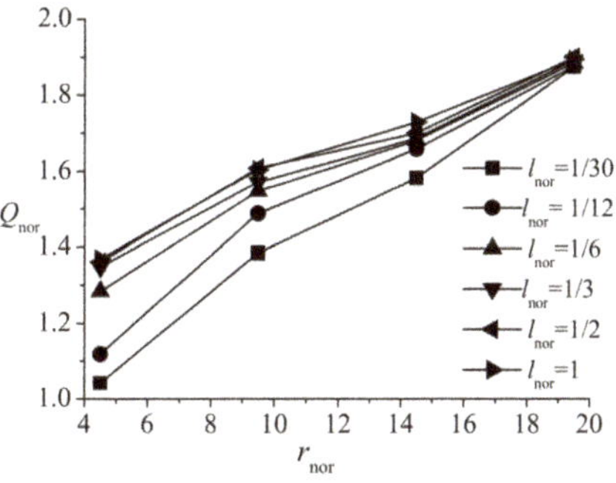

Figure 9. Relationship between seepage flux Q_{nor} and well radius r_{nor} for different degree of penetration l_{nor}, based on the results of the sand-tank experiments.

4. Numerical Simulation Results

4.1. Relationships between Seepage Flux, Well Length and Radius

The results of numerical simulations of the relationships between the seepage flux, well length and radius for different values of the boundary hydraulic head are shown in Figures 10 and 11. The relationship between the seepage flux and the degree of penetration can be described using an exponential function given by Equation (1) with $R^2 > 0.99$. The seepage flux increases with the increase in the well radius according to a power law relationship given by Equation (2) with $R^2 = 1$. The results of numerical simulations correspond to those from the laboratory experiments. Figures 10 and 11 also show that the seepage flux increases with the increase in the degree of penetration, and asymptotically reaches the maximum value for $l_{nor} \leq \zeta_c$. When the degree of penetration is greater than the critical value of ζ_c, the seepage flux reaches the maximum value and remains practically constant as the degree of penetration continues to increase. The critical degree of penetration ζ_c decreases with the increase in the well radius, and increases with the increase of the boundary head.

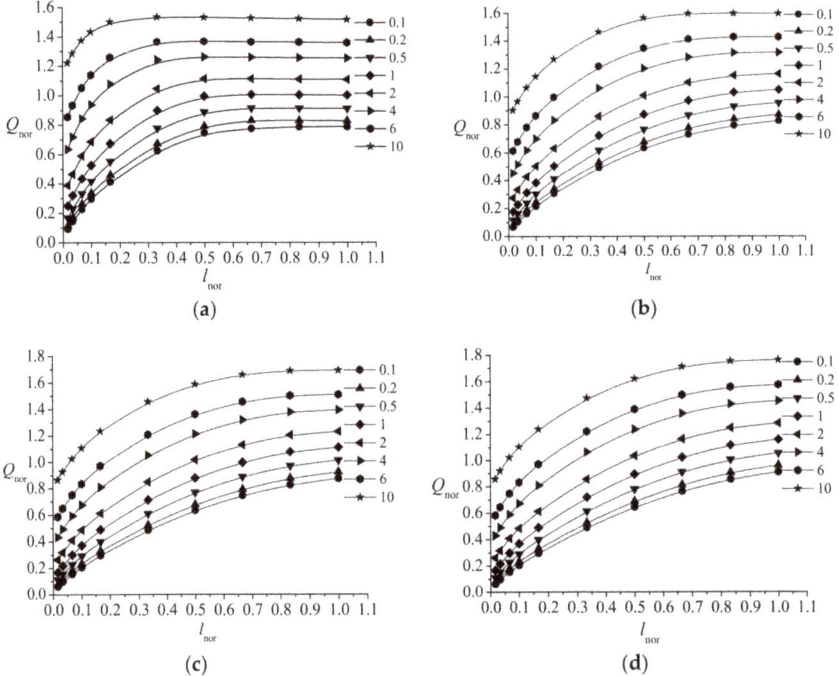

Figure 10. Relationships between the seepage flux Q_{nor} and the degree of well penetration l_{nor} for different boundary conditions: (a) $H_{nor} = 1$, (b) $H_{nor} = 2$, (c) $H_{nor} = 3$, and (d) $H_{nor} = 4$.

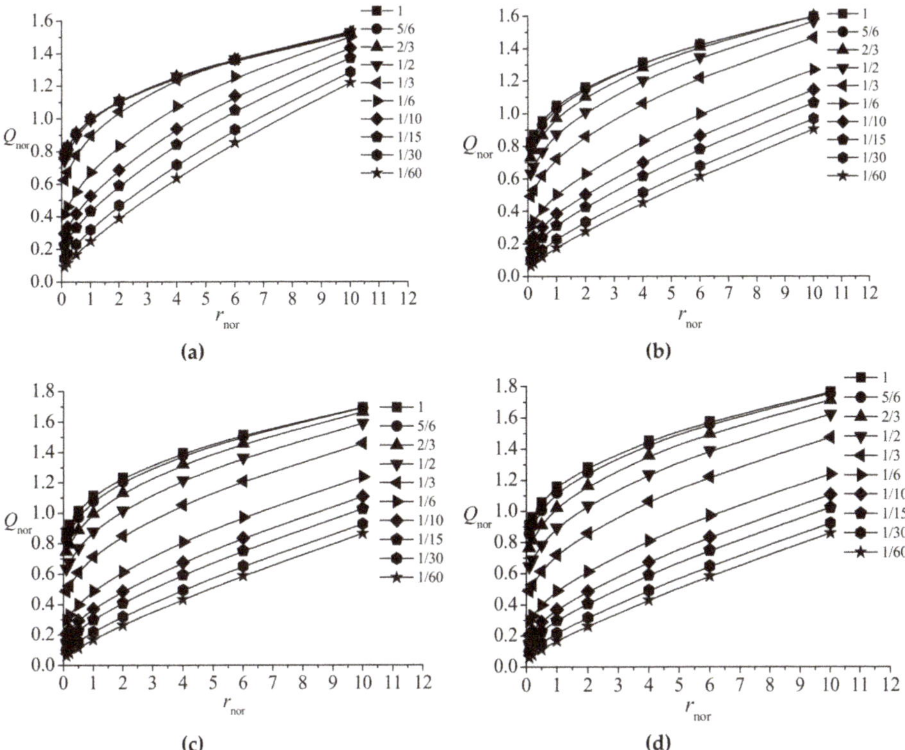

Figure 11. Relationships between the seepage flux Q_{nor} and the well radius r_{nor} for different boundary conditions: (**a**) $H_{nor} = 1$, (**b**) $H_{nor} = 2$, (**c**) $H_{nor} = 3$, (**d**) $H_{nor} = 4$.

4.2. Seepage Characteristics

The water piezometric head p_{nor} distributions above the well for the case $H_{nor} = 1$ and $r_{nor} = 1.0$ in an unconfined aquifer are shown in Figure 12. For the partially penetrating ARW (for l_{nor} from 1/60 to $\frac{1}{2}$), shown in the subfigures (1)–(7) of Figure 12, the lines of the piezometric head of zero (i.e., indicating the water level) are located above the well screen (at $x = 0$), which correspond to the degree of penetration being less than the critical value of ζ_c. For the partially penetrating wells (for $l_{nor} = 2/3$ and 5/6), shown in subfigures (8) and (9) and a fully penetrating well, shown in subfigure (10) of Figure 12, the lines of the zero piezometric head are below the top of the well, which correspond to conditions exceeding the critical value of ζ_c.

When the well length penetration exceeds the critical length l_c, the seepage flux of a partially penetrating ARW is the same as that of a fully penetrating well generating a maximum seepage flux. A similar pattern of the seepage flux was determined for a descending partially penetrating pumping well [57]. The piezometric head distribution characteristics determined using numerical modeling are consistent with the experimental results in the sand-tank described above in Section "Results of Laboratory Sand-Tank Tests."

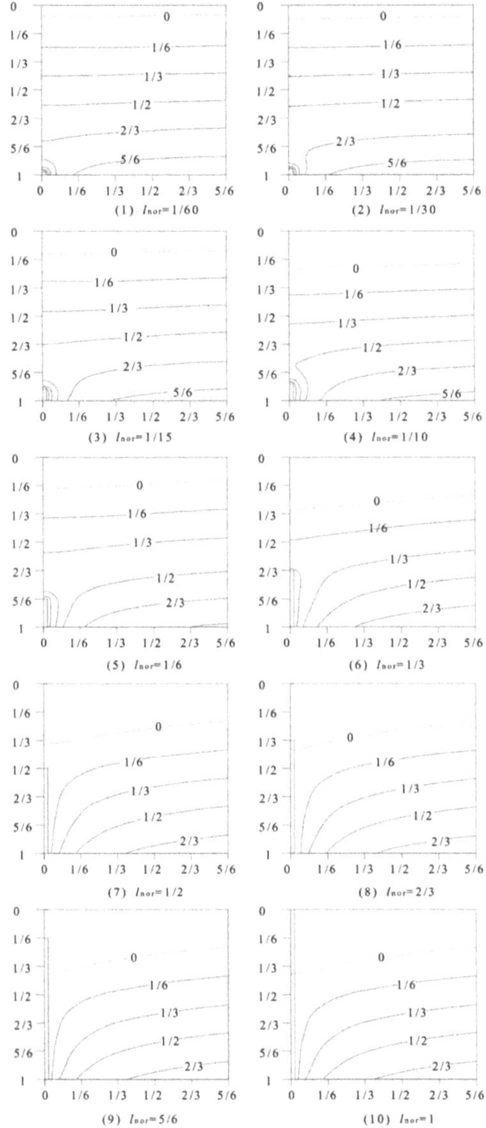

Figure 12. Results of simulations of the water piezometric head p_{nor} distribution around the ARW in the unconfined aquifer for different well length l_{nor}: (1) l_{nor} = 1/60, (2) l_{nor} = 1/30, (3) l_{nor} = 1/15, (4) l_{nor} = 1/10, (5) l_{nor} = 1/6, (6) l_{nor} = 1/3, (7) l_{nor} = 1/2, (8) l_{nor} = 2/3, (9) l_{nor} = 5/6, (10) l_{nor} = 1 (for the case of H_{nor} = 1 and r_{nor} = 1.0). The x-axis is the normalized radial distance from the well.

Figure 13 depicts the flow net composed of streamlines and equipotential lines for the boundary head of H_{nor} = 1 and r_{nor} = 1.0. The flow net indicates that the equipotential lines above the ARW are concaved in the vicinity above the ARW. The curvature of the arched shape equipotential lines gradually increases with the well length increase, and when the equipotential line intersects the water table line, the equipotential line bifurcates into two lines, symmetrically distributed around the ARW. When the ARW's degree of penetration of exceeds the critical value of ζ_c, the equipotential lines are symmetrically distributed around the ARW. The streamlines shown in Figure 13 indicate that a single ARW can drain the entire aquifer regardless of the degree of penetration, and even for the degree of the penetration less than ζ_c, groundwater flow from the entire aquifer thickness is directed into the ARW.

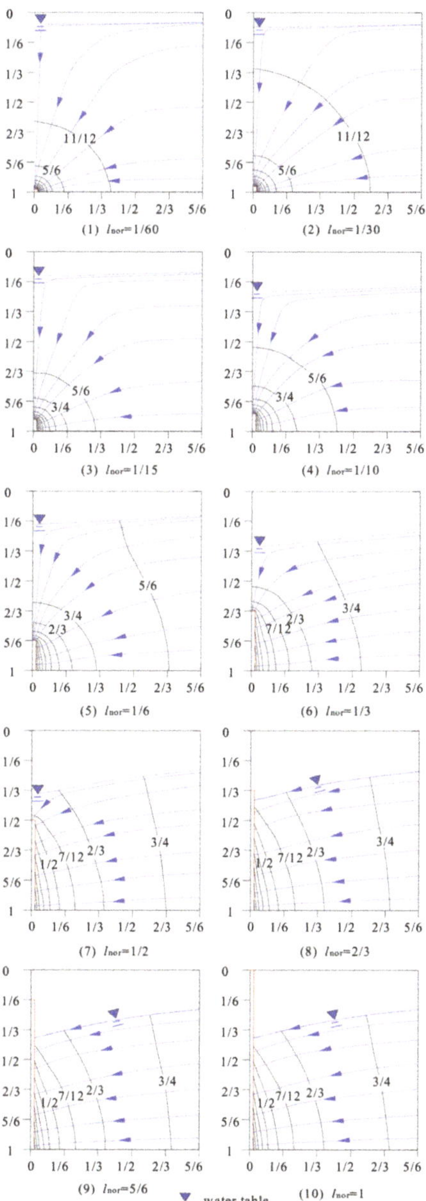

Figure 13. The flow net around the ARW in the unconfined aquifer for different well length l_{nor}: (**1**) l_{nor} = 1/60, (**2**) l_{nor} = 1/30, (**3**) l_{nor} = 1/15, (**4**) l_{nor} = 1/10, (**5**) l_{nor} = 1/6, (**6**) l_{nor} = 1/3, (**7**) l_{nor} = 1/2, (**8**) l_{nor} = 2/3, (**9**) l_{nor} = 5/6, (**10**) l_{nor} = 1 (for the case H_{nor} = 1 and r = 1.0). The x-axis is the radial distance from the well.

4.3. ARW's Critical Degree of Penetration

A summary of the calculations from the results of numerical simulation scenarios is presented in Table A2 and Figure 14. Figure 14 shows that ζ_c increases with H_{nor} for different r_{nor}, according to the relationship given by:

$$\zeta_c = (-0.0488 \times r_{nor} + 0.98) + (0.0273 r_{nor} + 0.1118) \times \ln(H_{nor} + 0.0569 r_{nor} - 1.0214) \tag{3}$$

where r_{nor} is the ARW normalized well radius, $H_{nor} = H/M$ for the case of a confined aquifer, and $H_{nor} = H/H = 1$ for an unconfined aquifer.

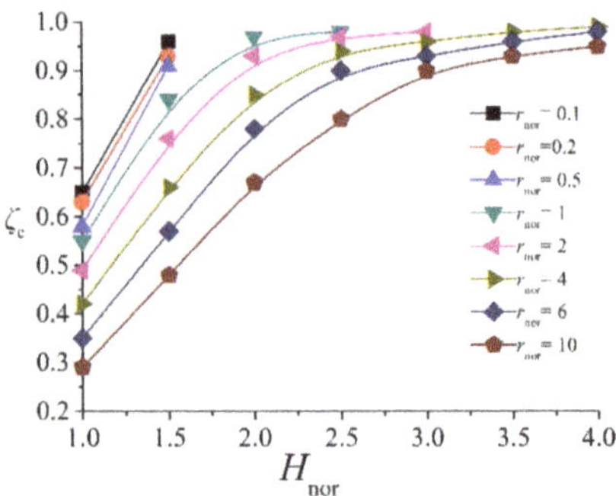

Figure 14. Relationship between ζ_c and H_{nor} for different well radii.

For large H_{nor} and a small well radius, for example, when $H_{nor} > 2.0$ and $r_{nor} < 0.2$, even for a fully penetrating well, the piezometric head above the well still exceeds zero, and the maximum seepage flux is not reached, as shown in Figure 14. Figure 15 presents a contour map of ζ_c for different H_{nor} and r_{nor} based on the results of numerical simulations. These results can be used for designing optimal ARWs to gain maximum seepage flux with minimum drilling lengths.

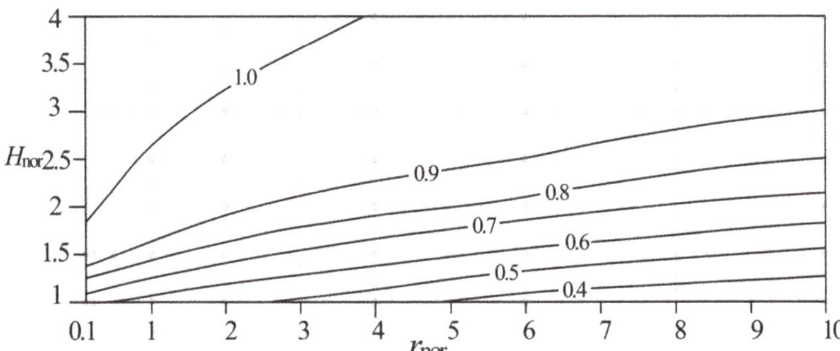

Figure 15. Type curves of ζ_c for as a function of H_{nor} and r_{nor}.

4.4. Model Validation

Validation of the numerical model was conducted based on a comparison of the simulation results, performed using the code MIDAS GTS NX and calculations using the Dupuit formulae for a fully penetrating well in an unconfined aquifer, to the results of laboratory experiments, which are shown in Figure 16. The results of calculations of errors are listed in Table A1. Based on the results of calculations, the absolute relatively errors of the numerical simulations and Dupuit formulae calculations compared to the results of laboratory experiments are <7.65% and <4.49%, respectively.

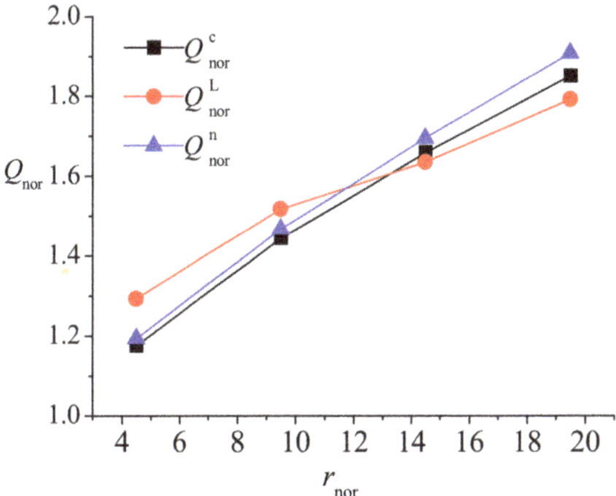

Figure 16. Comparison of the relationships between the seepage flux Q_{nor} with $l_{nor} = 1$ and $r_{nor} = 4.5$, 9.5, 14.5 and 19.5 from numerical simulations, Dupuit formula and laboratory sand-tank experiments.

5. Modified Dupuit and Dupuit–Thiem Formulae for a Single ARW

5.1. Unconfined Aquifer

Table A3 presents the results of simulations of the seepage flux from numerical simulations (Q_n) and calculations using the Dupuit formula (Q_s) for a single partially penetrating ARW in an unconfined aquifer, shown in Figure 1a, for $r_{nor} = 1.0$ and the boundary head $H_{nor} = 1$. One can see that the seepage flux calculated using the Dupuit equation is smaller than that determined from numerical simulations. For $l_w < l_c$, the effective water level drawdown above the ARW can be determined as the value $s_{eff} = s+l_w$, and a modified version of the Dupuit equation for the seepage flux Q_r can be given by:

$$Q_r = 1.366\frac{K(2H - s_{eff})s_{eff}}{\lg\frac{R}{r_w}} = 1.366\frac{K(2H - s_{eff})s_{eff}}{\lg\frac{10s_{eff}\sqrt{K}}{r_w}} \tag{4}$$

with parameters identified on Figure 1a.

The seepage flux Q_{nor}^{r} calculated using the modified Dupuit Equation (4) is also given in Table A3, and it is practically the same as that from numerical simulations, as shown in Figure 17a. Note that for $l_w \geq l_c$, the ARW seepage flux can be calculated from the Dupuit equation, as for a fully penetrating pumping well.

5.2. Confined Aquifer

For the case of a partially penetrating ARW in a confined aquifer, shown in Figure 1b, the seepage flux can be calculated using a modified version of the Dupuit–Thiem formula given by:

$$Q_r = 1.366\frac{KMs_{eff}}{\lg\frac{R}{r_w}} = 1.366\frac{KMs_{eff}}{\lg\frac{10s_{eff}\sqrt{K}}{r_w}} \tag{5}$$

where the value of $s_{eff} = s+l_w$ is the effective piezometric head drawdown.

As an example, the results of calculations of the seepage flux for $r_{nor} = 1.0$ and the boundary head $H_{nor} = 2$ for different values of the well penetration are summarized in Table A4 and Figure 17b. Table A4 includes the seepage flux Q_{nor}^{n} from numerical simulations, Q_{nor}^{s} calculated using the Dupuit–Thiem formula, and Q_{nor}^{r} from Equation (5). Table A4 also shows that the relative error of calculations Q_{nor}^{r}

using Equation (5) compared to the Q^n_{nor} is from -24.88% to 1.1%, while the relative error of using the Dupuit–Thiem formula is from -33.17% to -51.87%.

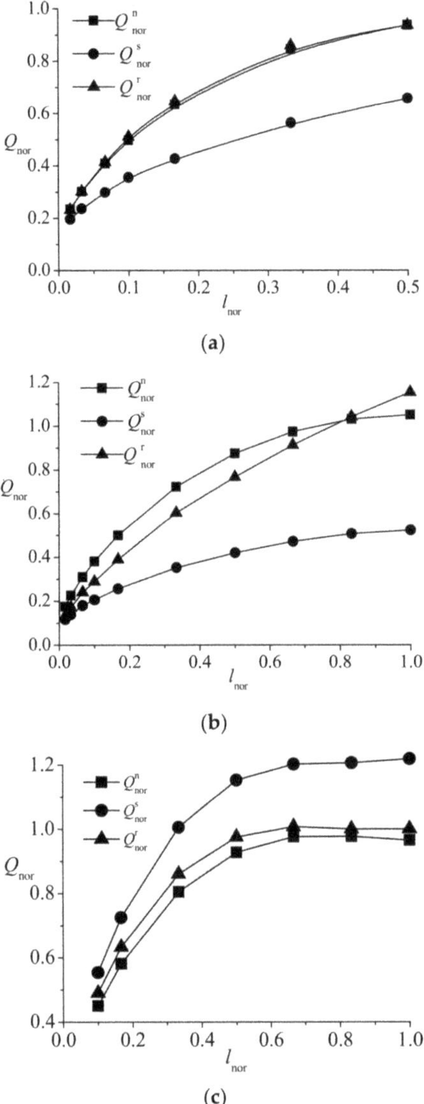

Figure 17. Seepage flux calculations using the results of numerical modeling (Q_n), original Dupuit (Q_s) and modified Dupuit formula (Q_r) for the cases of: (**a**) an unconfined aquifer, (**b**) confined aquifer, and (**c**) confined-unconfined aquifer.

5.3. Confined-Unconfined Aquifer

For the case of a partially penetrating ARW in a confined-unconfined aquifer, shown in Figure 1c, the results of simulations of the seepage flux can be described using a modified version of the Dupuit–Thiem formula given by:

$$Q_{\mathrm{r}} = 1.366 \frac{K(2HM - M^2 - h_{\mathrm{w}}^2)}{\lg \frac{R}{r_w}} = 1.366 \frac{K(2HM - M^2 - h_{\mathrm{w}}^2)}{\lg \frac{10 s_{\mathrm{eff}} \sqrt{K}}{r_w}} \tag{6}$$

As an example, the results of numerical simulations, calculations using the Dupuit–Thiem formula, and formula (7), for $r_{\mathrm{nor}} = 1.0$ and the boundary head $H_{\mathrm{nor}} = 7/6$, are summarized in Table A5 and shown in Figure 17c. The relative error of calculations of Q_{r} based on Equation (6), compared to numerical simulations, ranges from 2.37% to 8.87%.

5.4. Comparison of Modified Dupuit–Thiem Formulae with Results of Laboratory Sand-Tank Experiments

A comparison of calculations of the seepage flux based on the original and modified Dupuit formulae and sand-tank experiments are summarized in Figure 18 and Table A6. Figure 18 indicates that the seepage flux calculated using the modified Dupuit formula (Equation 6) is closer to the results of laboratory experiments. The seepage flux calculated from Dupuit formula is less than that from a modified Dupuit formula, with larger values being from laboratory experiments. However, the relative errors of $Q_{\mathrm{nor}}^{\mathrm{s}}$ compared to $Q_{\mathrm{nor}}^{\mathrm{r}}$ depend on the well length and the water table drawdown. As the well degree of penetration increases and the water table drawdown decreases, the relative error of $Q_{\mathrm{nor}}^{\mathrm{s}}$ compared to $Q_{\mathrm{nor}}^{\mathrm{r}}$ is increasing, as shown in Table A6. The data given in Tables A4–A6 confirm the validity of the application of a modified Dupuit–Thiem formula for calculations of the seepage flux for a single partially penetrating ARW.

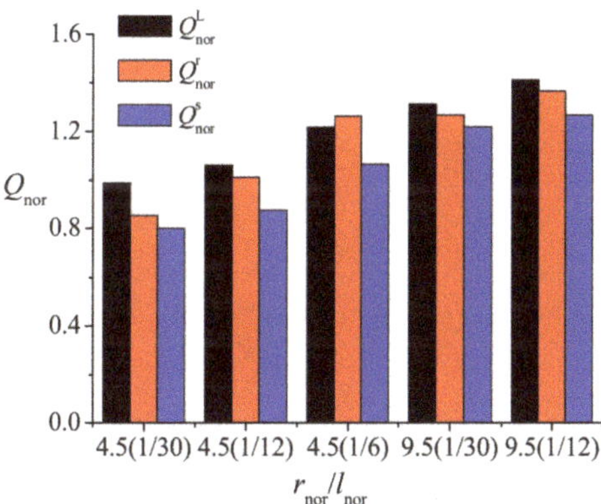

Figure 18. Comparison of the results of calculations of the seepage flux from the laboratory sand-tank experiments, original and modified Dupuit formulae.

6. Concluding Remarks

Laboratory sand-tank experiments and numerical simulations were conducted to assess the seepage characteristics of single partially and fully penetrating ARWs in homogeneous, isotropic unconfined, confined, and unconfined-confined aquifers. The seepage characteristics of the ARWs are affected by the seepage face boundary condition along an outflow boundary of the ARW, because the interior of the well casing remains open to the ambient air and is affected by the atmospheric pressure, so that water is discharging under gravity through the open bottom of the well casing. The experimental and modeling results indicate that (1) the relationships between the seepage flux and the degree of penetration, and seepage flux and the well radius, can be described using the exponential function (Equation 1) and the power function (Equation 2), respectively; (2) for the ARW's degree of penetration less than the critical value ζ_c, the piezometric head above the well is greater than zero with a piezometric head descent zone formed near the well.

The modeling results were used to develop (a) a series of type curves to evaluate the effect of the critical degree of penetration for different well radii and different aquifer water levels; and (b) the modified Dupuit and Dupuit–Thiem formulae by replacing the drawdown s by an effective drawdown $s_{eff} = s + l_w$. It is shown that for the ARWs in an unconfined aquifer with the degree of penetration exceeding a critical value, the seepage flux can be calculated based on the original Dupuit formula, and for confined or unconfined-confined aquifers—based on the Dupuit–Thiem formulae for a fully penetrating well.

Author Contributions: Conceptualization, W.W., T.J., J.D., and B.F.; Data curation, W.W.; Formal analysis, W.W., B.F. and J.D.; Funding acquisition, W.W.; Experimental investigations, W.W. and Y.L.; Methodology, W.W., B.F. and T.J.; Project administration, W.W.; Resources, W.W. and Y.L.; Software, W.W. and T.J.; Supervision, B.F.; Validation, W.W., B.F., J.D. and Y.L.; Writing—original draft, W.W. and J.D.; Writing—review and editing, W.W., B.F., T.J. and J.D. All authors have read and agreed to the published version of the manuscript.

Funding: Wenxue Wang: 41602298; Wenxue Wang: 2020HYTP014; Boris Faybishenko: DE-AC02-05CH11231.

Acknowledgments: Wang, W.X. would like to acknowledge the support of the National Natural Science Foundation of China under Grant No. 41.602.298 and the Young Talent Promotion Project of Henan Province (2020HYTP014). B. Faybishenko was supported by the U.S. Department of Energy, Office of Science, Office of Advanced Scientific Computing under Contract No. DE-AC02-05CH11231.

Appendix A

Table A1. Results of calculations of the errors of the seepage flux Q_{nor} (for $l_{nor} = 1$ and $r_{nor} = 4.5$, 9.5, 14.5 and 19.5) from numerical simulations and Dupuit formula in comparison to the laboratory sand-tank experiments.

Well Radius/r_{nor}	Seepage Flux Q_{nor}			Errors (%)	
	Lab Experiments	Dupuit Formula	Numerical Simulations		
	Q_{nor}^L	Q_{nor}^s	Q_{nor}^n	Errors of Q_{nor}^n	Errors of Q_{nor}^s
4.5	1.294	1.16	1.195	−7.65	3.02
9.5	1.517	1.425	1.467	−3.30	2.95
14.5	1.636	1.636	1.696	3.67	3.67
19.5	1.792	1.826	1.908	6.47	4.49

Table A2. For different well radii and the boundary head.

r_{nor} \\ H_{nor}	1	1.5	2.0	2.5	3.0	3.5	4.0
0.1	0.65	0.96					
0.2	0.63	0.93					
0.5	0.58	0.91					
1	0.55	0.84	0.97	0.98			
2	0.49	0.76	0.93	0.97	0.98		
4	0.42	0.66	0.85	0.94	0.96	0.98	0.99
6	0.35	0.57	0.78	0.90	0.93	0.96	0.98
10	0.29	0.48	0.67	0.80	0.9	0.93	0.95

Table A3. Seepage flux Q_{nor} calculations for an unconfined aquifer with $r_{nor} = 1.0$ and $H_{nor} = 1$.

l_{nor}	s_{nor}	$s_{nor}+l_{nor}$	Seepage Flux Q_{nor}			Errors (%)	
			Q_{nor}^n	Q_{nor}^r	Q_{nor}^s	Error of Q_{nor}^r	Error of Q_{nor}^s
1/2	0.362	0.862	0.937	0.936	0.657	−0.10	−29.88
1/3	0.282	0.615	0.846	0.860	0.563	1.60	−33.41
1/6	0.186	0.353	0.635	0.647	0.427	1.93	−32.73
1/10	0.143	0.243	0.497	0.511	0.356	2.82	−28.45
1/15	0.111	0.178	0.409	0.414	0.299	1.22	−27.04
1/30	0.080	0.113	0.302	0.303	0.237	0.39	−21.37
1/60	0.061	0.078	0.235	0.233	0.197	−0.79	−16.04

Notes: Q_{nor}^s represents the results of calculations using the Dupuit formula, Q_{nor}^n represents the results of numerical simulations, and Q_{nor}^r represents the results of calculations using the modified Dupuit formula.

Table A4. Parameters used for calculations of the seepage flux for a confined aquifer with $r_{nor} = 1.0$ and $H_{nor} = 2$.

l_{nor}	s_{nor}	$s_{nor}+l_{nor}$	Q_{nor}			Errors (%)	
			Q_{nor}^n	Q_{nor}^r	Q_{nor}^s	Error of Q_{nor}^r	Error of Q_{nor}^s
1	0.652	1.652	1.050	1.153	0.523	9.83	−50.18
5/6	0.628	1.461	1.031	1.042	0.507	1.10	−50.81
2/3	0.575	1.242	0.972	0.913	0.472	−6.10	−51.46
1/2	0.499	0.999	0.874	0.767	0.421	−12.20	−51.87
1/3	0.402	0.736	0.723	0.605	0.354	−16.35	−51.11
1/6	0.269	0.436	0.504	0.390	0.257	−22.72	−49.02
1/10	0.203	0.303	0.384	0.289	0.206	−24.75	−46.30
1/15	0.170	0.237	0.312	0.241	0.181	−22.87	−42.19
1/30	0.120	0.153	0.227	0.172	0.139	−24.47	−38.88
1/60	0.096	0.112	0.176	0.132	0.118	−24.88	−33.17

Table A5. Parameters used for calculations of the seepage flux for a confined-unconfined aquifer with $r_{nor} = 1.0$ and $H_{nor} = 7/6$.

l_{nor}	s_{nor}	h_{nor}^w	$s_{nor}+l_{nor}$	Q_{nor}			Errors (%)	
				Q_{nor}^n	Q_{nor}^r	Q_{nor}^s	Error of Q_{nor}^r	Error of Q_{nor}^s
1	0.391	0.000	0.779	0.965	1.000	1.219	3.64	26.28
5/6	0.411	0.000	0.789	0.977	1.000	1.207	2.37	23.50
2/3	0.405	0.095	0.786	0.977	1.007	1.202	3.16	23.09
1/2	0.350	0.317	0.758	0.928	0.976	1.153	5.16	24.25
1/3	0.279	0.555	0.723	0.805	0.860	1.006	6.84	24.88
1/6	0.192	0.808	0.679	0.581	0.633	0.725	8.87	24.73
1/10	0.147	0.920	0.657	0.451	0.490	0.554	8.82	22.90

Table A6. Comparison of seepage flux calculations using data from the laboratory experiments, original and modified Dupuit formulae. Errors of original and modified Dupuit formulae are calculated in comparison to the results of laboratory sand-tank experiments.

$r_{nor}(l_{nor})$.	s_{nor}	Q_{nor}			Errors (%)	
		Lab Experiments	Modified Dupuit Formula	Dupuit Formula		
		Q_{nor}^{L}	Q_{nor}^{r}	Q_{nor}^{s}	Errors of Q_{nor}^{r}	Errors of Q_{nor}^{s}
4.5(1/30)	0.20	0.988	0.853	0.801	−13.66	−18.93
4.5(1/12)	0.23	1.061	1.011	0.875	−4.71	−17.53
4.5(1/6)	0.33	1.217	1.264	1.066	3.86	−12.41
9.5(1/30)	0.37	1.313	1.267	1.220	−3.50	−7.08
9.5(1/12)	0.40	1.411	1.366	1.267	−3.19	−10.21

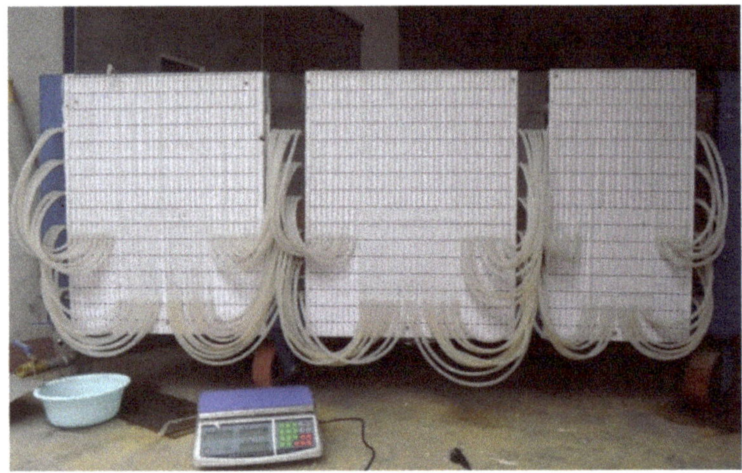

Figure A1. Photograph of the monitoring piezometer tubes.

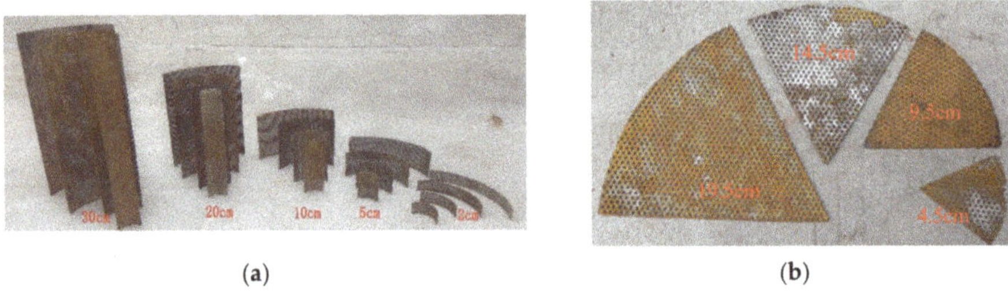

(a) (b)

Figure A2. Photographs of seepage well screens (**a**) and well covers (**b**) used in the sand-tank experiments.

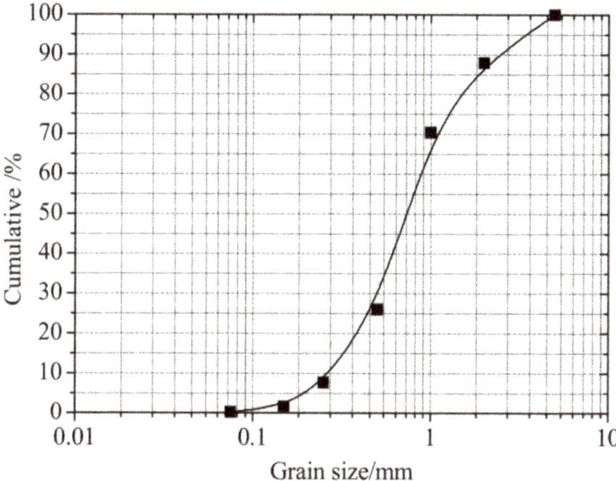

Figure A3. Particle size distribution curve of a sand used for the sand-tank experiments.

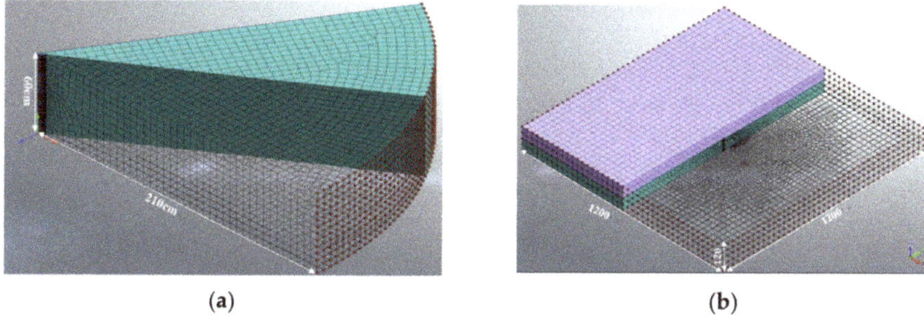

(a)   (b)

Figure A4. 3-D numerical discretization of the flow domains for simulations of: (**a**) the sand-tank experiments, and (**b**) the confined aquifer.

References

1. Schubert, J.P. *Reducing Water Leakage into Underground Coal Mines by Aquifer Dewatering*; No. ANL/LRP-CP-9; CONF-7809134-1; Argonne National Lab.: Argonne, IL, USA, 1978.
2. Sgambat, J.P.; LaBella, E.A.; Roebuck, S. *Effects of Coal Mining on Ground-Water in the Eastern United States*; EPA 60017-80-120; U.S. Environmental Protection Agency: Washington, DC, USA, 1980; p. 183.
3. Niskovskiy, Y.; Vasianovich, A. Investigation of possibility to apply untraditional and ecologically good methods of coal mining under sea bed. In Proceedings of the Sixth International Offshore and Polar Engineering Conference, Los Angeles, CA, USA, 26–31 May 1996.
4. Wu, Q.; Zhou, W.F. Prediction of inflow from overlying aquifers into coalmines: A case study in Jinggezhuang Coalmine, Kailuan, China. *Environ. Geol.* **2008**, *55*, 775–780. [CrossRef]
5. Zhou, N.Q.; Vermeer, P.A.; Lou, R.; Tang, Y.; Jiang, S. Numerical simulation of deep foundation pit dewatering and optimization of controlling land subsidence. *Eng. Geol.* **2010**, *114*, 251–260. [CrossRef]
6. Struzina, M.; Müller, M.; Drebenstedt, C.; Mansel, H.; Jolas, P. Dewatering of Multi-aquifer Unconsolidated Rock Opencast Mines: Alternative Solutions with Horizontal Wells. *Mine Water Environ.* **2011**, *30*, 90–104. [CrossRef]
7. Pujades, E.; López, A.; Carrera, J.; Vázquez-Suñé, E.; Jurado, A. Barrier effect of underground structures on aquifers. *Eng. Geol.* **2012**, *145*, 41–49. [CrossRef]
8. Wu, Y.X.; Shen, S.L.; Yuan, D.J. Characteristics of dewatering induced drawdown curve under blocking effect of retaining wall in aquifer. *J. Hydrol.* **2016**, *539*, 554–566. [CrossRef]

9. Brownlow, J.W.; James, S.C.; Yelderman, J.C. Influence of Hydraulic Fracturing on Overlying Aquifers in the Presence of Leaky Abandoned Wells. *Groundwater* **2016**, *54*, 781–792. [CrossRef] [PubMed]

10. Wang, W.X.; Sui, W.H.; Faybishenko, B.; Stringfellow, W.T. Permeability variations within mining-induced fractured rock mass and its influence on groundwater inrush. *Environ. Earth Sci.* **2016**, *75*, 1–15. [CrossRef]

11. Xu, J.P.; Chen, M.Y.; Hong, H.; Xu, X.Q.; Liu, B. Differences of hydraulic conductivities from pumping and draining tests with intensity effects. *J. Hydrol.* **2019**, *579*, 124165. [CrossRef]

12. Cartwright, K.; Hunt, C.S. *Hydrogeologic Aspects of Coal Mining in Illinois: An Overview. Champaign, III;* Illinois State Geological Survey: Champaign, IL, USA, 1981; p. 19.

13. Hill, J.G.; Price, D.R. The impact of deep mining on an overlying aquifer in western Pennsylvania. *Groundw. Monit. Remediat.* **1983**, *3*, 138–143. [CrossRef]

14. Younger, P.L.; Wolkersdorfer, C. Mining impacts on the fresh water environment: Technical and managerial guidelines for catchment scale management. *Mine Water Environ.* **2004**, *23*, 2–80.

15. Zhang, J.C. Investigations of water inrushes from aquifers under coal seams. *Int. J. Rock Mech. Min. Sci.* **2005**, *42*, 350–360. [CrossRef]

16. Booth, C.J. Groundwater as an environmental constraint of longwall coal mining. *Environ. Geol.* **2006**, *49*, 796–803. [CrossRef]

17. Hang, Y.; Zhang, G.L.; Yang, G.Y. Numerical simulation of dewatering thick unconsolidated aquifers for safety of underground coal mining. *Min. Sci. Technol.* **2009**, *19*, 312–316. [CrossRef]

18. Barnes, M.R.; Vermeulen, P.D. Guide to groundwater monitoring for the coal industry. *Water SA* **2012**, *38*, 831–836. [CrossRef]

19. Newman, C.; Agioutantis, Z.; Leon, G.B.J. Assessment of potential impacts to surface and subsurface water bodies due to longwall mining. *Int. J. Min. Sci. Technol.* **2017**, *27*, 57–64. [CrossRef]

20. Meiri, D. Unconfined groundwater flow calculation into a tunnel. *J. Hydrol.* **1985**, *82*, 69–75. [CrossRef]

21. Liu, X.X.; Shen, S.; Xu, Y.S. Analytical approach for time-dependent groundwater inflow into shield tunnel face in confined aquifer. *Int. J. Numer. Anal. Methods Geomech.* **2018**, *42*, 655–773. [CrossRef]

22. Hwang, J.H.; Lu, C.C. A semi-analytical method for analyzing the tunnel water inflow. *Tunn. Undergr. Space Technol.* **2007**, *22*, 39–46. [CrossRef]

23. Pulido-Velazquez, D.; Llopis-Albert, C.; Peña-Haro, S.; Pulido-Velazquez, M. Efficient conceptual model for simulating the effect of aquifer heterogeneity on natural groundwater discharge to rivers. *Adv. Water Res.* **2011**, *34*, 1377–1389. [CrossRef]

24. Mahdi Rasouli, M. Groundwater Seepage Rate (GSR); a new method for prediction of groundwater inflow into jointed rock tunnels. *Tunn. Undergr. Space Technol.* **2018**, *71*, 505–517. [CrossRef]

25. Goodman, R. Groundwater inflows during tunnel driving. *Eng. Geol.* **1965**, *2*, 39–56.

26. Shizhong, L. An Analytical Solution for Steady Flow into a Tunnel. *Groundwater* **1999**, *37*, 23–26.

27. El Tani, M. Circular tunnel in a semi-infinite aquifer. *Tunn. Undergr. Space Technol.* **2003**, *18*, 49–55. [CrossRef]

28. Kolymbas, D.; Wagner, P. Groundwater ingress to tunnels–The exact analytical solution. *Tunn. Undergr. Space Technol.* **2007**, *22*, 23–27. [CrossRef]

29. Corps of Engineers US Army (USACE). *Design, Construction, and Maintenance of Relief Wells;* Engineer Manual: Washington, DC, USA, 1994.

30. Tammetta, P. Estimation of the Change in Storage Capacity above Mined Longwall Panels. *Groundwater* **2016**, *54*, 646–655. [CrossRef] [PubMed]

31. Shamsai, A.; Narasimhan, T.N. A numerical investigation of free surface-seepage face relationship under steady state flow conditions. *Water Resour. Res.* **1991**, *27*, 409–421. [CrossRef]

32. Butler, J.J.; Healey, J.M. Relationship between Pumping-Test and Slug-Test Parameters: Scale Effect or Artifact? *Groundwater* **1998**, *36*, 305–312. [CrossRef]

33. Tsoflias, G.P.; Halihan, T.; Sharp, J.M. Monitoring pumping test response in a fractured aquifer using ground-penetrating radar. *Water Resour. Res.* **2001**, *37*, 1221–1229. [CrossRef]

34. Simpson, M.J.; Clement, T.P.; Gallop, T.A. Laboratory and Numerical Investigation of Flow and Transport near a Seepage-Face Boundary. *Groundwater* **2003**, *41*, 690–700. [CrossRef]

35. Li, W.; Englert, A.; Cirpka, O.A.; Vereecken, H. Three-dimensional geostatistical inversion of flowmeter and pumping test data. *Groundwater* **2008**, *46*, 193–201. [CrossRef]

36. Sethi, R. A dual-well step drawdown method for the estimation of linear and nonlinear flow parameters and wellbore skin factor in confined aquifer systems. *J. Hydrol.* **2011**, *400*, 187–194. [CrossRef]

37. Ji, S.H.; Koh, Y.K. Nonlinear groundwater flow during a slug test in fractured rock. *J. Hydrol.* **2015**, *520*, 30–36. [CrossRef]

38. Sun, H. A semi-analytical solution for slug tests in an unconfined aquifer considering unsaturated flow. *J. Hydrol.* **2016**, *532*, 29–36. [CrossRef]

39. Wu, Y.X.; Shen, J.S.; Chen, W.C.; Hino, T. Semi-analytical solution to pumping test data with barrier, wellbore storage, and partial penetration effects. *Eng. Geol.* **2017**, *226*, 44–51. [CrossRef]

40. Dupuit, J.É.J. *Études Théoriques et Pratiques sur le Mouvement des Eaux Dans les Canaux Découverts et à Travers les Terrains Perméables: Avec des Considérations Relatives au Régime des Grandes Eaux, au Débouché à leur Donner, et à la Marche des Alluvions dans les Rivières à Fond Mobile*; Dunod: Paris, France, 1863.

41. Thiem, G. *Hydrologische Methoden: Dissertation zur Erlangung der Wurde eines*; JM Gebhardt: Leipzig, Germany, 1906.

42. Boulton, N.S. The drawdown of the water-table under non-steady conditions near a pumped well in an unconfined formation. *Proc. Inst. Civ. Eng.* **1954**, *3*, 564–579. [CrossRef]

43. Streltsova, T. Unsteady radial flow in an unconfined aquifer. *Water Resour. Res.* **1972**, *8*, 1059–1066. [CrossRef]

44. Neuman, S.P. Effect of partial penetration on flow in unconfined aquifers considering delayed gravity response. *Water Resour. Res.* **1974**, *10*, 303–312. [CrossRef]

45. Neuman, S.P. Theory of flow in unconfined aquifers considering delayed response of the water table. *Water Resour. Res.* **1972**, *8*, 1031–1045. [CrossRef]

46. Ruud, N.C.; Kabala, Z.J. Response of a partially penetrating well in a heterogeneous aquifer: Integrated well-face flux vs. uniform well-face flux boundary conditions. *J. Hydrol.* **1997**, *194*, 76–94. [CrossRef]

47. Cassiani, G.; Kabala, Z.J. Hydraulics of a partially penetrating well: Solution to a mixed-type boundary value problem via dual integral equations. *J. Hydrol.* **1998**, *211*, 100–111. [CrossRef]

48. Hemker, C.J. Transient well flow in vertically heterogeneous aquifers. *J. Hydrol.* **1999**, *225*, 1–18. [CrossRef]

49. Chang, C.C.; Chen, C.S. A flowing partially penetrating well in a finite-thickness aquifer: A mixed-type initial boundary value problem. *J. Hydrol.* **2003**, *271*, 101–118. [CrossRef]

50. Perina, T.; Lee, T.C. General well function for pumping from a confined, leaky, or unconfined aquifer. *J. Hydrol.* **2006**, *317*, 239–260. [CrossRef]

51. Feng, Q.G.; Zhan, H.B. On the aquitard–aquifer interface flow and the drawdown sensitivity with a partially penetrating pumping well in an anisotropic leaky confined aquifer. *J. Hydrol.* **2015**, *521*, 74–83. [CrossRef]

52. Mishra, P.K.; Kuhlman, K.L. Unconfined aquifer flow theory: From Dupuit to present. In *Advances in Hydrogeology*; Springer: New York, NY, USA, 2013; pp. 185–202.

53. Charnyi, I.A. A rigorous derivation of Dupuit's formula for unconfined seepage with seepage surface. *Dokl. Akad. Nauk Mosc.* **1951**, *79*, 937–940. (In Russian)

54. Shercliff, J.A. Seepage flow in unconfined aquifers. *J. Fluid Mech.* **1975**, *71*, 181–192. [CrossRef]

55. Bhunya, P.K.; Anjaneyulu, B. Study of flow analysis in semiconfined aquifer using a sand tank model. *ISH J. Hydraul. Eng.* **2001**, *7*, 13–21. [CrossRef]

56. Faybishenko, B.A. Hydraulic behavior of quasi-saturated soils in the presence of entrapped air: Laboratory experiments. *Water Resour. Res.* **1995**, *31*, 2421–2435. [CrossRef]

57. Faybishenko, B.A.; Javandel, I.; Witherspoon, P.A. Hydrodynamics of the Capture Zone for Contaminant Transport with Partially Penetrating Well in a Confined Aquifer. *Water Resour. Res.* **1995**, *31*, 859–866. [CrossRef]

Article

Dynamic Variation Characteristics of Seawater Intrusion in Underground Water-Sealed Oil Storage Cavern under Island Tidal Environment

Yutao Li, Bin Zhang *, Lei Shi and Yiwei Ye

School of Engineering and Technology, China University of Geosciences (Beijing), Beijing 100083, China; liyutaomail@163.com (Y.L.); shilei1990@cugb.edu.cn (L.S.); wzeg20090317@hotmail.com (Y.Y.)
* Correspondence: sc_zhb@cugb.edu.cn; Tel.: +86-131-2670-2993

Received: 16 December 2018; Accepted: 8 January 2019; Published: 12 January 2019

Abstract: In the case of constructing underground water-sealed oil storage caverns in island environments, the groundwater seepage characteristics are more complicated under the influence of seawater and tidal fluctuations. It also faces problems such as seawater intrusion. This research is based on multi-physical field coupling theory and analyzed the influence of tidal fluctuation and water curtain systems on the temporal-spatial variations of seawater intrusion in an island oil storage cavern in China using the finite element method. The results show that the operation of an underground water-sealed oil storage cavern in an island environment has a risk of inducing seawater intrusion. The tidal fluctuation has a certain degree of influence on the seepage field of the island. The water curtain system can decrease seawater intrusion and reduce the influence of tidal fluctuation on the seepage field inside the island. The research results provide a theoretical basis for the study of seawater intrusion in underground oil storage caverns under island tidal environments.

Keywords: underground water-sealed oil storage cavern; seawater intrusion; island tidal environment; vertical water curtain system; multi-physical field coupling

---

1. Introduction

Compared with other oil storage methods, underground water-sealed oil storage caverns have the advantages of using less space, high safety and good economic efficiency, and have become the preferred method of oil storage in the world [1,2]. Most of the underground water-sealed oil storage caverns in the world are built in freshwater areas with good rock integrity and keep a certain distance from the shoreline [3–5]. However, sites suitable for the construction of underground water-sealed oil storage caverns are limited [6]. The construction of underground water-sealed oil storage caverns on islands with suitable geological conditions not only provides a new possibility for underground oil storage but also promotes the utilization of island resources. In addition, island environments are convenient for the arrival and departure of oil tankers, facilitating oil transport. At present, there are few cases of building underground water-sealed oil storage caverns in island environments.

The construction of underground water-sealed oil storage caverns in an island environment will face the problem of seawater intrusion [7]. Surrounded by seawater, the excavation of an underground water-sealed oil storage cavern in an island environment will affect the natural seepage field of the research area, and the groundwater in the surrounding rock will flow into the underground caverns [8]. When the water inflow of the underground caverns exceeds the underground freshwater recharge, the freshwater head will fall. When the freshwater head is lower than the seawater head, the freshwater-seawater interface will push into the island, thus causing seawater intrusion (as shown in Figure 1) [9]. Seawater intrusion will not only corrode the structure of the underground caverns (steel, concrete, etc.) but also affect the quality of the crude oil and underground freshwater on the island [10]. Therefore, seawater intrusion will affect the long-term stability and safety of underground water-sealed oil storage caverns. Worse still, it may even destroy the original ecological balance of the island [11–13]. Therefore, more attention should be paid to seawater intrusion in the construction of underground water-sealed oil storage caverns in island environments.

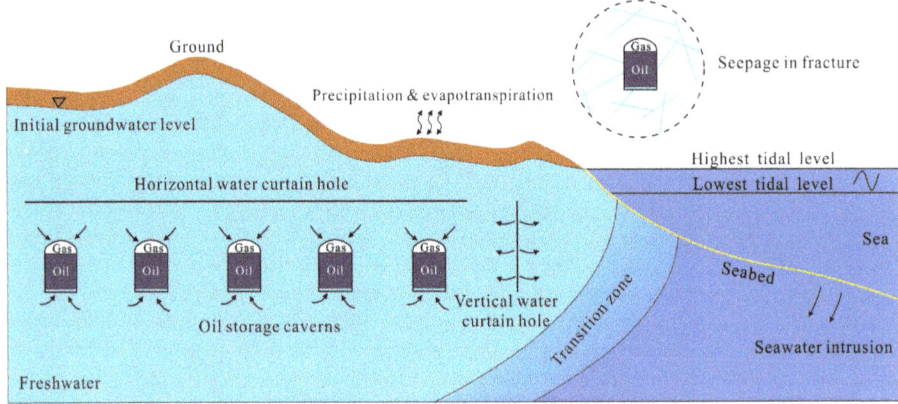

Figure 1. Model diagram of an underground water-sealed oil storage cavern.

Numerical simulation has become one of the most effective methods for studying seawater intrusion [14]. According to the groundwater aquifer, the numerical models of seawater intrusion can be divided into a porous media model and fracture-karst media model. Research on the porous media model is relatively mature [9]. Fracture-karst media models can be divided into equivalent porous media model, discrete fracture network model, fractured porous media model and discrete-continuum media model [15–18].

The groundwater seepage field of the island is affected by the tidal fluctuation, but the seepage characteristics are more complicated [19,20] and the groundwater level changes periodically [21,22]. The fluctuation of the groundwater level will affect the hydraulic head differences between the freshwater and seawater in the island environment, which will affect the development and rate of seawater intrusion [23–26]. Moreover, the fluctuation of groundwater level will affect the water-sealed safety of underground oil storage caverns. However, the amplitudes of the tidal waves are increasingly damped with horizontal distance from the shoreline, and there exists a phase difference at different locations [27,28]. Therefore, the extent of tidal influence is worth studying.

The prevention measures of seawater intrusion are mainly as follows: (a) rational exploitation of groundwater, (b) artificial recharge of groundwater, and (c) setting up barrier wells along the coast [29,30]. Barrier wells located along the coast can effectively decrease seawater intrusion into the underground caverns [31]. In general, a water curtain system needs to be set up to ensure the water-sealed reliability of an underground water-sealed oil storage cavern [32–35]. The water curtain system is mainly horizontal. Vertical water curtain systems have not been paid much attention because of the problem of excessive water inflow [36]. Vertical water curtain systems can form a water pressure wall on both sides of the underground caverns, with a role similar to barrier wells. Therefore, the effect of vertical water curtain systems should be studied in the process of building an underground water-sealed oil storage cavern in an island environment.

In this paper, the equivalent porous media model was adopted. The influence of fracture anisotropy on groundwater seepage field was not considered. The objective of this paper is to obtain the dynamic characteristics of seawater intrusion in an underground water-sealed oil storage cavern under an island tidal environment using a transient simulation based on the multi-physical field coupling method. By comparing the temporal-spatial variations of seawater intrusion under four conditions (un-excavated, excavated, horizontal water curtain system, vertical water curtain system), the effect of the water curtain system on seawater intrusion was studied. Then, the tidal monitoring data were fitted to a sinusoidal curve, and the influence of the tidal fluctuation on the seawater intrusion was analyzed. Finally, the influence of the water curtain system on the tidal wave propagation to the island was discussed. The research results provide a theoretical basis for the construction of underground water-sealed oil storage caverns under island tidal environments in the future.

2. Research Area

2.1. Project Overview

The island is located in Zhejiang Province, China, with an area of approximately 2.53 km^2 (as shown in Figure 2). The storage capacity of the proposed underground water-sealed oil storage cavern is approximately 750×10^4 m^3, consisting of 16 caverns. According to the island topography, the length of the main caverns can be divided into four types: 1000 m, 930 m, 620 m and 465 m. The main cavern spacing is divided into four cases of 30 m, 50 m, 70 m and 95 m. The underground oil storage cavern cross-section has a horseshoe shape, 20 m wide, 30 m high. The elevation of the top of the cavern is −45 m, and the axis direction of the caverns is NE 80°.

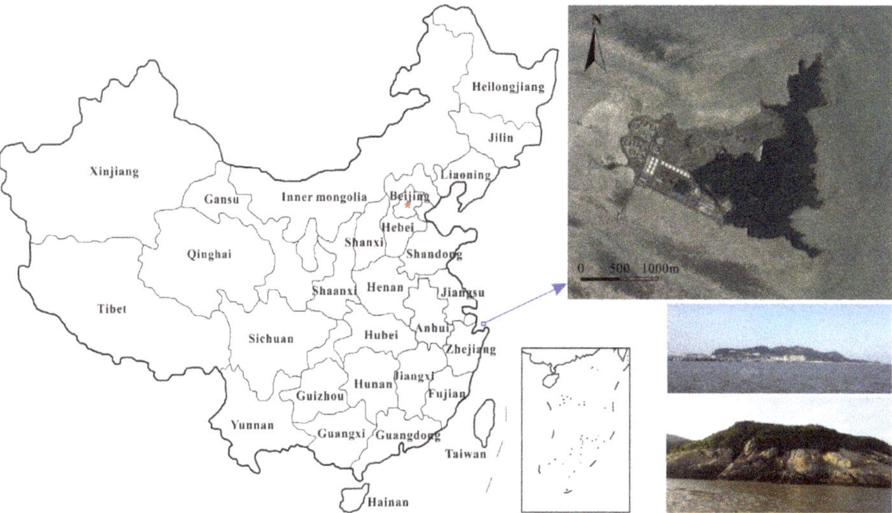

Figure 2. Location map of the research area.

2.2. Engineering Geology

Geologically, the stratum of the research area (as shown in Figure 3a) comprises metamorphic rocks (schist, gneiss and plagioclase amphibolite) of the Mesoproterozoic Chencai Group Mo Jiu Wan Formation (Ptd(AnDch^a)), medium-acidic and acidic pyroclastic rocks of the Lower Cretaceous System Chawan Group (K$_1$c), granite of the 2nd stage intrusions of Late Yanshanian ($\gamma_5^{3(2)}$) and quaternary strata (as shown in Figure 3b). According to the field exploration results, the lithology of the research area along the vertical direction can be divided into the following four types: residual sandy clay, intense weathered granite, intermediary weathered granite, and weak weathered granite. The average thickness of the residual candy clay is 2.80 m, the average thickness of the intense weathered granite is 9.13 m, the average thickness of the intermediary weathered granite is 23.03 m, and the rest is weak weathered granite. All underground caverns are located in the weak weathered granite. The rock mass of the research area has good integrity, and a total of 8 fault zones are found, mainly in NE direction, followed by the NW direction and nearly EW direction. Joint attitude statistics show that the main directions of the joints are NE 60° and NW 306°.

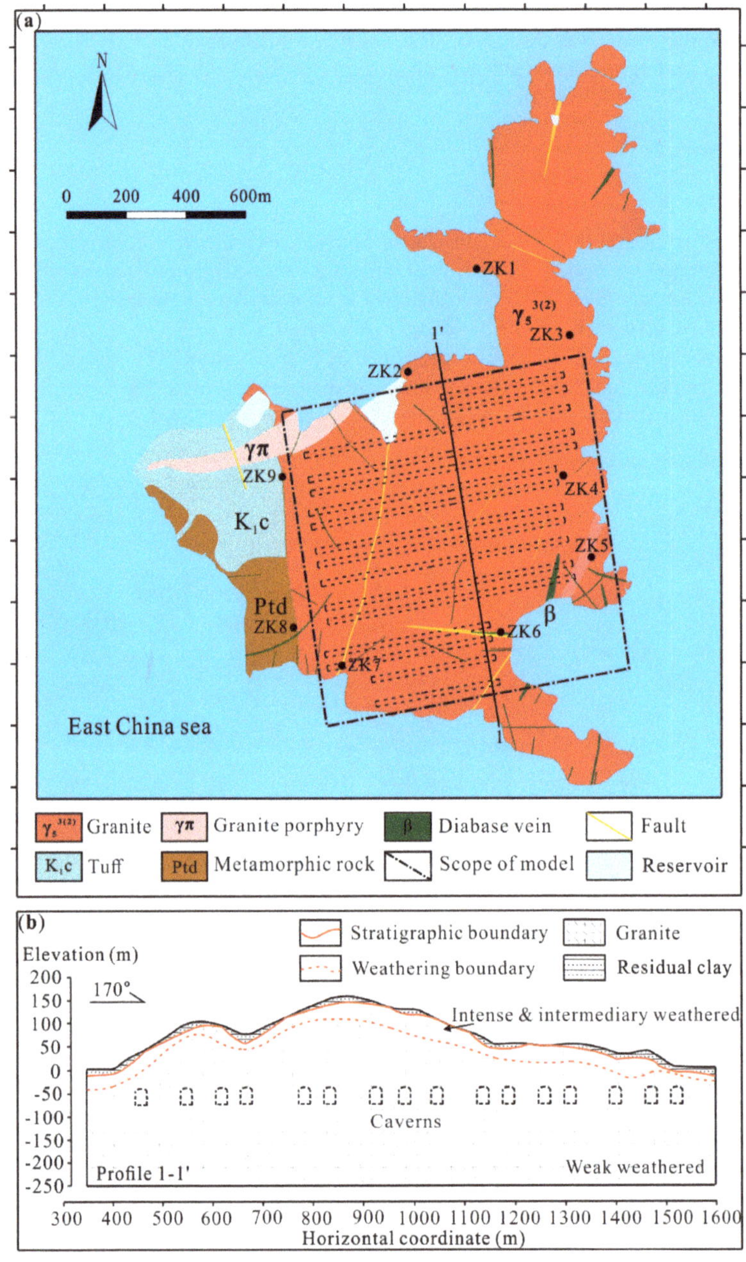

Figure 3. Geological map of the study area: (**a**) plane map, ZK1-ZK9 are the number of groundwater table monitoring boreholes; (**b**) profile map.

2.3. Hydrogeology

The groundwater in the research area can be divided into 2 types: loose-rock pore water and bedrock fracture water. There are 9 groundwater table monitoring holes arranged in the research area (as shown in Figure 3). By analyzing the monitoring data of the groundwater table, it can be seen that: the groundwater depth is between 1.4 and 28.5 m, and the average groundwater depth is 13.3 m (as shown in Figure 4). According to the research results of relevant scholars and our sensitivity analysis, the hydraulic conductivity and the hydrodynamic dispersion coefficient are the main parameters influencing seawater intrusion. The hydraulic conductivity of rock mass is the most sensitive. The hydrodynamic dispersion coefficient is less sensitive. According to the rock mass permeability test, the hydraulic conductivity of the rock mass in the research area was obtained. The hydraulic conductivity of the rock mass decreases gradually as the depth increases. According to the variation law of hydraulic conductivity of granitic rock masses [37], curve fitting can be carried out, and the fitting formula is as follows:

$$k = 2.72 \times 10^{-8} \times e^{h/22.96} + 2.24 \times 10^{-9} \tag{1}$$

where k is the hydraulic conductivity (m/s), and h is the depth (m).

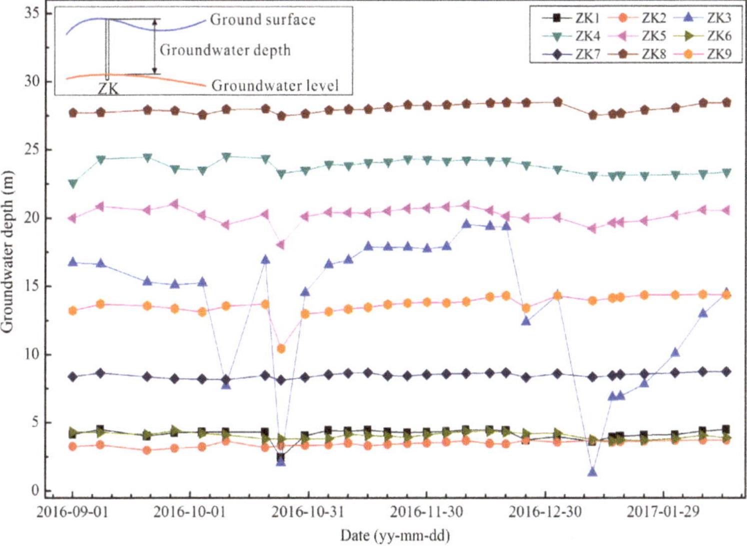

Figure 4. Variation in the groundwater depth.

According to the sampling analysis of the seawater around the island, it can be seen that: the density of seawater is 1025 kg/m^3. The most abundant ions in the seawater are chloride ions and sodium ions. The concentration of chloride ion in the seawater is 0.31 mol/kg, the concentration of sodium ion in the seawater is 0.22 mol/kg.

The research area is subjected to an irregular semidiurnal tide [38]. Affected by the relative positions of the Earth, the Moon and the Sun, the tidal cycles show the phenomena of diurnal inequality, semi-menstrual inequality and annual inequality [39]. According to the tidal monitoring data, which are shown in Figure 5a (the monitoring data came from the Navigation Guarantee Department of Chinese Navy Headquarters), the characteristics of the tidal fluctuations include: the highest tidal level is 2.42 m, the lowest tidal level is −2.28 m, two high tidal levels and two low tidal levels occurred in one day, with a maximum tidal fluctuation range of 4.34 m and a minimum range of 0.28 m.

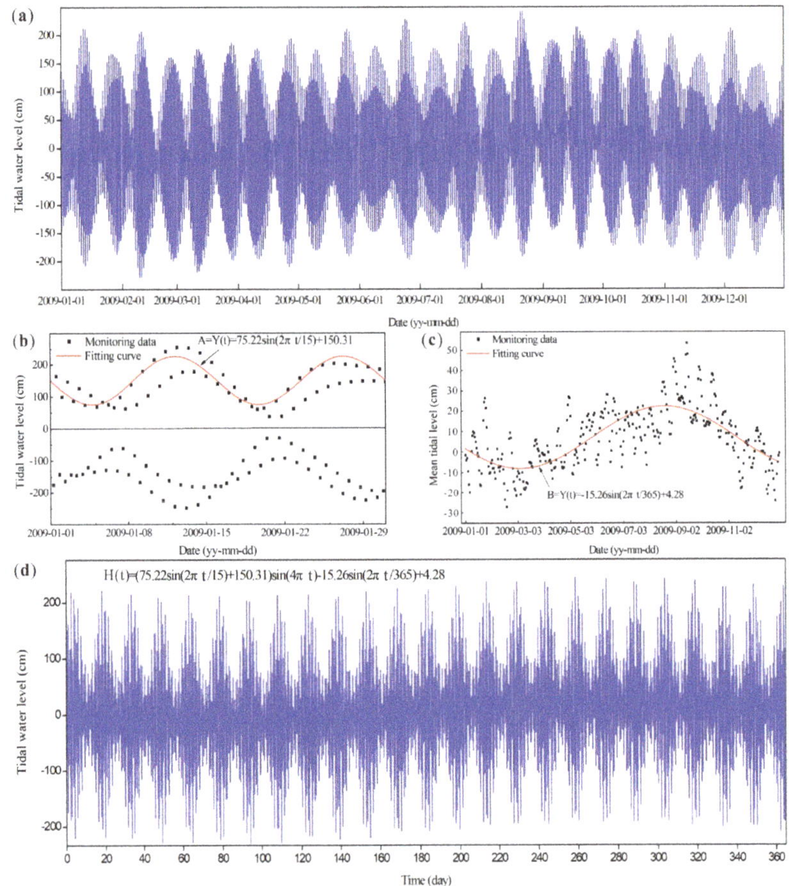

Figure 5. Tidal monitoring data and the fitting curve: (**a**) tidal monitoring data; (**b**) tidal amplitude and its fitting curve; (**c**) sinusoidal setover and its fitting curve; (**d**) fitting curve. Data source: The Navigation Guarantee Department of Chinese Navy Headquarters.

The tidal monitoring data are taken as four values per day, namely higher high water, higher low water, lower high water and lower low water. Therefore, tidal data are discontinuous. To apply a tidal fluctuation boundary to the model, a sinusoidal fitting is performed for the tidal monitoring data. It can be seen from the analysis of the tidal data: (1) the tidal water level fluctuates sinusoidally and the fluctuation period is half a day, (2) the amplitude of the tidal fluctuation also varies sinusoidally, and the fluctuation period is one month, and (3) the deviation of the tidal fluctuation also fluctuates sinusoidally with a period of one year [40]. The fitting equation is as follows.

The equation for the average tidal water level (deviation) is (as shown in Figure 5c):

$$B = -15.26 \times \sin\frac{2\pi t}{365} + 4.28 \tag{2}$$

The equation for the tidal amplitude is (as shown in Figure 5b):

$$A = 75.22 \times \sin\frac{2\pi t}{15} + 150.31 \tag{3}$$

The equation for the tidal water level is:

$$H_t = A \times \sin(4\pi t) + B \tag{4}$$

Substituting Equations (2) and (3) into Equation (4) gives (as shown in Figure 5d):

$$H_t = \left(75.22 \times \sin\frac{2\pi t}{15} + 150.31\right) \times \sin(4\pi t) - 15.26 \times \sin\frac{2\pi t}{365} + 4.28 \tag{5}$$

where t is the time (day), B is the average tidal level (cm), A is the amplitude of the tidal fluctuation (cm), and H_t is the tidal level (cm).

3. Simulation Method

3.1. Numerical Model

In this paper, the finite element software COMSOL Multiphysics (version 5.3, manufactured by COMSOL Inc. in Stockholm, Sweden) is used to establish a three-dimensional numerical model (as shown in Figure 6). The numerical model set the Y-axis as geographic north, the vertical direction as the Z-axis and the axis of the main caverns and X-axis formed an angle of $10°$. The model size is 2080 m $\times$ 2730 m $\times$ 250 m. The simulation is divided into the following four conditions: the underground caverns are un-excavated, the underground caverns are excavated, the underground caverns are excavated and a horizontal water curtain system is set above the caverns, and the underground caverns are excavated and a vertical water curtain system is arranged on both sides of the caverns. The horizontal water curtain system is set with the horizontal water curtain holes in a parallel arrangement. The bottom elevation of the water curtain hole is -15 m, the diameter is 0.1 m, the separation distance is 10 m, and the water injection pressure is 0.3 MPa. The vertical water curtain system is distributed on both sides of the main cavern. The top elevation of the water curtain hole is -15 m, the hole length is 75 m, the diameter is 0.1 m, and the injection pressure is 0.3 MPa. To facilitate the analysis of the simulation results, profile 1-1' is set along the direction vertical to the axis of the cavern. The four models are divided into tetrahedral meshes. The computational grid and the size resolution of the four models are shown in Table 1.

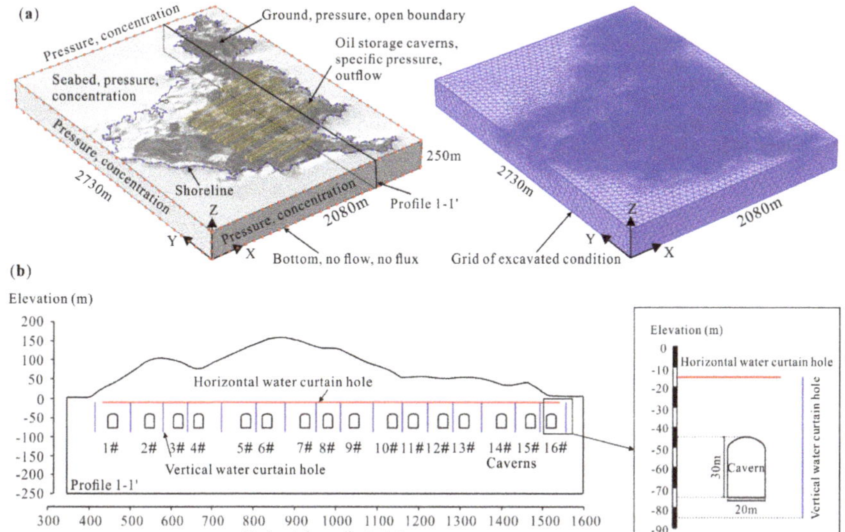

Figure 6. Numerical model: (**a**) three-dimensional numerical model; (**b**) numerical model profile.

Table 1. The computational grid and the size resolution of the four models.

Items	Four Conditions			
	Un-Excavated	Excavated	Horizontal Water Curtain	Vertical Water Curtain
Number of elements	440,710	970,463	1,083,967	3,054,415
Minimum element size (m)	9.45	4.26	1.54	1.29
Maximum element size (m)	94.75	91.65	80.72	92.452
Average element size (m)	51.22	46.98	44.39	44.59

3.2. Governing Equation

The study of the dynamic variation characteristics of seawater intrusion is based on the advection-dispersion model. Freshwater and seawater are miscible. Under the influence of hydrodynamic dispersion in the process of seawater intrusion, there is no freshwater–seawater sharp interface. Instead, there is a transition zone. Therefore, the advection-dispersion equation of seawater intrusion is composed of two parts. One part is the Darcy equation and the other part is the solute transport equation. In the numerical simulation, the multi-physical field coupling method is used to couple these two equations.

(1) The underground seepage field is subject to the seepage continuity equation (Equation (6)) and Darcy's law (Equation (7)) and [41]:

$$\left[\frac{\partial}{\partial t}(n\rho_w) + \nabla\cdot(\rho_w v)\right] = Q_m \tag{6}$$

$$v = -\frac{k}{\rho_w g}(\nabla p + \rho_w g \nabla z) \tag{7}$$

$$\frac{\partial \rho_w}{\partial C_w} = \frac{\rho_{sw} - \rho_{fw}}{C_{sw}} \tag{8}$$

where k is the hydraulic conductivity (m/s), n is the porosity, Q_m is the rate of recharge to the water table per unit volume of the aquifer (kg/s), v is the Darcy's flux (m/s), p is the pore water pressure (Pa), z is the vertical coordinate (m), g is the gravitational acceleration (m/s^2), ρ_w is the density of water (kg/m^3), ρ_{fw} is the density of freshwater (kg/m^3), ρ_{sw} is the density of seawater (kg/m^3), C_w is the concentration of chlorine ions in water (kg/m^3) and C_{sw} is the concentration of chlorine ions in seawater (kg/m^3).

(2) In the three-dimensional numerical model, the solute transport equation can be described as follows [42]:

$$\frac{\partial}{\partial t}(nC_w) - \nabla(nD\nabla C_w) + \nabla(C_w v) = Q_m C_q \tag{9}$$

$$D = \begin{bmatrix} D_{xx} & D_{xy} & D_{xz} \\ D_{yx} & D_{yy} & D_{yz} \\ D_{zx} & D_{zy} & D_{zz} \end{bmatrix} \tag{10}$$

$$D_{ii} = D^* + \alpha_T u + (\alpha_L - \alpha_T)\frac{u_i^2}{u}, \quad i = x,\, y,\, z \tag{11}$$

$$D_{ij} = D^* + (\alpha_L - \alpha_T)\frac{u_i u_j}{u}, \quad i,\, j = x,\, y,\, z \text{ and } i \neq j \tag{12}$$

$$u = \sqrt{u_x^2 + u_y^2 + u_z^2} \tag{13}$$

where C_w is the concentration of chlorine ions (mol/kg), C_q the concentration of solute in recharged water (mol/kg), x is the flow direction, and y and z are vertical to the flow direction, D is the hydrodynamic dispersion coefficient tensor (m^2/s), D_{ii} and D_{ij} are the components of the hydrodynamic dispersion coefficient tensor (m^2/s), D^* is the diffusion coefficient (m^2/s), α_L is the longitudinal dispersity (m), α_T is the transversal dispersity (m), u_x, u_y and u_z are the three components of the seepage velocity (m/s).

3.3. Model Parameters

The numerical model parameters are selected from the Engineering Geological Investigation Report provided by the Zhejiang Engineering Geochemical Prospecting Academy of China. The diffusion coefficient, longitudinal dispersity and transversal dispersity are selected based on engineering experience [43]. The detailed model parameters are shown in Table 2.

Table 2. The model parameters.

Granite Density ρ_g (kg/m^3)	Freshwater Density ρ_{fw} (kg/m^3)	Seawater Density ρ_{sw} (kg/m^3)	Porosity n	Diffusion Coefficients D^* (m^2/s)	Longitudinal Dispersity α_L (m)	Transversal Dispersity α_T (m)
2600	1000	1025	0.015	1.7×10^{-11}	21.3	4.2

3.4. Boundary and Initial Conditions

The seawater intrusion model is divided into the Darcy seepage field and solute transport field, so there are two sets of boundary and initial conditions.

1   There are several types of boundary and initial conditions for the transient seepage of groundwater:

(1)   Initial conditions:

$$H^*(x,\, y,\, z,\, 0) = H_0^*(x,\, y,\, z) \tag{14}$$

(2)   The first boundary condition is:

$$H^*(x,\, y,\, z,\, t) = H_1^*,\, (x,\, y,\, z) \in \Gamma_1 \tag{15}$$

(3)   The second boundary condition is:

$$q_\Gamma = q^*(x,\, y,\, z,\, t),\, (x,\, y,\, z) \in \Gamma_2 \tag{16}$$

where H^* is the hydraulic head (m), q^* is the quantity of flow, and Γ is the boundary.

The boundary around the model and the top surface covered by seawater are affected by tidal fluctuation, the pressure value is $P_1 = \rho_w g\,(H_t - z)$. The tidal monitoring data given in this paper are within 1 year. In order to simulate the seawater intrusion of the research area in 50 years, we repeated the one-year monitoring data 50 times. The rest of the top surface of the model is set as the pressure boundary. According to the initial groundwater depth in the research area, the value is $P_2 = \rho_w g\,(z_0 - 13.3 - z)$, z_0 is the ground elevation over the position. The bottom of the model is set as the second boundary condition, as a no-flow boundary. The underground cavern is set as the zero-pressure boundary.

2   The boundary conditions of the solute transport field mainly include the following:

(1)   Initial conditions:

$$C_\Gamma(x,\, y,\, z,\, 0) = C_0(x,\, y,\, z) \tag{17}$$

(2)   The first boundary condition, also known as the Dirichlet boundary condition, where the solute concentration is known:

$$C_\Gamma(x,\, y,\, z,\, t) = C_1(x,\, y,\, z,\, t) \tag{18}$$

(3)   The second boundary condition, also called Neumann boundary condition, which describes the change rate of the solute concentration in normal direction at the boundary:

$$-n\boldsymbol{D}\nabla C_w = q_1(x,\ y,\ z,\ t) \tag{19}$$

(4)   Input boundary condition. Under the effect of groundwater flow and hydrodynamic dispersion, the solute flux across the boundary is known:

$$-n\boldsymbol{D}\nabla C_w + C_w u = q_2(x,\ y,\ z,\ t) \tag{20}$$

where q is the flux through the boundary ($\mathrm{kg \cdot m^{-2} \cdot s^{-1}}$).

According to the analysis of the borehole water samples, it can be seen that there is no obvious seawater intrusion under the initial condition, so the initial concentration of chloride ion in the model is set to 0. The boundary around the model and the top surface covered by seawater are set as the first boundary, according to the hydrogeological conditions of the research area, the chloride ion concentration is 0.31 mol/kg. The underground cavern is set as the outflow boundary, that is, chloride ions can flow into the cavern with the groundwater. The rest of the top surface is set as the open boundary, the chloride ions concentration outside the surface is 0. The bottom of the model is set as a no-flux boundary.

4. Results

4.1. Validation of the Numerical Model

The model is established according to the terrain of the research area. Numerical model parameters are obtained by field monitoring and testing. Boundary conditions are selected according to hydrogeological conditions of the research area. In this paper, the model is verified by the variation of the groundwater level with tidal fluctuation. The amplitude attenuation and phase lag of the groundwater table are observed in the process of tidal fluctuation propagation to the island [28,44,45]. In the un-excavated condition, the groundwater table fluctuation is increasingly damped with the increase of the horizontal distance from the coast. The attenuation is faster in the early stage and slower in the later stage (as shown in Figure 7a). At different locations within the island, the fluctuation phase of the groundwater table is also different. As the distance from the coast increases, the fluctuation phase of the groundwater table changes (as shown in Figure 7b). It can be concluded that the model we have established is reliable.

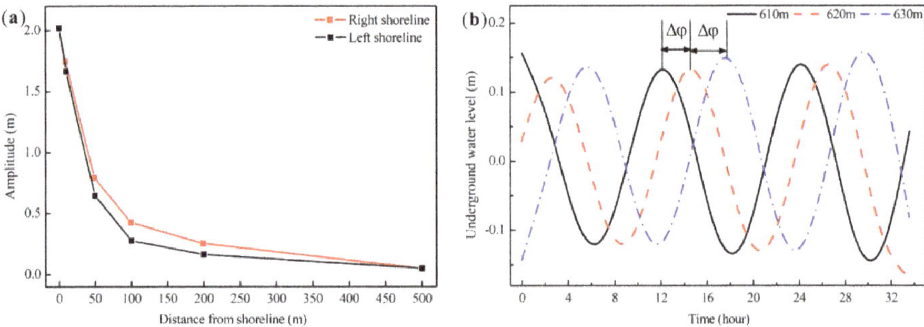

Figure 7. The amplitude attenuation and phase lag of the groundwater table in the un-excavated condition: (**a**) the fluctuation amplitude; (**b**) the fluctuation phase, $\Delta\varphi$ phase difference.

4.2. Characteristics of Groundwater Seepage

4.2.1. Groundwater Seepage Direction

The seepage direction of the groundwater on profile 1-1' was analyzed. Figure 8 shows the groundwater seepage direction at the maximum tidal water level. Before the excavation of the underground cavern (as shown in Figure 8a), groundwater seepage from the island to the coast occurred under the natural recharge effect. After the excavation of the underground cavern (as shown in Figure 8b), the groundwater in the surrounding rocks seeped into the cavern. On the coast, closer to the underground cavern, the water begins to flow into the island. However, the direction of groundwater seepage is not changed significantly on the coast far from the underground cavern. In the case of setting a horizontal water curtain system (as shown in Figure 8c), the seepage field between the underground cavern and the water curtain hole is supplied by the horizontal water curtain system. Compared with the condition in which no water curtain system is present, the direction of groundwater seepage has not changed significantly. The vertical water curtain system (as shown in Figure 8d) obviously changes the seepage direction of the groundwater around the cavern, that is, the vertical water curtain system greatly changes the seepage field in the underground cavern's surrounding rock.

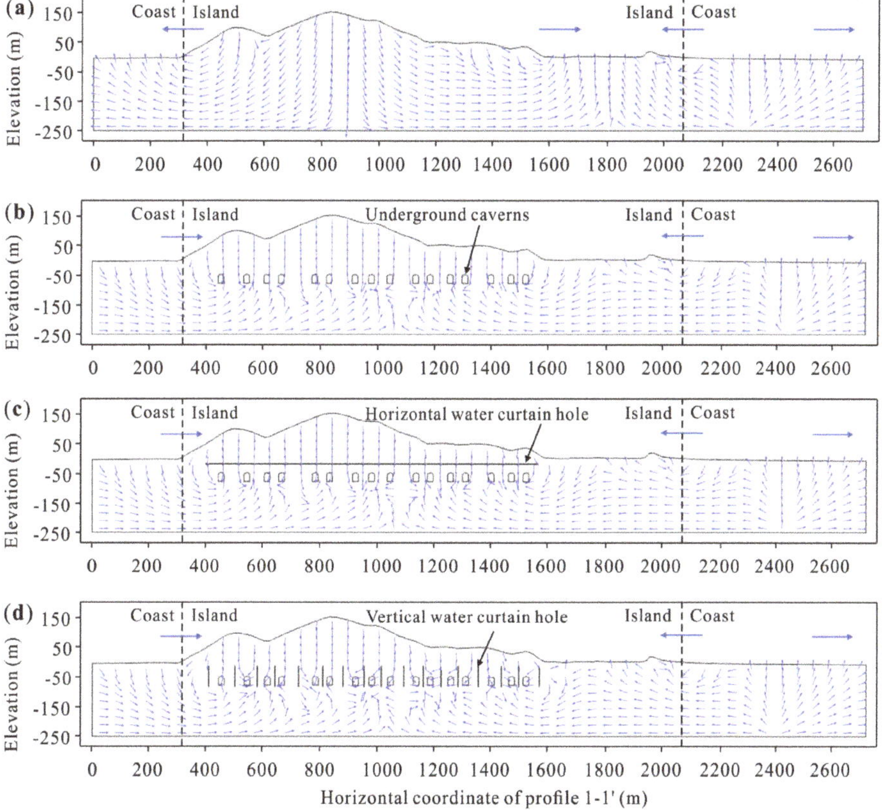

Figure 8. Groundwater seepage direction: (**a**) un-excavated; (**b**) excavated; (**c**) horizontal water curtain system; (**d**) vertical water curtain system.

4.2.2. Groundwater Seepage Velocity

Six horizontal measuring lines were arranged at different elevations (−20 m, −40 m, −60 m, −80 m, −100 m and −120 m) of profile 1-1'. Figure 9 shows the groundwater seepage velocity at the maximum tidal water level. The seepage velocities of the groundwater on the lines were analyzed. Before the excavation of the underground caverns, due to the influence of the terrain, the groundwater seepage velocity inside the island is large, while the groundwater seepage velocity outside the shoreline is relatively small. After the excavation of the underground caverns, the groundwater seepage velocity increases inside the island, especially in the cavern's area. However, the groundwater seepage velocity outside the shoreline is relatively stable. After setting up the horizontal and vertical water curtain system, the groundwater seepage velocity in the cavern area is further increased as a result of the water curtain system recharge. Within the elevation of 40 m to 80 m, the groundwater seepage velocity of the vertical water curtain system condition is higher than that of the un-excavated condition, the excavated condition and the horizontal water curtain system condition. In the vertical direction, the closer to the storage cavern, the larger fluctuation of the seepage velocity.

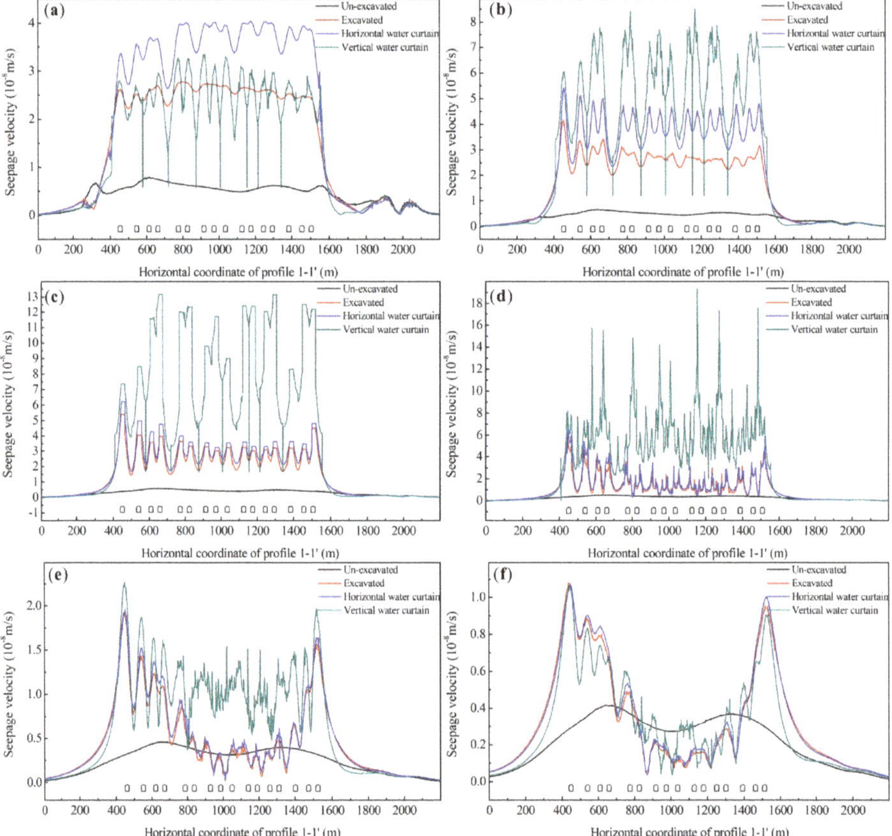

Figure 9. Groundwater seepage velocity: (**a**) elevation of −20 m; (**b**) elevation of −40 m; (**c**) elevation of −60 m; (**d**) elevation of −80 m; (**e**) elevation of −100 m; (**f**) elevation of −120 m.

At the elevation of −20 m (as shown in Figure 9a), the groundwater seepage velocities of the 4 conditions are less than 4×10^{-8} m/s. Above the underground caverns, the groundwater seepage velocity of the horizontal water curtain model is the highest. At the elevation of −40 m (as shown in

Figure 9b), the groundwater seepage velocities of the four conditions are less than 8.5×10^{-8} m/s. Compared with the elevation of -20 m, the groundwater seepage velocity under the vertical water curtain system condition increases the most. At the elevation of -60 m (as shown in Figure 9c), the measuring line passes through the underground caverns. The groundwater seepage velocities of the four conditions are less than 13.2×10^{-8} m/s. The seepage velocity of the groundwater in the location of the underground cavern fluctuates sharply, and the local maximum value of the groundwater seepage velocity is obtained at the location of the underground cavern. Compared with the other 3 conditions, the groundwater seepage velocity of the vertical water curtain system increased significantly. At the elevation of -80 m (as shown in Figure 9d), the measuring line is located below the underground caverns. The groundwater seepage velocity of the un-excavated condition, the excavated condition and the horizontal water curtain system condition decrease slightly compared with -60 m. On the contrary, under the influence of the vertical water curtain, the groundwater seepage velocity increases with a maximum value of 19.3×10^{-8} m/s. Figure 9e,f show the groundwater seepage velocity on the measuring line at the elevation of -100 m and -120 m respectively. The maximum groundwater seepage velocities decrease to 2.27×10^{-8} m/s and 1.1×10^{-8} m/s respectively. And at the elevation of -120 m, the groundwater seepage velocity of the un-excavated condition is higher than the other three conditions between the 4# cavern and the 14# cavern.

Take the excavated condition as an example. The monitoring points of the groundwater seepage velocity were set on profile 1-1′ at distances of 10 m, 50 m, 100 m, 200 m and 300 m from the shoreline. Under the influence of tidal fluctuation, the groundwater seepage velocity will also fluctuate (as shown in Figure 10a). As the distance from the shoreline increases, the fluctuation range of the groundwater seepage velocity decreases. At the same time, the phase of fluctuation at different positions is also different. By comparing the maximum amplitude of the groundwater seepage velocity under four different conditions, it is found that the fluctuation amplitude of the groundwater seepage velocity decreases exponentially with the increasing distance from shoreline (as shown in Figure 10b). At 200 m from the shoreline, the fluctuation amplitude of the groundwater seepage velocity decreases to 4.3% of the shoreline.

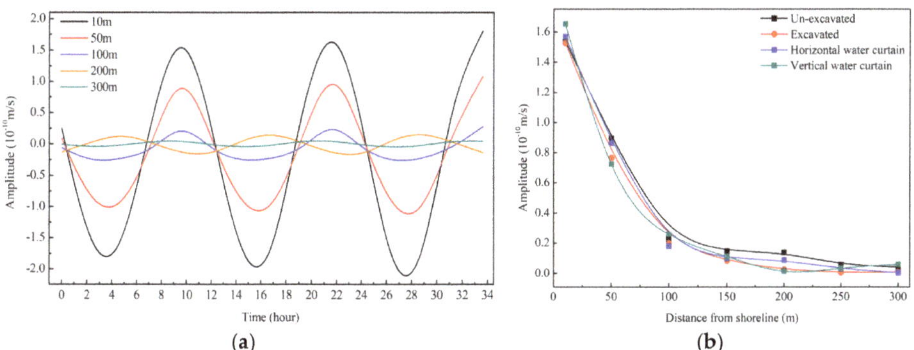

Figure 10. Seepage velocity fluctuation amplitude: (**a**) different distance from shoreline under excavated condition; (**b**) fluctuation amplitude damping.

4.2.3. Groundwater Level

Groundwater level is one of the driving factors of seawater intrusion. The distribution of the groundwater level under four conditions (un-excavated, excavated, horizontal water curtain system, vertical water curtain system) were compared. Taking the maximum tidal water level as the boundary condition, the distribution of the groundwater level under the steady state is calculated (as shown in Figure 11). The variation of the groundwater level is affected by the groundwater seepage direction and the groundwater seepage velocity. Before the underground cavern is excavated (as shown in

Figure 11a), the groundwater level is slightly lower than the initial groundwater level. After the underground oil storage cavern is excavated (as shown in Figure 11b), the groundwater level drops sharply within the range of the cavern due to the flow of groundwater from surrounding rocks into the cavern, and a large cone of depression can be formed above the underground cavern. After the horizontal water curtain system is set (as shown in Figure 11c), due to the recharge of the horizontal water curtain system, the groundwater level above the cavern rises. However, the pore water pressure under the cavern shows little change when compared with the excavated condition. After the vertical water curtain system is set (as shown in Figure 11d), the groundwater level above the underground cavern is high, which is similar to the condition where the horizontal water curtain system is set. The pore water pressure under the cavern is higher than the excavated condition, but it is still significantly smaller than the un-excavated condition.

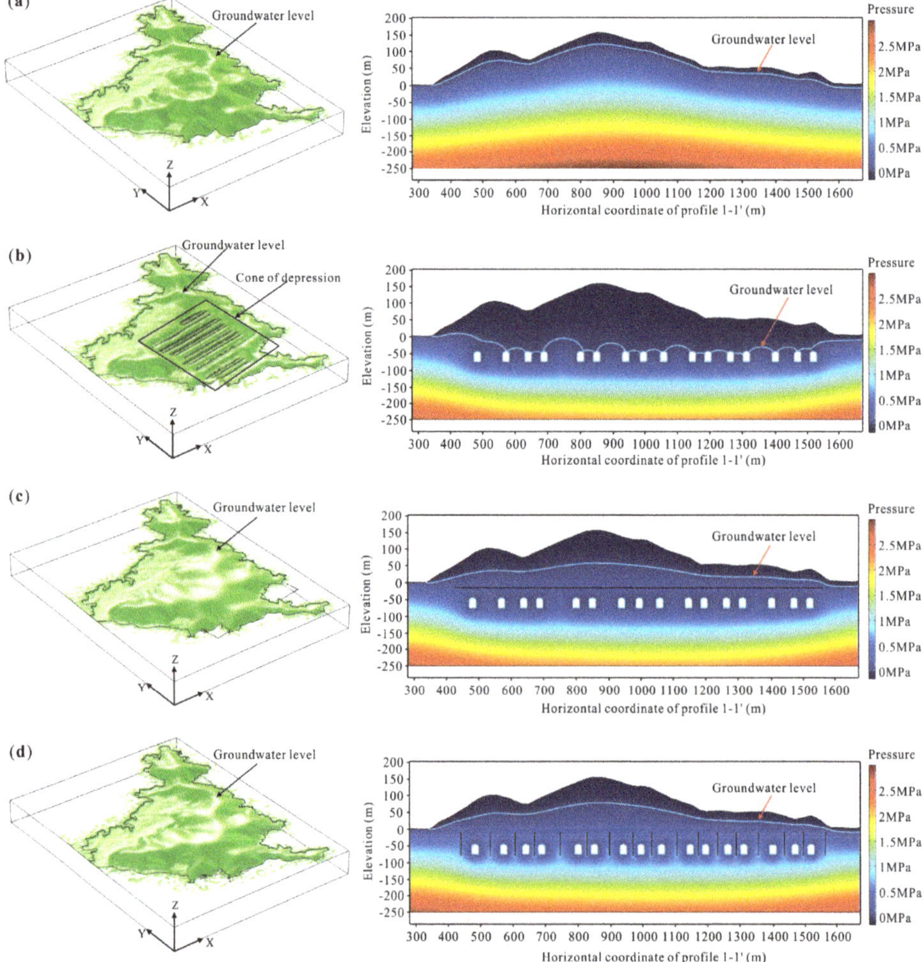

Figure 11. Groundwater level: (**a**) un-excavated; (**b**) excavated; (**c**) horizontal water curtain system; (**d**) vertical water curtain system.

Under the influence of tidal fluctuation, the groundwater level of the island fluctuates regularly. The tides fluctuate sinusoidally. At different distances from the shoreline, the fluctuating phase of the groundwater table varied (as shown in Figure 12). With the increase of the distance from the shoreline, the fluctuating phase of the groundwater table lags. From shoreline to the island, the maximum amplitude of groundwater fluctuation decreases exponentially (as shown in Figure 13). At 200 m from the shoreline, the maximum fluctuation amplitude of the groundwater table decreases to 9.1% of the shoreline. At the same position, the difference in the groundwater fluctuation amplitude under the four conditions is small, and the maximum amplitude ranges from large to small are excavated, un-excavated, the horizontal water curtain system and the vertical water curtain system.

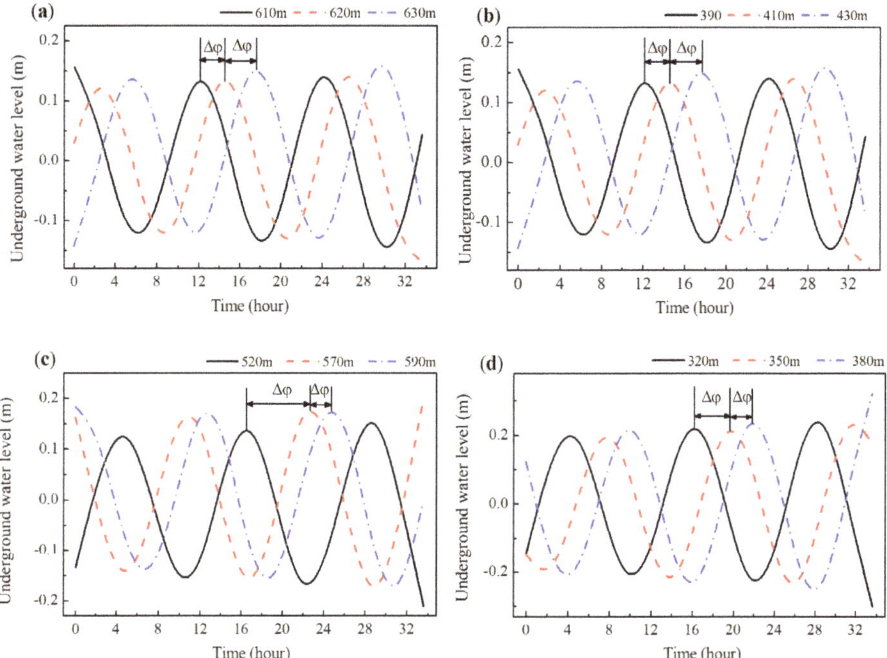

Figure 12. Groundwater level fluctuation with different distance from shoreline: (**a**) un-excavated; (**b**) excavated; (**c**) horizontal water curtain; (**d**) vertical water curtain; $\Delta\varphi$ phase difference.

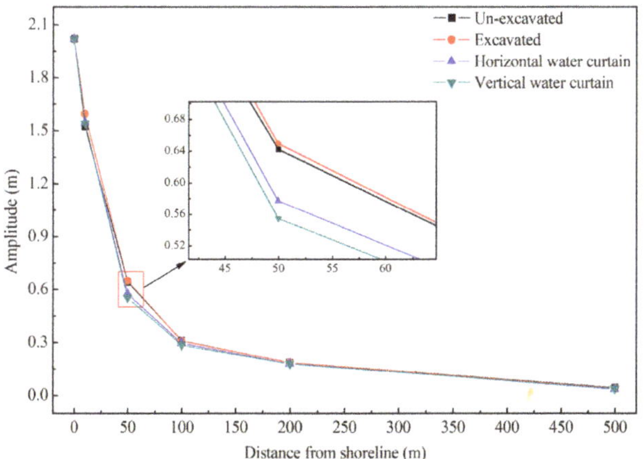

Figure 13. Tidal water table fluctuation amplitude damping.

4.3. Characteristics of Seawater Intrusion

4.3.1. Temporal-Spatial Variations of Seawater Intrusion

After the underground cavern is excavated, the seawater will gradually intrude into the island. The temporal-spatial variations of seawater intrusion in 50 years after underground cavern excavation were simulated. The simulation results are shown in Figure 14. The temporal-spatial variations of seawater intrusion are different at different locations within the island. Closer to the underground cavern, the seawater intrusion velocity is higher. In terms of this project, the seawater intrusion on the right side and the front of the island develops rapidly. This characteristic is related to the distribution of groundwater seepage velocity in the island. Because the density of seawater is greater than that of freshwater, seawater will intrude into the interior of the underground cavern from below.

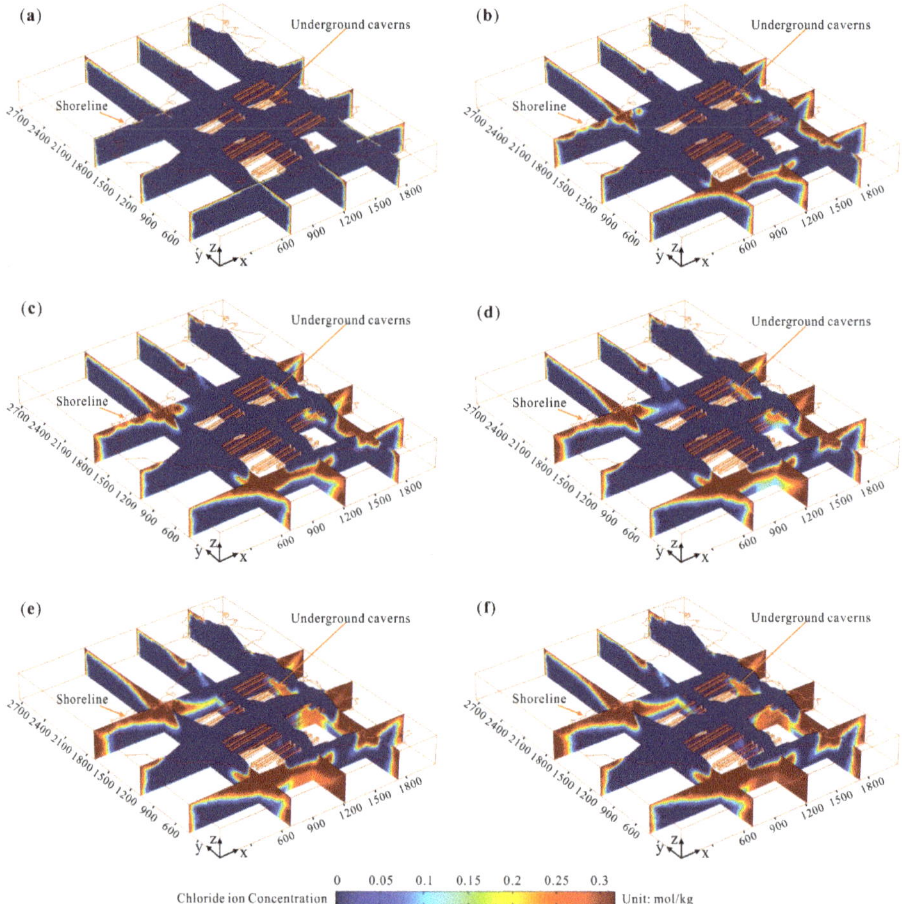

Figure 14. Temporal-spatial variations of seawater intrusion: (**a**) 0 year; (**b**) 10 years; (**c**) 20 years; (**d**) 30 years; (**e**) 40 years; (**f**) 50 years.

4.3.2. Influence of Water Curtain System on Seawater Intrusion

The seawater intrusion on the left side of profile 1-1′ in the horizontal direction was compared in the four tested conditions (as shown in Figure 15). Before the excavation of the underground cavern, the seawater intrusion velocity is quite low and has no tendency to increase. After the excavation of the underground cavern, the seawater intrusion velocity gradually increases with time. After the horizontal water curtain system is set, the rate of seawater intrusion in the early stage changes little. However, within the horizontal water curtain range, the seawater intrusion velocity has significantly reduced. After the vertical water curtain system is set, the seawater intrusion velocity decreases significantly during the entire simulation time.

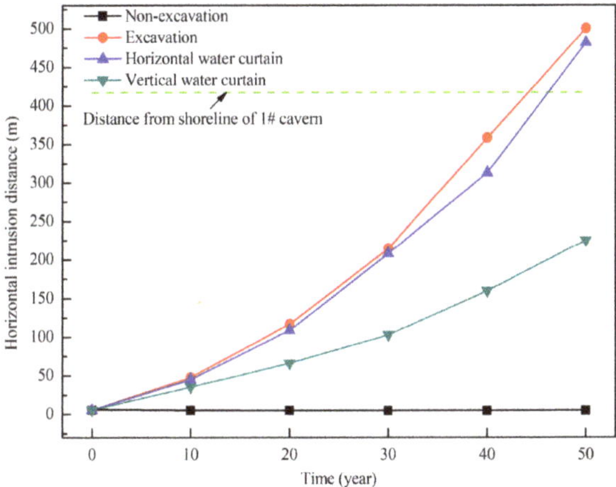

Figure 15. Horizontal distance seawater intrusion.

The seawater intrusion at the end of the simulation (50 years) was compared and analyzed. Before the excavation of the underground cavern (as shown in Figure 16a), the seawater will hardly intrude into the island. After the underground cavern is excavated (as shown in Figure 16b), the groundwater seepage field insides the island will be disturbed, and the groundwater will seep from the surrounding rock to the underground cavern. A large extent of seawater intrusion has occurred around the caverns. The underground cavern is closer to the coast, and the seawater intrusion is obviously greater. In addition, seawater intrudes into the underground cavern from the bottom of the cavern. After the horizontal water curtain system is set above the underground cavern (as shown in Figure 16c), the degree of seawater intrusion between the cavern and the horizontal water curtain system decreases. However, there is no significant change in the seawater intrusion outside the range of the underground cavern and horizontal water curtain system. After the vertical water curtain system is set (as shown in Figure 16d), the degree of seawater intrusion around the cavern and the vertical water curtain system is reduced compared to that of the excavated condition.

4.3.3. Tidal Influence on Seawater Intrusion

According to the above analysis, after the underground cavern is excavated, the seawater will gradually intrude from the shoreline to the interior of the island regardless of whether the water curtain system is set or not. Under the four conditions, concentration monitoring points are set at an elevation of −100 m, 10 m away from the shoreline on profile 1-1′. The simulation results of the four conditions are shown in Figure 17a. Before the excavation of the underground cavern, the concentration of chlorine ions in the groundwater decreases. After the excavation of the underground cavern, the chloride ion concentration gradually increases. In the case where a horizontal water curtain system was set, the variation in the chloride ion concentration is similar to the excavated condition, and the concentration of chloride ion at this position is slightly less than the excavated condition at the same time. After the vertical water curtain system is set, the chloride ion concentration gradually increases, but the rate of increase is obviously lower than the excavated condition and the horizontal water curtain system condition.

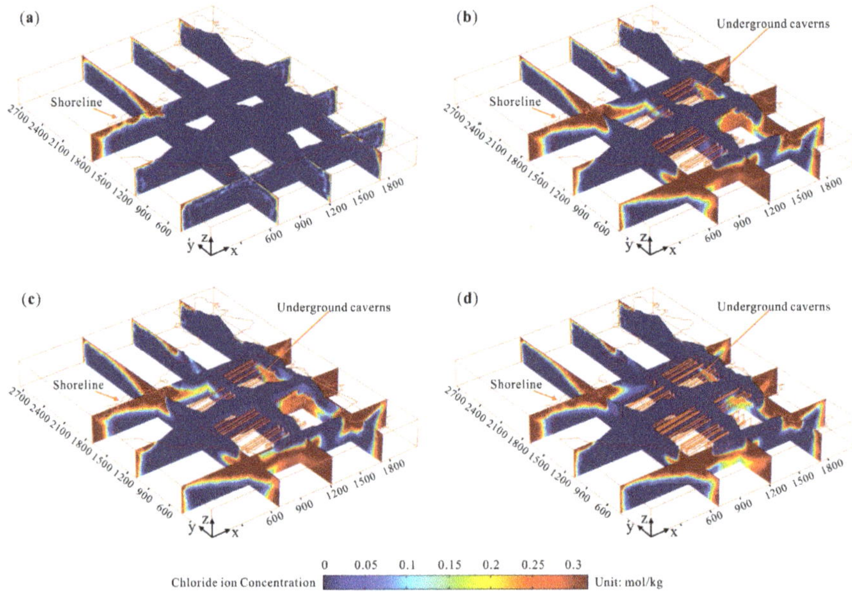

Figure 16. Contrast of seawater intrusion in fifty years: (**a**) un-excavated; (**b**) excavated; (**c**) horizontal water curtain system; (**d**) vertical water curtain system.

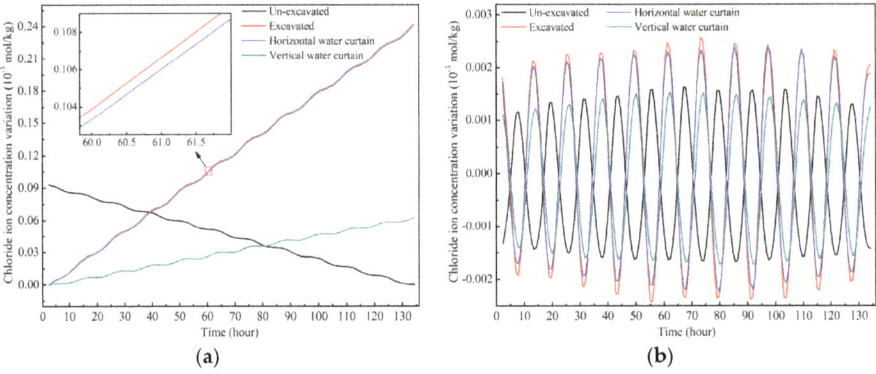

Figure 17. Chloride concentration curve: (**a**) chloride concentration varies over time; (**b**) chloride concentration fluctuation.

In the process of seawater intrusion, the chloride ion concentration will gradually increase, but be affected by the tidal dynamic fluctuation, and the chloride ion concentration will also fluctuate slightly. To research its fluctuation characteristics, the curve of the chloride ion concentration was filtered (as shown in Figure 17b). The maximum fluctuation amplitude of the chlorine ion concentration of the excavated condition and the horizontal water curtain system condition is approximately 0.0025 mol/m^3. The maximum fluctuation amplitude of the chlorine ion concentration of the un-excavated condition and the vertical water curtain system condition is approximately 0.0015 mol/m^3. Therefore, the excavation of the underground cavern will increase the influence of tidal fluctuation on seawater intrusion, and a vertical water curtain system can better decrease the influence of the tidal fluctuation on seawater intrusion than horizontal water curtain system.

5. Discussion

Based on the above analysis, it can be seen that the tidal fluctuation causes the groundwater in the island to show fluctuation characteristics. The water curtain system can reduce the seawater intrusion and weaken the influence of tidal fluctuation on seawater intrusion. Characteristics analysis is as follows:

(1)   Under the island tidal environment, the groundwater level within the island is affected by tidal fluctuations, which show fluctuation characteristics. When the underground water-sealed oil storage cavern is built in the island, the fluctuation amplitude damping rate of the groundwater level can be successively sorted from low to high: excavated, un-excavated, with a horizontal water curtain system and with a vertical water curtain system (as shown in Figure 13). The tidal wave is transmitted from the shoreline to the island, and the wave energy decreases continually in the process of transmission, so the amplitude damping is generated. After the underground cavern is excavated, the groundwater seeps into the underground cavern, which accelerates the velocity of groundwater flow. Therefore, the amplitude damping velocity of the groundwater level is smaller than that of the un-excavated condition. After the water curtain system is set, the water curtain system recharges the groundwater, and the groundwater seepage around the island is restrained, so the amplitude of the groundwater level damps faster than that without the water curtain system. The amplitude damping velocity under the influence of a vertical water curtain system is greater than that of a horizontal water curtain system, which indicates that the vertical water curtain system has a greater barrier effect on the seepage flow of groundwater into the island. This phenomenon reveals the advantages of vertical water curtain systems in the construction of underground water-sealed oil storage caverns in island tidal environments.

(2)   Before the excavation of the underground cavern (as shown in Figure 17a), the chloride ion concentration at the monitoring point shows a decreasing trend. According to the governing equation, seawater intrusion is affected by the groundwater flow and the hydrodynamic dispersion of solute ions. In the case of an un-excavated island, groundwater seeps from the island to the shoreline (as shown in Figure 8). In contrast, solute diffusion is affected by its concentration, from high concentration to low concentration, that is, it diffuses from the shoreline to the island interior. It can be seen from the results that at the location of the monitoring point, the influence of the groundwater flow on seawater intrusion is greater than that of hydrodynamic dispersion.

This paper is a preliminary study on the seawater intrusion in the construction of the underground water-sealed oil storage cavern under the island environment. There are limitations such as limited field-monitoring data. The model built in this research covers the whole island and has a larger size. Therefore, the equivalent hydraulic conductivity method is adopted to study the rock fracture mass as an isotropic medium with a larger hydraulic conductivity. However, in practice, the rock mass of the research area contains complex fracture networks. The groundwater seepage and seawater intrusion also mainly develop along the penetrating fissures [46,47]. In the following research, the fracture should be added into the model for more accurate simulation.

6. Conclusions

Based on the multi-physical field coupling theory, this paper researched the seawater intrusion characteristics of the underground water-sealed oil storage cavern under an island tidal environment using the transient simulation method. Some conclusions were obtained.

Seawater intrusion is mainly controlled by groundwater seepage characteristics (groundwater seepage direction, groundwater seepage velocity and groundwater level). The direction of groundwater seepage is the dominant factor of the direction of seawater intrusion. The influence of hydrodynamic dispersion on the direction of seawater intrusion is small. The velocity of groundwater seepage determines the velocity of seawater intrusion.

The amplitudes of groundwater seepage speed and groundwater level decrease exponentially along the horizontal direction, so the influence of tide on the groundwater seepage field in the island is limited. At 200 m from the shoreline, the fluctuation amplitude of the groundwater seepage velocity and the groundwater level decrease to 4.3% and 9.1% of the shoreline, respectively.

The water curtain system can reduce seawater intrusion caused by underground water-sealed oil storage cavern excavation. In addition, the water curtain system can also reduce the influence of tidal fluctuations on the underground seepage field of the island. In both of these aspects, the effect of a vertical water curtain system is better than that of a horizontal water curtain system.

Author Contributions: This paper was completed under the guidance of B.Z. Y.L. established the numerical model, analyzed the results and completed the writing of most of the manuscripts. L.S. contributed to the concept, thought and organization of the study. Writing—review and editing, Y.Y.

Funding: This research was funded by the National Natural Science Foundation of China (No. 41572301, No. 40902086) and the Fundamental Research Funds for the Central Universities of China (No. 2018CDJSK04XK09, No. 2-9-2017-089).

Acknowledgments: Thanks are due to senior engineer Zhenhua Peng and Junyan Li of Petrochemical Engineering Co., Ltd. for supplying geological data of research area.

Conflicts of Interest: The authors declare no conflict of interest.

References

1. Morfeldt, C.O. Storage of petroleum products in man-made caverns in Sweden. *Bull. Eng. Geol. Environ.* **1983**, *28*, 17–30. [CrossRef]
2. Lu, M. Finite element analysis of a pilot gas storage in rock cavern under high pressure. *Eng. Geol.* **1998**, *49*, 353–361. [CrossRef]
3. Makita, T.; Miyanaga, Y.; Iguchi, K.; Hatano, T. Underground oil storage facilities in japan. *Eng. Geol.* **1993**, *35*, 191–198. [CrossRef]
4. Li, Y.; Chen, Y.F.; Zhang, G.J.; Liu, Y.; Zhou, C.B. A numerical procedure for modeling the seepage field of water-sealed underground oil and gas storage caverns. *Tunn. Undergr. Space Technol.* **2017**, *66*, 56–63. [CrossRef]
5. Shi, L.; Zhang, B.; Wang, L.; Wang, H.X.; Zhang, H.J. Functional efficiency assessment of the water curtain system in an underground water-sealed oil storage cavern based on time-series monitoring data. *Eng. Geol.* **2018**, *239*, 79–95. [CrossRef]
6. Xue, Y.G.; Li, S.C.; Qiu, D.H.; Wang, Z.C.; Li, Z.Q.; Tian, H.; Su, M.X.; Yang, W.M.; Lin, C.J.; Zhu, J.Y. A new evaluation method for site selection of large underground water-sealed petroleum storage depots. *Sci. China Technol. Sci.* **2015**, *58*, 967–978. [CrossRef]
7. Lim, J.W.; Lee, E.; Moon, H.S.; Lee, K.K. Integrated investigation of seawater intrusion around oil storage caverns in a coastal fractured aquifer using hydrogeochemical and isotopic data. *J. Hydrol.* **2013**, *486*, 202–210. [CrossRef]
8. Lee, J.Y.; Cho, B.W. Submarine groundwater discharge into the coast revealed by water chemistry of man-made undersea liquefied petroleum gas cavern. *J. Hydrol.* **2008**, *360*, 195–206. [CrossRef]
9. Huyakorn, P.S.; Andersen, P.F.; Mercer, J.W.; White, H.O. Saltwater intrusion in aquifers: Development and testing of a three-dimensional finite element model. *Water Resour. Res.* **1987**, *23*, 293–312. [CrossRef]
10. Lee, J.; Kim, J.H.; Kim, H.M.; Chang, H.W. Statistical approach to determine the salinized ground water flow path and hydrogeochemical features around the underground LPG cavern, Korea. *Hydrol. Process.* **2010**, *21*, 3615–3626. [CrossRef]
11. Mas-Pla, J.; Rodríguez-Florit, A.; Zamorano, M.; Roqué, C.; Menció, A.; Brusi, D. Anticipating the effects of groundwater withdrawal on seawater intrusion and soil settlement in urban coastal areas. *Hydrol. Process.* **2013**, *27*, 2352–2366. [CrossRef]
12. Werner, A.D.; Bakker, M.; Post, V.E.A.; Vandenbohede, A.; Lu, C.H.; Ataie-Ashtiani, B.; Simmons, C.T.; Barry, D.A. Seawater intrusion processes, investigation and management: Recent advances and future challenges. *Adv. Water Resour.* **2013**, *51*, 3–26. [CrossRef]

13. Boschetti, T.; Awaleh, M.O.; Barbieri, M. Waters from the Djiboutian Afar: A Review of Strontium Isotopic Composition and a Comparison with Ethiopian Waters and Red Sea Brines. *Water* **2018**, *10*, 1700. [CrossRef]
14. Sreekanth, J.; Datta, B. Review: Simulation-optimization models for the management and monitoring of coastal aquifers. *Hydrogeol. J.* **2015**, *23*, 1155–1166. [CrossRef]
15. Nocchi, M.; Salleolini, M. Prediction of pollutant remediation in a heterogeneous aquifer in Israel: Reducing uncertainty by incorporating lithological, head and concentration data. *J. Hydrol.* **2018**, *564*, 651–666. [CrossRef]
16. Neuman, S.P. Trends, prospects and challenges in quantifying flow and transport through fractured rocks. *Hydrogeol. J.* **2005**, *13*, 124–147. [CrossRef]
17. Sebben, M.L.; Werner, A.D.; Graf, T. Seawater intrusion in fractured coastal aquifers: A preliminary numerical investigation using a fractured Henry problem. *Adv. Water Resour.* **2015**, *85*, 93–108. [CrossRef]
18. Arfib, B.; Marsily, G.D. Modeling the salinity of an inland coastal brackish karstic spring with a conduit-matrix model. *Water Resour. Res.* **2004**, *40*, 417–427. [CrossRef]
19. Liu, W.C.; Liu, H.M. Assessing the Impacts of Sea Level Rise on Salinity Intrusion and Transport Time Scales in a Tidal Estuary, Taiwan. *Water* **2014**, *6*, 324–344. [CrossRef]
20. Petti, M.; Bosa, S.; Pascolo, S. Lagoon Sediment Dynamics: A Coupled Model to Study a Medium-Term Silting of Tidal Channels. *Water* **2018**, *10*, 569. [CrossRef]
21. Oberle, F.K.J.; Swarzenski, P.W.; Storlazzi, C.D. Atoll Groundwater Movement and Its Response to Climatic and Sea-Level Fluctuations. *Water* **2017**, *9*, 650. [CrossRef]
22. Taniguchi, M. Tidal effects on submarine groundwater discharge into the ocean. *Geophys. Res. Lett.* **2002**, *29*. [CrossRef]
23. Li, L.; Barry, D.A.; Parlange, J.Y.; Pattiaratchi, C.B. Beach water table fluctuations due to wave run-up: Capillarity effects. *Water Resour. Res.* **1997**, *33*, 817–824. [CrossRef]
24. Jeng, D.S.; Li, L.; Barry, D.A. Analytical solution for tidal propagation in a coupled semi-confined/phreatic coastal aquifer. *Adv. Water Resour.* **2002**, *25*, 577–584. [CrossRef]
25. Teo, H.T.; Jeng, D.S.; Seymour, B.R.; Barry, D.A.; Li, L. A new analytical solution for water table fluctuations in coastal aquifers with sloping beaches. *Adv. Water Resour.* **2003**, *26*, 1239–1247. [CrossRef]
26. Huang, F.K.; Chuang, M.H.; Wang, G.S.; Yeh, H.D. Tide-induced groundwater level fluctuation in a u-shaped coastal aquifer. *J. Hydrol.* **2015**, *530*, 291–305. [CrossRef]
27. Levanon, E.; Shalev, E.; Yechieli, Y.; Gvirtzman, H. Fluctuations of fresh-saline water interface and of water table induced by sea tides in unconfined aquifers. *Adv. Water Resour.* **2016**, *96*, 34–42. [CrossRef]
28. Nielsen, P. Tidal dynamics of the water table in beaches. *Water Resour. Res.* **1990**, *26*, 2127–2134. [CrossRef]
29. Siarkos, I.; Latinopoulos, D.; Mallios, Z.; Latinopoulos, P. A methodological framework to assess the environmental and economic effects of injection barriers against seawater intrusion. *J. Environ. Manag.* **2017**, *193*, 532–540. [CrossRef]
30. Huang, P.S.; Chiu, Y.C. A simulation-optimization model for seawater intrusion management at pingtung coastal area, Taiwan. *Water* **2018**, *10*, 251. [CrossRef]
31. Lee, E.; Lim, J.W.; Moon, H.S.; Lee, K.K. Assessment of seawater intrusion into underground oil storage cavern and prediction of its sustainability. *Environ. Earth Sci.* **2015**, *73*, 1179–1190. [CrossRef]
32. Ivars, D.M. Water inflow into excavations in fractured rock—A three-dimensional hydro-mechanical numerical study. *Int. J. Rock Mech. Min.* **2006**, *43*, 705–725. [CrossRef]
33. Li, S.C.; Wang, Z.C.; Ping, Y.; Zhou, Y.; Zhang, L. Discrete element analysis of hydro-mechanical behavior of a pilot underground crude oil storage facility in granite in china. *Tunn. Undergr. Space Technol.* **2014**, *40*, 75–84. [CrossRef]
34. Wang, Z.C.; Li, S.C.; Qiao, L.P.; Zhang, Q.S. Finite element analysis of the hydro-mechanical behavior of an underground crude oil storage facility in granite subject to cyclic loading during operation. *Int. J. Rock Mech. Min.* **2015**, *73*, 70–81. [CrossRef]
35. Ren, F.; Ma, G.W.; Wang, Y.; Fan, L.F. Pipe network model for unconfined seepage analysis in fractured rock masses. *Int. J. Rock Mech. Min.* **2016**, *88*, 183–196. [CrossRef]
36. Wang, Z.C.; Li, W.; Li, S.C.; Qiu, W.G.; Ding, W.T. Development of an optimum forepole spacing (OFS) determination method for tunnelling in silty clay with a case study. *Tunn. Undergr. Space Technol.* **2018**, *74*, 20–32. [CrossRef]
37. Louis, C. *Rock Hydraulics*; Springer Vienna: New York, NY, USA, 1972; p. 59. ISBN 978-3-7091-4109-0.

38. Navigation Guarantee Department of Chinese Navy Headquarters. *Tide Table: 2009, East China Sea*; China Navigation Publications Press: Beijing, China, 2008; pp. 100–102. ISBN 978-7-80224-554-9.

39. Chen, Z.Y. *Ocean Tide*; Science Press: Beijing, China, 1979; pp. 18–24. ISBN 13031-946.

40. Chun, J.A.; Lim, C.; Kim, D.; Kim, J.S. Assessing Impacts of Climate Change and Sea-Level Rise on Seawater Intrusion in a Coastal Aquifer. *Water* **2018**, *10*, 357. [CrossRef]

41. Bear, J. Dynamics of fluids in porous media. *Eng. Geol.* **1972**, *7*, 174–175. [CrossRef]

42. Zhu, X.Y.; Xie, C.H. *Transport in Subsurface Flow*; China Architecture & Building Press: Beijing, China, 1990; pp. 214–219. ISBN 7-112-01092-6.

43. Gelhar, L.W.; Welty, C.; Rehfeldt, K.R. A critical review of data on field-scale dispersion in aquifers. *Water Resour. Res.* **1992**, *28*. [CrossRef]

44. Harrison, W.; Fang, C.S.; Wang, S.N. Groundwater flow in a sandy tidal beach: One dimensional finite element analysis. *Water Resour. Res.* **1971**, *7*, 1313–1322. [CrossRef]

45. Li, L.; Barry, D.A.; Pattiaratchi, C.B. Numerical modelling of tide-induced beach water table fluctuations. *Coast. Eng.* **1997**, *30*, 105–123. [CrossRef]

46. Liu, H.J.; Zhang, X.H.; Lu, X.B.; Liu, Q.J. Study on flow in fractured porous media using pore-fracture network modeling. *Energies* **2017**, *10*, 17. [CrossRef]

47. Xu, S.Y.; Zhang, Y.B.; Shi, H.; Wang, K.; Geng, Y.P.; Chen, J.F. Physical Simulation of Strata Failure and Its Impact on Overlying Unconsolidated Aquifer at Various Mining Depths. *Water* **2018**, *10*, 650. [CrossRef]

Article

Comparative Study of Methods for Delineating the Wellhead Protection Area in an Unconfined Coastal Aquifer

Yue Liu [1,2,*], Noam Weisbrod [2] and Alexander Yakirevich [2]

[1] School of Civil Engineering, University of Queensland, Brisbane QLD 4072, Australia
[2] Zuckerberg Institute for Water Research, The Jacob Blaustein Institutes for Desert Research, Ben-Gurion University of the Negev, Sede Boqer Campus, Midreshet Ben-Gurion 84990, Israel; weisbrod@bgu.ac.il (N.W.); alexy@bgu.ac.il (A.Y.)
* Correspondence: yue.liu5@uqconnect.edu.au

Received: 6 May 2019; Accepted: 31 May 2019; Published: 4 June 2019

Abstract: Various delineation methods, ranging from simple analytical solutions to complex numerical models, have been applied for wellhead protection area (WHPA) delineation. Numerical modeling is usually regarded as the most reliable method, but the uncertainty of input parameters has always been an obstacle. This study aims at examining the results from different WHPA delineation methods and addressing the delineation uncertainty of numerical modeling due to the uncertainty from input parameters. A comparison and uncertainty analysis were performed at two pumping sites—a single well and a wellfield consisting of eight wells in an unconfined coastal aquifer in Israel. By appointing numerical modeling as the reference method, a comparison between different methods showed that a semi-analytical method best fits the reference WHPA, and that analytical solutions produced overestimated WHPAs in unconfined aquifers as regional groundwater flow characteristics were neglected. The results from single well and wellfield indicated that interferences between wells are important for WHPA delineation, and thus, that only semi-analytical and numerical modelling are recommended for WHPA delineation at wellfields. Stochastic modeling was employed to analyze the uncertainty of numerical method, and the probabilistic distribution of WHPAs, rather a deterministic protection area, was generated with considering the uncertain input hydrogeological parameters.

Keywords: groundwater protection; wellhead protection area (WHPA); uncertainty analysis

---

1. Introduction

Worldwide, groundwater provides essential and valuable water resources for drinking water production, agricultural irrigation [1] and industrial processes [2]. Particularly in arid and semi-arid areas where ample surface water sources do not exist, groundwater is often the major, or even the sole, drinking water resource. However, human activities, due to increasing livelihood and development demands, have contributed to the deterioration of groundwater quality. Experience in the last decades has shown that once groundwater is contaminated by chemical, biological or radiological agents, it is always difficult to clean up these contaminants, and the remediation involves high costs [3]. Therefore, protecting groundwater from pollution is a major task for sustainable water management. A central prerequisite for this endeavor is the delineation of wellhead protection area (WHPA) [4].

WHPA delineation is a concept that refers to the adequate management of the areas surrounding groundwater pumping wells where contamination could potentially exist [5,6]. Once WHPA delineation has been conducted, subsequent land-use restrictions and remediation techniques can be applied. WHPA delineation is an effective means of protecting groundwater resources; hence, it is necessary and important to obtain thorough knowledge on how to delineate such an area accurately and precisely.

The very early groundwater resource protection with safety area being defined around pumping wells should be dated back to the 1950s in Europe [5], in the purpose of eliminating pathogenic bacteria and viruses from entering groundwater [7]. Later, in the United States, the Amendments to the Safe Drinking Water Act (SDWA) established the first nationwide program to protect groundwater resources [4]. A series of different delineation solutions were then proposed by the U.S. Environmental Protection Agency (EPA) [8]. These methods can be summarized into five major categories, namely, arbitrary and simplified variable shapes, analytical solutions, hydrogeological mapping, semi-analytical method, and numerical modeling [9,10]. Different methods vary greatly in their degrees of complexity and precision, as well as producing different types of protection zones such as the zone of contribution (ZOC), which is the land surface area of the aquifer volume containing all the groundwater that may eventually flow toward the pumping well, or the zone of travel (ZOT), which is defined with a given time of travel (TOT) value for the water to reach the well [8].

Traditional practices determine WHPA using the arbitrary or easily calculated analytical solutions, which ignore the temporal and spatial variability of the local hydrogeological conditions. On the other hand, applying the three-dimensional numerical model gives rise to the possibility of integrating more hydrogeologic characteristics to suit the highly anisotropic and heterogeneous settings [11,12]. The commonly used model in delineating WHPA is MODFLOW [13,14]. Coupled with MODPATH, it provides numerical solution to tracking the pathlines of the particles moving within groundwater flow in a forward or backward manner [15]; thus, the area in which particles could move to can be delineated as the WHPA. However, despite their powerful simulation efficiency, the application of numerical models for WHPA delineation requires a great deal of hydrogeological data from field investigation, through which a complete knowledge of aquifer properties has never been fully achieved. Therefore, the uncertainty of required input data has consistently been an obstacle for obtaining the numerical modeling results [16,17]. Some studies have been conducted to assess the influence of hydrogeological properties on WHPA delineation results, and hydraulic conductivity was identified as the most significant parameter in determining the dimensions of WHPA [18,19]. The principle idea of analyzing the uncertainty of WHPA simulation is to select the input parameters randomly, then run the stochastic modeling over the numerical model to delineate WHPA [6,19].

Reviewing all of the WHPA delineation methods, the simple analytical solution allows us to obtain the protection area through quick calculations, but the results are usually the least accurate, since the real local aquifer conditions are often oversimplified. Numerical modeling has been regarded as the most accurate WHPA delineation method to represent complex aquifer conditions, but the model construction requires the performance of a considerable intensive field investigation and the integration of a large number of hydrogeological parameters, necessitating careful consideration of the modeling uncertainty. Many countries (e.g., the United States, the United Kingdom, Australia, Israel) had listed different methods for defining WHPA, and most of them adopted analytical solutions to define different levels of WHPAs by using ZOTs with different travel times for the national-wide groundwater protection programs [20]. However, a detailed comparative study of using different delineation methods and the comparison between delineation results are still missing in most of the countries. From a practical point of view, the most appropriate method for WHPA delineation should simplify the groundwater flow systems as much as possible for the ease of nationwide program implementation, while still preserving the essential geological and hydrologic characteristics for effective groundwater protection [21].

Therefore, a comparison between the common types of WHPA delineation methods is necessary to ensure a proper protection zone. It is also particularly important for proposing recommendations on the choice of a WHPA delineation method based on the comparison. The major objectives of this study are: (1) to present a comparative study on different methods for delineating WHPAs to examine their performances and to reveal the essential factors that impact WHPA delineation results; (2) to address WHPA delineation uncertainty with numerical modeling by stochastic simulation; and (3) to suggest a general WHPA delineation method selection guideline for conducting groundwater protection program. This study was conducted using two pumping sites in the unconfined coastal aquifer in Israel as an example.

2. Material and Methods

2.1. Delineation Methods Selection

Various delineation methods have been introduced and been applied based on different delineation criteria: distance to the water source, drawdown range of pumping well, time of travel (TOT) of contaminant with flow, assimilative capacity of contaminant, and boundaries which control groundwater flow [8,20,22]. Given the fact that delineating WHPA based on contaminant travel time allows defining WHPA into different vulnerability levels, most of the countries incorporated different TOT values into delineation methods to legislate groundwater protection programs [20,23–26]. Therefore, in this study, TOT was selected as the delineation criterion. Five commonly used delineation methods were compared: the calculated fixed radius (CFR) method [8], the uniform flow equation method [8,27,28], the WhAEM2000 model [29], the HYBRID method [30] and the MODFLOW–MODPATH model [13,15]. All of the selected methods have been thoroughly explained in the literature, so only brief introductions are provided here. Table 1 summarized the equations for calculating the protection areas by each of the analytical methods.

Table 1. Equations of different analytical WHPA delineation methods.

Analytical Methods	Equations	Note
CFR	$r = \sqrt{\dfrac{Qt}{\pi nH + N\pi t}}$	r (L): radius of the circular WHPA, Q (L^3/T): pumping rate, t (T): time of travel, n (-): aquifer porosity, H (L): aquifer thickness, N (L/T): infiltration rate.
Uniform flow equation	$X_L = -\dfrac{Q}{2\pi Kbi}$ $Y_L = \pm\dfrac{Q}{2Kbi}$ $t_x = \dfrac{n}{Ki}\left[r_x - \left(\dfrac{Q}{2\pi Kbi}\right)\ln\left\{1 + \left(\dfrac{2\pi Kbi}{Q}\right)r_x\right\}\right]$	X_L (L): down-gradient flow boundary, Y_L (L): max. width of up-gradient zone, t_x (T): time of travel, $r_x(\pm)$ (L): distance to up-gradient boundary (+), or to down-gradient boundary (−), K (L/T): hydraulic conductivity, b (L): aquifer thickness, i (-): hydraulic gradient, n (-): aquifer porosity, Q (L^3/T): pumping rate.
HYBRID	$t_x = \dfrac{n}{Ki}\left[r_x - \left(\dfrac{Q}{2\pi Kbi}\right)\ln\left\{1 + \left(\dfrac{2\pi Kbi}{Q}\right)r_x\right\}\right]$ $\dfrac{d}{2} = \dfrac{r_x(+) + r_x(-)}{2}$ $\pi\dfrac{w}{2}\dfrac{d}{2} = \dfrac{tQ}{bn}$	w/2 (L): vertical dimension of ellipse, d/2 (L): horizontal dimension of ellipse, other parameters are the same as the parameters from uniform flow equation method.

The CFR method is based on the assumption that water pumped out from the well comes from the porous aquifer media around the well screen, which is distributed as a cylinder, and the infiltrated water is contributed by precipitation. By using CFR method, a fixed radius around the pumping well will be calculated in terms of a specified TOT value. The uniform flow equation calculates the down-gradient flow boundary X_L, the maximum width of the up-gradient zone Y_L and the distance to up-gradient boundary r_x (with regard to the specified TOT) to delineate the envelope-shaped protection zone. WhAEM2000 is a semi-analytical model developed by U.S. EPA for facilitating capture zone and protection area mapping. WhAEM2000 includes the CFR and uniform flow methods, but also has its own semi-analytical solution to model steady pumping wells and can solve the influence of hydrological boundaries. HYBRID method combines some of the analytical equations from the CFR and uniform flow equation method. By following the five steps illustrated by Paradis and Martel [30], this method is relatively straight forward. MODFLOW-MODPATH is a widely used groundwater numerical model. All of the WHPA delineation analysis was based on the data compiled from past annual average values which covered both the dry and wet seasons.

Application of the selected delineation methods requires that TOT values be specified for different dimensions of protection areas. In Israel, WHPAs are classified into three protection levels, as in many other countries. According to the regulations from the Israel Ministry of Health [31], a Level-A protection zone is a circle closely located to the pumping well with a radius of 10 m, a Level-B protection zone is delineated by a 100-day TOT, and a Level-C protection zone is delineated by a 400-day TOT. The Level-B and Level-C protection zone results from selected methods were compared, since all the Level-A zones are identical.

2.2. Reference WHPA

In order to compare the protection zones produced by different methods, a reference WHPA is necessary. We chose the WHPA delineated by MODFLOW-MODPATH as the reference WHPA because numerical modeling most completely accounts for hydrological information of the study area. Referring to the work from Paradis [21] and Miller [32] who tried to assess the differences between WHPAs, we used a comparative index C_i to quantify the difference between the reference WHPA and the WHPAs produced by other methods. C_i index represents how the WHPA from the tested method fits the WHPA from the reference method. The principle of C_i can be illustrated by Figure 1 and C_i is calculated by Equation (1):

$$C_i = \frac{(S_{CA})}{(S_{CA} + S_{NPA} + S_{OPA})} \times 100\% \tag{1}$$

where S_{CA} is the common area between the reference and tested WHPA, S_{NPA} is the non-protected area by the tested method, and S_{OPA} is the over-protected area by the tested method.

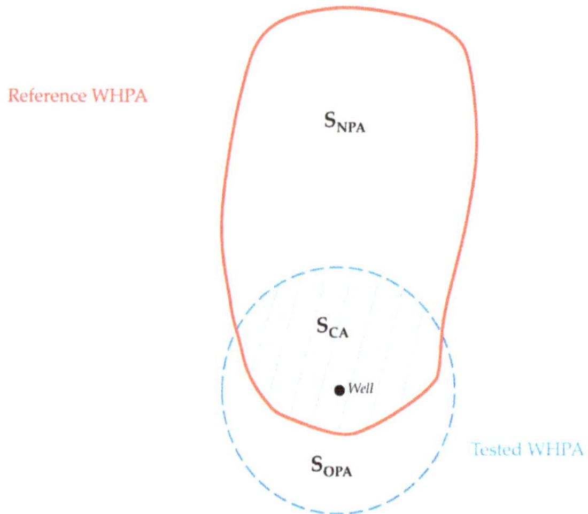

Figure 1. Illustration of comparative index C_i (modified according to Paradis et al. [21]).

According to the definition of the comparative index, a higher C_i value indicates a better fitting to the WHPA delineated by reference numerical modeling method, and vice versa.

2.3. Stochastic Modeling for Uncertainty Analysis

In order to address the impacts of uncertain input parameters of the MODFLOW-MODPATH numerical model on the results of delineation, an uncertainty analysis was conducted by Latin hypercube sampling (LHS) based stochastic modeling. An uncertainty analysis is an evaluation that takes a set of randomly chosen input values and passes them through the testing model to obtain the distribution of the outputs [33]. LHS, rather the common Monte Carlo sampling, was used to avoid clustering samples and to produce fewer iterations of random input number series. By using LHS method, the probability distribution of the sample parameter firstly being stratified into equal probability intervals on the cumulative probability scale, then a random value is taken from each interval. The number of intervals equals the number of iterations. Hydraulic conductivity (K) was chosen as the randomized parameter as it has been examined by many studies on the importance of determining the size of WHPA and its large uncertainty [6,16,34]. Other parameters, such as aquifer effective porosity, can also affect the WHPA delineation. However, the effects of its spatial variability can be smaller than that of the hydraulic conductivity [19]. By giving the minimum, mean, and maximum values of K, as well as the distribution pattern of K, a series of random K values were produced. The random K values were then introduced into the numerical WHPA delineation model for each study area to get the series of WHPAs with different probabilities.

2.4. Study Area

The study was conducted at two pumping sites in Israel: a single groundwater pumping well in the Nahariya city and a multi-well pumping field in the Rehovot area (Figure 2a). Both pumping sites are municipal drinking water supplying wells located on the unconfined coastal aquifer. The overall relief of the coastal plain aquifer increases gradually from east to west, since a series of mountains are located to the east and in the west is the Mediterranean Sea. The coastal plain aquifer of Israel can be represented by a typical cross-section in Figure 2b [35]. The coastal sediments have a thickness up to 200 m in the west and pinch-out to the east. Calcareous sandstone is the major component of the sediments, and is also the major component of the aquifer storing groundwater [36,37]. Several silt and clay aquicludes exist within the sandstone, which divide the overall aquifer into distinct sub-aquifers, significantly improving the heterogeneity of the local aquifer [38]. Due to the local topography, groundwater moves from the east and is discharged into the sea.

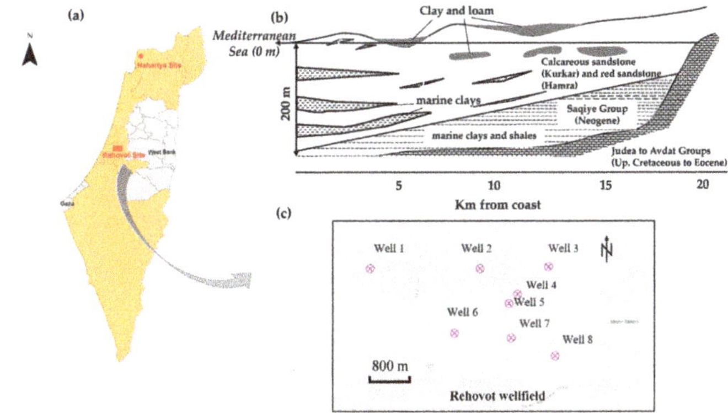

Figure 2. Location, representative aquifer cross-section of the two study areas and the well distribution map of the multi-well pumping site in Rehovot: (**a**) location of two study areas; (**b**) example hydrogeological cross-section of coastal aquifer in Israel; (**c**) distribution pattern of eight pumping wells in the Rehovot wellfield.

The aquifer properties of two sites were compiled from past studies. In the single well pumping site (the Nahariya Site), aquifer recharge due to natural precipitation is approximately 250 mm/year [39], local hydraulic gradient was estimated to be ranging between 0.002 and 0.0032, transmissivity is between 775 and 4650 m^2/day, and hydraulic conductivity is 50–300 m/day. The porosity of the supplying aquifer is estimated as 30% [40]. The Nahariya pumping well has a nearly constant pumping rate of 6600 m^3/day [41]. In the multi-well pumping field (the Rehovot Site), the local aquifer recharge rate is around 208 mm/year with a hydraulic conductivity of 10.5 m/day [35]. Eight pumping wells in the wellfield are distributed as shown in Figure 2c. The average pumping rate of Well 1 is 3000 m^3/day, with 1500 m^3/day of Well 2 and 2500 m^3/day of Well 6; the remaining wells have the same pumping rate, i.e., 1000 m^3/day.

3. Results

For the tested analytical methods (CFR method, uniform flow equation method and HYBRID method), the input parameters are the average values of the data compiled from past field records listed in Section 2. For the semi-analytical model and the reference numerical modeling method, input hydrogeological parameters may have been calibrated to fit the model.

3.1. WHPA Delineation of the Nahariya Site (Single Well)

Table 2 shows the hydrogeological parameters used by different WHPA delineation methods on the Nahariya and the Rehovot pumping sites. The application of CFR method produced circular protection zones around pumping wells (Figure 3). With a pumping rate of 6600 m^3/day and the saturated thickness of well is 15.5 m, the radii of Level-B (TOT = 100 days) and Level-C (TOT = 400 days) protection zone was 211 m and 421 m, respectively. The surface area of Level-B zone was 0.14 km^2 and the area of the Level-C zone was 0.56 km^2. Applying the uniform flow equation method, an average hydraulic gradient of 0.0025 and hydraulic conductivity of 250 m/day [41] were used. The calculated distances from well to the down-gradient boundary (X_L) and the maximum width (Y_L) were 108 m and 341 m. X_L and Y_L values were the same for both Level-B and Level-C protection zone. The size of WHPA depends on distance to the up-gradient boundary (r_x), which is the travel distance of contaminants with the up-gradient groundwater flow to enter pumping well. r_x of Level-B zone was 365 m and was 1098 m for Level-C zone. Thus, two envelope-shaped WHPAs with different dimensions that extend to the up-gradient direction were produced by uniform flow method. The WHPAs delineated by HYBRID method were represented by two ellipses with different horizontal dimensions (d/2) and vertical dimensions (w/2). Surface area of Level-B and Level-C protection zones were 0.14 km^2 and 0.55 km^2.

Table 2. Hydrogeological parameters used in different methods for the Nahariya and Rehovot pumping sites.

Method	Pumping Rate (m^3/day)	Saturated Thickness (m)	Porosity (%)	Recharge Rate (mm/year)	Conductivity (m/day)	Hydraulic Gradient
Nahariya site						
CFR	6600	15.5	30	250	-	-
Uniform flow equation	6600	15.5	30	-	250	0.0025
WhAEM2000	6600	15.5	30	250	250	-
HYBRID	6600	15.5	30	250	250	0.0025
MODFLOW-MODPATH	6600	15.5	30	250	250	-
Rehovot site						
CFR	1000–3000	4.9–18.5	30	208	-	-
WhAEM2000	1000–3000	85	30	208	10.5	-
HYBRID	1000–3000	4.9–18.5	30	208	10.5	0.0019–0.0068
MODFLOW-MODPATH	1000–3000	4.9–18.5	30	0.1–475	0.006–30	-

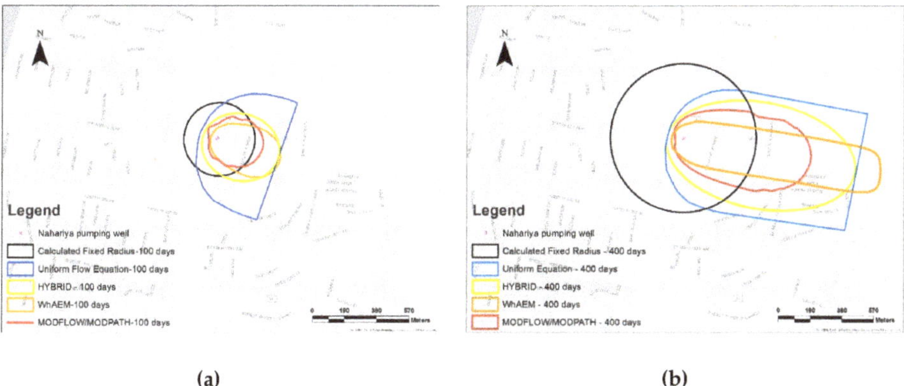

(a)   (b)

Figure 3. Level-B and Level-C WHPAs of the Nahariya pumping site delineated by different methods: (**a**) Level-B (TOT = 100 days) WHPAs; (**b**) Level-C (TOT = 400 days) WHPAs.

Since the aquifer at the Nahariya pumping well is less heterogeneous than that at the Rehovot site, and groundwater was extracted from the substantial sandstone underneath a thin loam layer on the ground surface, the hydrogeological WhAEM2000 model was constructed as a simple unconfined sandy aquifer with an average thickness of 15.5 m. The groundwater flow direction and magnitude were assigned by model calibration with the groundwater heads data from three observation wells [41]. As shown in Figure 3, a smaller Level-B WHPA with the surface area of 0.097 km^2 and a Level-C WHPA with the surface area of 0.37 km^2 were delineated.

The reference WHPAs delineation was based on the simulation conducted by the Groundwater Modeling System (GMS) 10.2 based on MODFLOW and MODPATH. A 2D single sandy layer model was constructed as shown in Figure 4. The no-flow boundary condition was set for the two boundaries parallel to the groundwater flow, and specific head boundary condition was set for the other two boundaries at the west and east. The same recharge rate of 250 mm/year was used as the analytical methods. WHPAs were outlined with regard to the simulation range of backward particle tracking. The reference Level-B protection zone was 0.0073 km^2 and the Level-C protection zone was 0.29 km^2. Reference WHPAs aligned with the groundwater flow direction.

Figure 4. 2D numerical model domain and boundary conditions of the Nahariya site: yellow lines represent the simulated groundwater heads.

3.2. WHPA Delineation of the Rehovot Site (Multi-Well Field)

The same delineation methods were applied for the Rehovot multi-well pumping site. For the application of analytical methods, the WHPA of each pumping well was calculated separately. For the WhAEM2000 and MODFLOW-MODPATH modeling methods, groundwater flow model was constructed for the whole site. Hydrogeological parameter values used for each method were shown in Table 2. The pumping rate of each well in the Rehovot pumping site is much smaller than the pumping rate of the Nahariya well. Consequently, the WHPA of each pumping well would be very small if the TOT values were the same as the Israeli regulation. In order to project all of the Level-B and Level-C WHPAs of each pumping well on the same map to compare the differences, different time scale for delineating Level-B and Level-C protection areas was used for the Rehovot Site. A new TOT of 400 days was used for the Level-B protection zone, and a TOT of 15 years was used for the Level-C protection zone.

The calculation results of the protection area lateral dimensions and surface area of the analytical methods are listed in Table 3. The results of uniform flow equation methods showed great differences between X_L, Y_L and r_x. Therefore, the WHPAs of all pumping wells delineated by the uniform equation method expanded significantly in the direction that was perpendicular to the groundwater flow direction, which already deviated from the realistic protection areas for protecting groundwater resources. Owning to this, the uniform flow equation was not used for delineating WHPAs in the Rehovot wellfield and the surface area of WHPAs were not calculated. Since WhAEM2000 is incapable of simulating the aquicludes within aquifer, a homogeneous aquifer similar to the Nahariya site was built and eight pumping wells were distributed with different saturation thickness and pumping rates.

Table 3. Calculation results of Level-B and Level-C WHPAs in the Rehovot pumping site by analytical methods.

Well Number	Analytical Methods								
	CFR		Uniform Flow Equation				HYBRID		
	r (m)	S (km²)	X_L (m)	Y_L (m)	r_x (m)	S (km²)	w/2(m)	d/2 (m)	S (km²)
				Level-B WHPA					
Well 1	397	0.49	3002	9431	415	-	398	397	0.50
Well 2	335	0.35	1173	3686	367	-	334	336	0.35
Well 3	295	0.27	462	1450	361	-	292	298	0.27
Well 4	238	0.18	374	1176	291	-	236	241	0.18
Well 5	221	0.15	323	1014	274	-	219	224	0.15
Well 6	367	0.42	707	2223	433	-	365	370	0.42
Well 7	152	0.07	160	505	203	-	148	156	0.07
Well 8	231	0.17	289	907	296	-	227	235	0.16
				Level-C WHPA					
Well 1	1467	6.76	3002	9431	1715	-	1457	1477	6.76
Well 2	1236	4.80	1173	3686	1704	-	1120	1275	4.48
Well 3	1089	3.72	462	1450	2069	-	941	1260	3.72
Well 4	880	2.43	374	1176	1670	-	761	1018	2.43
Well 5	817	2.10	323	1014	1613	-	692	965	2.10
Well 6	1356	5.77	707	2223	2330	-	1229	1497	5.78
Well 7	560	0.98	160	505	1336	-	420	748	0.99
Well 8	852	2.28	289	907	1833	-	685	1060	2.28

Note: r is the radius of the circular WHPA by CFR method.

For the numerical simulations, a three-dimensional model (Figure 5) of saturated flow and particle transport was used [42]. Considering the more complex hydrogeological conditions at the Rehovot pumping site, five major materials composing the aquifer were defined: sand, silt, sandstone, clay, and gravel as shown by the cross-section A–A' in Figure 5b. The model was calibrated to the observed groundwater levels in observation wells (grey dots in Figure 5a) from 1975–1985 [42], and it was found that a significantly improved fit was obtained when the major water-bearing sandstone was divided into two sub-materials with different hydraulic properties which were located in the western and eastern domain of the model (Figure 5b). The recharge rate (Table 2) was produced through calibration as well and varied in space. The MODPATH module was operated by releasing particles at the groundwater table of each pumping well and running backward tracking. All the WHPAs delineated by different methods can be found in Figure 6. Level-C WHPAs delineated with a longer TOT got overlapped between a few of the wells.

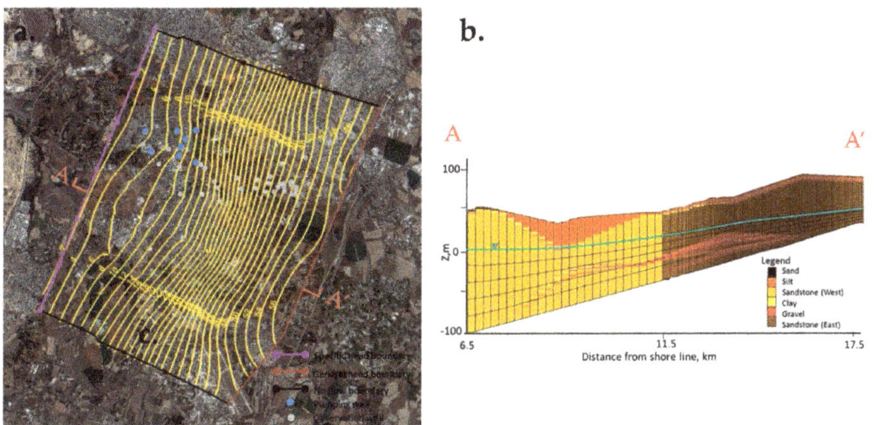

Figure 5. 3D numerical model domain of the Rehovot pumping field: (**a**) Model domain and boundary conditions; (**b**) Representative cross-section A-A′ of the model after calibration.

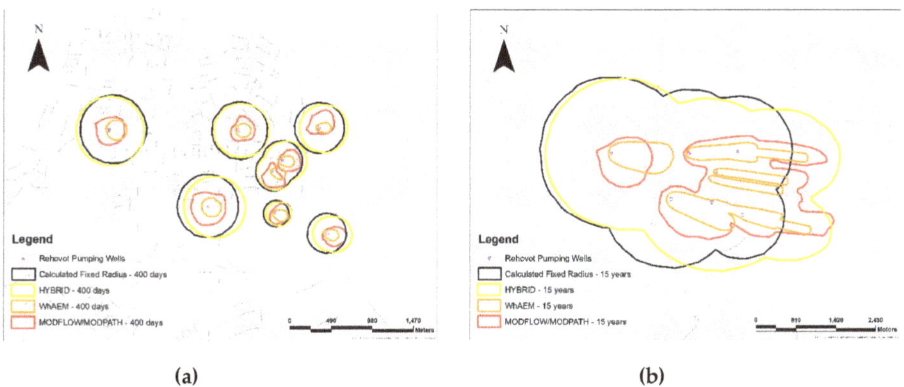

(a) (b)

Figure 6. Level-B and Level-C WHPAs of the Rehovot wellfield pumping site delineated by different methods: (**a**) Level-B (TOT = 400 days) WHPAs; (**b**) Level-C (TOT = 15 years) WHPAs.

3.3. Uncertainty Analysis Results of Numerical Modeling (MODFLOW-MODPATH) Method

Hydraulic conductivities of two pumping sites were randomized to produce a series of conductivity values. For the Nahariya pumping site, the numerical model was built as a homogeneous aquifer; thus only one hydraulic conductivity value, which represents the entire site, was randomized. For the Rehovot pumping site, the hydraulic conductivities of two major sandstone sub-aquifers were randomized. The statistical characteristics of hydraulic conductivities are listed in Table 4. For each pumping site, 64 iterations of parameter sampling were created.

Table 4. Statistical characteristics of the hydraulic conductivities to be randomly sampled at two pumping sites.

Parameters	Mean(Starting) Value	Value Range	Distribution Pattern
Nahariya Site			
K (m/day)	200	100–300	Lognormal
Rehovot Site			
K_{west} (m/day)	10.43	5.0–15.0	Lognormal
K_{east} (m/day)	7.30	3.5–10.5	Lognormal

Note: K_{west} is the hydraulic conductivity of the western part sandstone of Rehovot pumping site MODFLOW model; K_{east} is the hydraulic conductivity of the eastern part sandstone of Rehovot pumping site MODFLOW model.

The objective of the stochastic simulation was to generate the stochastic map of the possible areas to be protected and to evaluate the probability distribution of WHPAs that are caused by the uncertainty of input hydraulic conductivities. Here, only the stochastic map produced at the Rehovot multi-well pumping site is shown in Figure 7 since the Nahariya site is a single well and the stochastic modeling results can be represented by the results of the multi-well site. As shown in Figure 7, delineating WHPAs with considering uncertainty of hydraulic conductivity produced a series of protection zones with different possibilities around each pumping well. The darker color area with a higher probability represents a higher priority that this particular location should be protected, and the lighter color refers to a lower probability and a lower protection demand. From Figure 7a,b, it was found that more uncertainties of the generated protection zones, which were indicated by the wider distributed different probability zones, appeared on the southeast of the pumping site. The reason for this was that the groundwater was contributed to by the mountainous area in the southeast side. The heterogeneity of this direction was more influential for the entire groundwater system, as well as for the particle movements. Thus, more uncertainty was encountered in this direction.

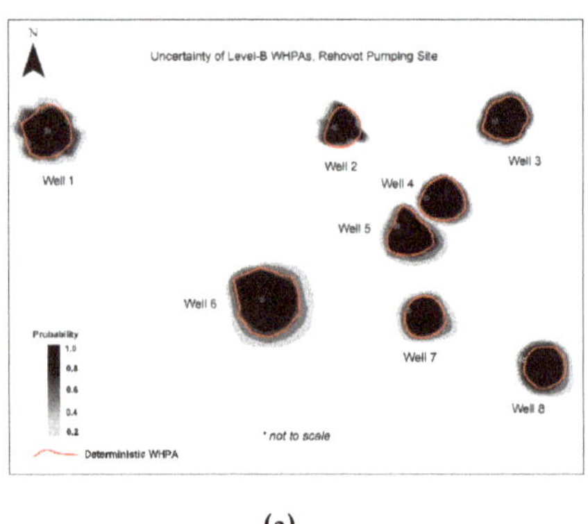

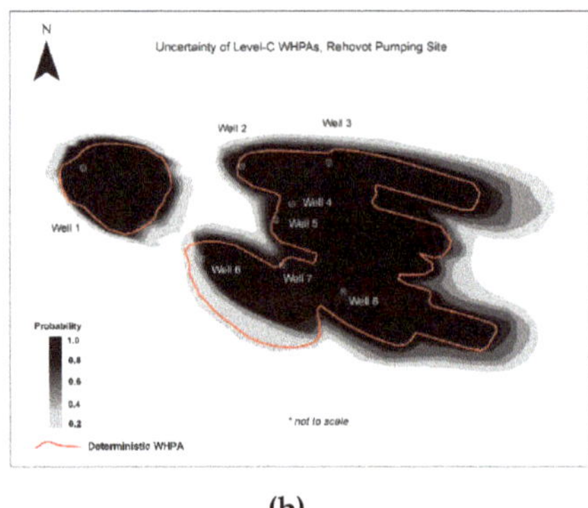

(a)   (b)

Figure 7. Probabilistic maps produced by stochastic modeling that reflects the uncertainty of WHPAs at the Rehovot wellfield pumping site: (**a**) Level-B probabilistic map; (**b**) Level-C probabilistic map. The red shapes represent the WHPAs from the normal deterministic numerical modeling method.

Comparing the uncertainty of Level-B WHPAs (Figure 7a) with the uncertainty of Level-C WHPAs (Figure 7b), more uncertainties were found in the larger protection zone delineation which was indicated by the more expanded protection areas with different probabilities around each well in Figure 7b. This is because confidently determining a larger area is more difficult than determining a relatively smaller area. Spatial variables can increase greatly with expanding area. The red shapes in Figure 7 represent the WHPAs for Rehovot pumping site generated by the deterministic MODFLOW-MODPATH numerical model without considering the uncertainty of the input hydraulic conductivities. By comparing the deterministic results with the stochastic modeling results, it was easy to find that the probabilistic maps basically covered all of the 100% to be protected zones and most of the 80% to be protected zones, which indicates that WHPAs delineated by the deterministic numerical modeling method can provide the most necessary protection for groundwater resources. However, there are still the uncovered protection areas due to the uncertainty of input parameters and this could result in possible risks on groundwater resources contamination.

4. Discussion and Recommendation

4.1. Comparison of WHPA Delineation Results

The results of WHPA delineation at the two study areas are shown in Figures 3 and 6. In order to compare the application of five different methods, the Level-B and Level-C WHPA results were analyzed in terms of the surface area, geometric property, and the C_i value.

4.1.1. Comparison of Level-B Protection Zones

The Level-B protection zones of the Nahariya and the Rehovot pumping site are shown in Figures 3a and 6a, respectively. WHPAs delineated by different methods were represented by different colors, and the red zones are the reference protection areas delineated by numerical model. The comparative index C_i values and the surface area of protection zone of each method are listed in Table 5.

The level-B protection zone delineation results from the two pumping sites showed that the differences among the results were significant when different methods were compared. In general, the protection area from the semi-analytical method—WhAEM2000 model, best fit the reference results in both pumping sites (C_i = 58.6% in the Nahariya site, Average C_i = 34.1% in the Rehovot site), while the delineation results from the uniform flow equation method (C_i = 25.0% in the Nahariya site) and the HYBRID method (C_i = 27.0% in the Rehovot site) least fit the reference results. From the perspective of the surface area (S) of the WHPAs being delineated, all of the analytical methods generated overestimated protection zones compared to the surface area of the WHPAs delineated by the MODFLOW-MODPATH model. Exceptionally, the WhAEM2000 model produced a similar scale of protection zones as the reference method.

The application of the CFR method was based on the water balance between the pumping extraction and the aquifer storage and recharge, although the recharge rate can be neglected, as it is usually a small value. In this study, the circular protection zones delineated by the CFR method was four times larger than the reference WHPAs at the Rehovot pumping site, and the C_i values at the two pumping sites were relatively small (C_i = 43.1% at the Nahariya site, and the average C_i = 28.8% at the Rehovot site), mainly due to the unrealistic shape of the WHPAs from the CFR method. The equation of the CFR method suggests that this method is only valid when the supplying aquifer around pumping well can be regarded as a cylinder, which means that the CFR method is most suitable for confined aquifers and unconfined aquifer with a nearly flat gradient (aquifer hydraulic gradient < 0.0015). In the Nahariya site, the well was located in an unconfined aquifer with a hydraulic gradient of 0.0025 (cannot be regarded as a flat water table); thus the reference protection area ought to elongate along the direction of the groundwater flow, which significantly differed from the circular WHPA determined by the CFR method. Similar results can be observed in the Rehovot site. The hydraulic gradient in this wellfield increased gradually from west to eat (from 0.0015–0.0070). Well 1 was located in the west with a nearly flat hydraulic gradient, therefore, the reference WHPA of well 1 tended to be in a similar shape as the WHPA determined by the CFR method. The WHPAs of the other wells in the east had an elongated shape towards the groundwater flow direction as the gradient increased, which can be seen in Figure 6a.

Implementing the uniform flow equation method requires a certain degree of sloping groundwater table (hydraulic gradient). According to the equations of this method, for a nearly flat aquifer with a small hydraulic gradient value, an extremely large width for the up-gradient zone would be produced.

In this study, the uniform flow equation method was only used at the Nahariya site as the pumping well has a relatively large pumping rate (6600 m^3/day), and it was located in an unconfined aquifer where a significant asymmetrical drawdown could be created, and the difference between X_L and Y_L values were appropriately small. However, the uniform flow method does not guarantee the water balance between extraction and storage as the CFR method does [30]. Therefore, the C_i value (25.0%) of the uniform flow method was still small at the Nahariya site.

Table 5. Level-B protection areas and comparison index C_i of different delineation methods at the Nahariya and the Rehovot pumping site.

Pumping Sites	Surface Area (S) Index (C_i)	CFR	Uniform Flow Equation	WhAEM2000	HYBRID	MODFLOW-MODPATH
		Nahariya Site (Level-B: 100 days)				
Nahariya well	S (km^2)	0.14	0.29	0.097	0.14	0.073
	C_i (%)	43.1	25.0	58.6	51.7	Reference
		Rehovot Site (Level-B: 400 days)				
Well 1	S (km^2)	0.49	-	0.05	0.51	0.09
	C_i (%)	18.4	-	43.3	17.7	Reference
Well 2	S (km^2)	0.35	-	0.02	0.33	0.06
	C_i (%)	17.1	-	33.3	18.2	Reference
Well 3	S (km^2)	0.27	-	0.02	0.25	0.07
	C_i (%)	25.9	-	28.6	28.00	Reference
Well 4	S (km^2)	0.18	-	0.02	0.18	0.05
	C_i (%)	27.8	-	40.0	27.8	Reference
Well 5	S (km^2)	0.15	-	0.02	0.15	0.05
	C_i (%)	33.3	-	16.7	33.3	Reference
Well 6	S (km^2)	0.42	-	0.04	0.41	0.13
	C_i (%)	31.0	-	30.8	31.7	Reference
Well 7	S (km^2)	0.07	-	0.02	0.06	0.05
	C_i (%)	51.2	-	40.0	33.3	Reference
Well 8	S (km^2)	0.17	-	0.02	0.15	0.05
	C_i (%)	29.4	-	40.0	33.3	Reference
Average	S (km^2)	-	-	-	-	-
	C_i (%)	28.8	-	34.1	27.03	Reference
Sum	S (km^2)	2.00	-	0.21	1.91	0.55
	C_i (%)	-	-	-	-	-

Note: since the uniform flow equation method was not used in the Rehovot site, surface area of each well was not calculated by this method. Neither the average surface area (S) nor the comparative index C_i values can be calculated to represent the whole Rehovot multi-well site.

The HYBRID method integrates the water balance factor and the boundary limits of the protection area. An ellipse-shaped WHPA is usually produced by this method. The C_i value of the HYBRID method delineated WHPA was 51.7%, indicating that this method was well suited to the reference method in this setting. However, the HYBRID method performed poorly in delineating WHPAs in the Rehovot site (average C_i = 27.0%). Several WHPAs determined by the HYBRID method at the Rehovot wellfield were similar to the WHPAs determined by the CFR method, particularly for the wells located in small hydraulic gradient zones.

The semi-analytical WhAEM2000 model had the highest accuracy (C_i = 58.6%) compared with the other analytical methods. At the Rehovot wellfield, although the WhAEM2000 model was still more accurate (C_i = 34.1%) than the other methods, the WHPAs from the WhAEM2000 model were much smaller than the reference WHPAs and the WHPAs from other methods. This was because the

WhAEM2000 model cannot represent the saturated thickness of an individual well. In the WhAEM2000 model, only a constant thickness (in the Rehovot case, saturated thickness = 85 m) can be given to all the well, which was much greater than the real saturated thickness of each well in the numerical model. Hence, under similar hydraulic conductivity values and hydraulic gradients, the WHPAs determined by the WhAEM2000 method had to be much narrower than the WHPAs determined by the numerical model, in order to ensure that the water flux to the well was equivalent to the pumping rate.

4.1.2. Comparison of Level-C Protection Zones

Level-C protection zones were delineated with a longer TOT value. The Level-C WHPAs of two pumping sites can be found in Figures 3b and 6b. The comparison data are listed in Table 6. Since the Level-C WHPA of the individual wells at the Rehovot wellfield overlapped each other, the protection zone information of the entire wellfield is presented. The comparison results of the two sites showed that the WhAEM2000 model still provided the best fitting WHPAs with the highest C_i values, followed by the HYBRID method then the CFR method.

Table 6. Level-C protection areas and comparison index C_i of different delineation methods at the Nahariya and Rehovot pumping sites.

Pumping Sites	Surface Area (S) / Index (C_i)	CFR	Uniform Flow Equation	WhAEM2000	HYBRID	MODFLOW-MODPATH
	Nahariya Site (Level-B: 100 days)					
Nahariya well	S (km^2)	0.56	0.75	0.37	0.55	0.29
	C_i (%)	25.1	38.6	54.3	53.2	Reference
	Rehovot Site (Level-B: 400 days)					
Entire wellfield	S (km^2)	14.38	-	2.62	16.04	5.66
	C_i (%)	30.2	-	40.6	35.3	Reference

Note: C_i value of the entire Rehovot wellfield shown in this table was the average C_i value of eight pumping wells in the site.

With the CFR methods, a much smaller C_i value (25.1%) was obtained in the Level-C WHPA at the Nahariya site. This was likely due to the fact that the analytical CFR method does not account for the heterogeneity of local aquifers. When the CFR method was implemented to delineate a relatively small area (Level-B zone), the local heterogeneity could be ignored. However, with the expansion of delineating the higher level protection zone, aquifer heterogeneity could no longer be ignored, and the uncertainty that accumulated with the heterogeneity finally resulted in the deviation of the WHPA. At the Rehovot site, the C_i value of the entire Level-C WHPAs was basically the same as the average C_i value of the Level-B protection zones. However, CFR method didn't produce the WHPAs that reflect the local groundwater flow condition. For instance, by the reference method, the WHPA of well 1 in the Rehovot site should be nearly circular and not overlapping other protection zones of other wells.

The HYBRID method was still more accurate than CFR method, since it considered the groundwater flow property. The WhAEM2000 model generally fits the reference WHPA well. Despite the fact that the WHPA of well 1 delineated by this model at the Rehovot site was more narrowed and elongated to the hydraulic gradient direction. This is because WhAEM2000 model can only handle the same hydrogeological conditions for all pumping wells, which means that the individual properties of the aquifer where each well was located were disregarded. This further proves that WhAEM2000 model is less appropriate for simulating the multi-well field, though this method provides a relatively accurate approximation and offers an easier alternative compared with building a numerical model.

The WHPAs generated by different methods at the Rehovot multi-well pumping site indicated that WHPAs in the wellfield can be overlapped when there are a few wells nearby (Figure 6), or when delineating WHPAs to a larger range. Comparing the C_i values (Tables 5 and 6) of both Level-B and Level-C WHPAs at the Nahariya site with the average C_i value at the Rehovot wellfield, smaller

C_i values were obtained in the Rehovot wellfield, indicating that the same methods performed less accurately in the multi-well field. This is likely due to the influence of well-interference. In a wellfield consisting of several pumping wells, the groundwater flow, as well as the particle transport with the groundwater flow system, can be greatly affected by adjacent wells. Other methods, except WhAEM2000 model and numerical modeling, do not account for the adjacent well interferences.

4.2. Recommendation for Selecting Delineation Method

The comparison between WHPAs, determined by different types of methods, showed significant differences. The WhAEM2000 model fits the reference method best, while all the analytical methods overestimated the protection zones. Analysis in detail indicated that it is important to apply the analytical methods to the appropriate hydrogeological setting.

The principle idea of CFR method is to ensure the water balance, so it can only be used for a confined aquifer or an unconfined aquifer with a nearly flat water table (local hydraulic gradient < 0.0015), and it can only provide a very rough estimate for WHPA delineation. The uniform flow equation is only suitable for a sloping aquifer with a hydraulic gradient higher than 0.0015, but this method aligns well to the groundwater flow direction which reflects the regional flow characteristics. The HYBRID method had better fitting compared to other analytical methods, which indicated that water balance and regional flow characteristics are two essential factors for delineating a proper WHPA. Water balance controls the general size of the protection zone, and the consideration of flow characteristics determines the shape of the protection zone. Furthermore, well interferences existed in the multi-well field could have significant impact on the accuracy of WHPAs delineation. Since the analytical methods are not able to consider the influences from adjacent wells, only WhAEM2000 model and numerical modeling methods are recommended. But the WhAEM2000 model is limited to handle the uniform aquifer condition, which means it's not appropriate to be used for WHPA delineation while the aquifer parameters vary greatly at each well location.

The uncertainty analysis through stochastic modeling produced the probabilistic distribution map of the protection zones (Figure 7). This map shows the different necessities of protecting the areas around the pumping well, and improved the deterministic delineation method which excludes the possibly threated zones. It is useful for land-use risk management around pumping wells to prevent potential contaminating activities. A general WHPA delineation method selection flow chart is depicted in Figure 8. This scheme is recommended to choose a suitable protection zone delineation method based on data availability and delineation expectations.

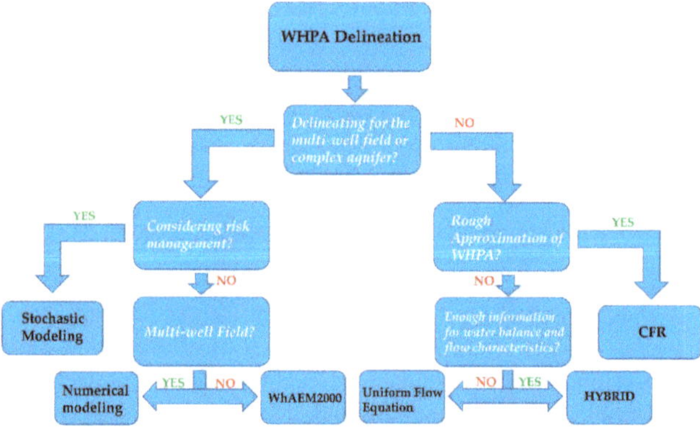

Figure 8. Recommended procedure for selecting a WHPA delineation based on data availability and delineation expectation.

5. Conclusions

Since the WHPA cannot be measured but only be calculated or simulated, it is necessary to justify the performance of the commonly applied WHPA delineation methods. In this comparative study, five different WHPA delineation methods were employed to define the protection zones for two pumping sites in the unconfined coastal aquifer in Israel. The MODFLOW-MODPATH numerical modeling method was selected as the reference method. By comparing the WHPA results with the reference WHPA, the semi-analytical WhAEM2000 model provides the best fitting, while all of the analytical methods overestimated the protection zones. The CFR method, which is based on water balance between aquifer extraction and storage, provided fast WHPA approximation. But it can only be used for the confined or the unconfined aquifer with a nearly flat water table. On the other hand, the uniform flow equation method preserved groundwater flow property and can only be applied at sloping aquifers. However, the latter does not consider the water balance as the CFR method does; therefore, it could produce unrealistic WHPAs. The better fitting to the reference method of the analytical HYBIRD method indicated that it is essential to ensure water balance calculation as well as to reflecting regional flow characteristics while defining the WHPA. Dealing with a complex hydrological condition and a multi-well field, only semi-analytical or numerical modeling should be applied. Stochastic modeling of both pumping sites generated a probability distribution map for WHPAs, which suggested the different necessities of protecting various areas around the pumping well and offered risk management into WHPA delineation. Therefore, selecting a WHPA delineation method should be based on data availability and delineation expectations. The recommended procedure for choosing the proper delineation method could facilitate this process and resolve the puzzle of finding the "best" method.

Author Contributions: This paper was completed under the supervision of A.Y. and N.W.; Y.L. conducted the calculation and modeling, analyzed the results and wrote the paper. N.W. contributed to the analysis discussion and helped with reviewing the manuscript. A.Y. guided building the concepts of this research, helped with analyzing the results and building up the numerical models, as well as reviewing this manuscript.

Funding: This research was funded by Maccabi Carasso, grant number 87382911.

Acknowledgments: This research was supported by China Scholarship Council and the Council for Higher Education, Israel. Thanks due to Noam Dvory for providing field data of the Nahariya site and to Mikhail Kuznetzov for offering modeling skills supervision.

Conflicts of Interest: The authors declare no conflict of interest.

References

1. Siebert, S.; Burke, J.; Faures, J.M.; Frenken, K.; Hoogeveen, J.; Döll, P.; Portmann, F.T. Groundwater use for irrigation—A global inventory. *Hydrol. Earth Syst. Sci.* **2010**, *14*, 1863–1880. [CrossRef]
2. Shah, T.; Burke, J.; Villholth, K.G.; Angelica, M.; Custodio, E.; Daibes, F.; Hoogesteger, J.; Giordano, M.; Girman, J.; Van Der Gun, J.; et al. *Groundwater: A Global Assessment of Scale and Significance*; Earthscan: London, UK; International Water Management Institute (IWMI): Colombo, Sri Lanka, 2007; pp. 395–423.
3. World Water Assessment Programme (United Nations). *Water: A Shared Responsibility*; the United Nations World Water Development Report 2; UN-HABITAT: Nairobi, Kenya, 2006.
4. EPA. *Guidelines for Delineation of Wellhead Protection Areas*; EPA-440/5-93-001; United States Environmental Protection Agency: Washington, DC, USA, 1987; p. 212.
5. Cleary, T.C.B.F.; Cleary, R.W. Delineation of Wellhead Protection Areas: Theory and Practice. *Water Sci. Technol.* **1991**, *24*, 239–250. [CrossRef]
6. Vassolo, S.; Kinzelbach, W.; Schafer, W. Determination of a well head protection zone by stochastic inverse modelling. *J. Hydrol.* **1998**, *206*, 268–280. [CrossRef]
7. Schleyer, R.; Milde, G.; Milde, K. Wellhead Protection Zones in Germany—Delineation, Research and Management. *J. Inst. Water Environ. Manag.* **1992**, *6*, 303–311. [CrossRef]
8. EPA. *Handbook—Groundwater and Wellhead Protection*; United States Environmental Protection Agency: Washington, DC, USA, 1994; p. 288.

9. Harter, T. *Delineating Groundwater Sources and Protection Zones*; Rollins, L., Ed.; University of California at Davis: Davis, CA, USA, 2002.

10. EPA. *Literature Review of Methods for Delineating Wellhead Protection Areas*; United States Environmental Protection Agency: Washington, DC, USA, 1998; p. 48.

11. Rock, G.; Kupfersberger, H. Numerical delineation of transient capture zones. *J. Hydrol.* **2002**, *269*, 134–149. [CrossRef]

12. Piccinini, L.; Fabbri, P.; Pola, M.; Marcolongo, E.; Rosignoli, A. Numerical modeling to well-head protection area delineation, an example in Veneto Region (NE Italy). *Rend. Online Soc. Geol. Ital.* **2015**, *35*, 232–235. [CrossRef]

13. Harbaugh, A.W. *MODFLOW-2005: The U.S. Geological Survey Modular Ground-Water Model—The Ground-Water Flow Process*; US Department of the Interior, US Geological Survey: Reston, VA, USA, 2005.

14. Forster, C.B.; Lachmar, T.E.; Oliver, D.S. Comparison of Models for Delineating Wellhead Protection Areas in Confined to Semiconfined Aquifers in Alluvial Basins. *Groundwater* **1997**, *35*, 689–697. [CrossRef]

15. Pollock, D.W. *User Guide for MODPATH Version 7—A Particle-Tracking Model for MODFLOW*; US Department of the Interior, US Geological Survey: Reston, VA, USA, 2016; p. 41.

16. Evers, S.; Lerner, D.N. How uncertain is our estimate of a wellhead protection zone? *Ground Water* **1998**, *36*, 49–57. [CrossRef]

17. Theodossiou, N.; Fotopoulou, E. Delineating well-head protection areas under conditions of hydrogeological uncertainty. A case-study application in northern Greece. *Environ. Process.* **2015**, *2*, 113–122. [CrossRef]

18. Fadlelmawla, A.A.; Dawoud, M.A. An approach for delineating drinking water wellhead protection areas at the Nile Delta, Egypt. *J. Environ. Manag.* **2006**, *79*, 140–149. [CrossRef] [PubMed]

19. Stauffer, F.; Guadagnini, A.; Butler, A.; Franssen, H.J.H.; Van den Wiel, N.; Bakr, M.; Riva, M.; Guadagnini, L. Delineation of source protection zones using statistical methods. *Water Resour. Manag.* **2005**, *19*, 163–185. [CrossRef]

20. Schmoll, O.; World Health Organization. *Protecting Groundwater for Health: Managing the Quality of Drinking-Water Sources*; IWA Pub.: London, UK, 2006.

21. Paradis, D.; Martel, R.; Karanta, G.; Lefebvre, R.; Michaud, Y.; Therrien, R.; Nastev, M. Comparative study of methods for WHPA delineation. *Ground Water* **2007**, *45*, 158–167. [CrossRef] [PubMed]

22. Landmeyer, J.E. Description and Application of Capture Zone Delineation for A Wellfield at Hilton Head Island, South Carolina. *Water-Resour. Investig. Rep.* **1994**, *94*, 4012.

23. ANWQMS. *National Water Quality Management Strategies: Guidelines for Groundwater Protection in Australia*; Agriculture and Resources Management Council of Australia and New Zealand: Canberra, Australia, 1995.

24. DoELG. *Groundwater Protection Schemes*; Environmental Protection Agency and Geological Survey of Ireland, Department of Environment and Local Government: Dublin, Ireland, 1999.

25. DVGW. *Code of Practice W101 for Drinking Water Protection Areas Part 1, Protective Areas for Groundwater*; German Association of Gas and Water Experts: Bonn, Germany, 1995.

26. Government of Oman. *Water Resources of the Sultanate of Oman*; Ministry of Water Resources: Muscat, Oman, 1991.

27. Bear, J.; Jacobs, M. On the movement of water bodies injected into aquifers. *J. Hydrol.* **1965**, *3*, 37–57. [CrossRef]

28. Todd, D.K.; Mays, L.W. 4.3 WELL IN A UNIFORM FLOW. In *Groundwater Hydrology*, 3rd ed.; Bill, Z., Ed.; John Wiley & Sons, Inc.: New York, NY, USA, 2004; p. 656.

29. Haitjema, H.M.; Wittman, J.; Kelson, V.; Bauch, N. *WhAEM: Program Documentation for the Wellhead Analytic Element Model*; EPA/600/R-94/210 (NTIS PB95-167373); U.S. Environmental Protection Agency: Washington, DC, USA, 1994.

30. Paradis, D.; Martel, R. *HYBRID: A Wellhead Protection Delineation Method for Aquifers of Limited Extent*; Geological Survey of Canada: Ottawa, ON, Canada, 2007; p. 5.

31. Israel Ministry of Health. *Public Health Regulations 2013—The Sanitary Quality of Drinking Water and Drinking Water Facilities*; Collection of Regulations 7262; Department of Environmental Health, Public Health Services, Ministry of Health: Jerusalem, Israel, 2013; p. 34.

32. Miller, C.; Chudek, P.; Babcock, S. A Comparison of wellhead Protection Area Delineation Methods for Public Drinking Water Systems in Whatcom County, Washington. *J. Environ. Health* **2003**, *66*, 17. [PubMed]

33. Loucks, D.P.; van Beek, E.; Stedinger, J.R.; Dijkman, J.P.M.; Villars, M.T. 9. Model Sensitivity and Uncertainty Analysis. In *Water Resources Systems Planning and Management: An Introduction to Methods, Models and Applications*; Springer International Publishing AG: Basel, Switzerland, 2005. [CrossRef]
34. Guha, H. A Stochastic Modeling Approach to Address Hydrogeologic Uncertainties in Modeling Wellhead Protection Boundaries in Karst Aquifers1. *JAWRA J. Am. Water Resour. Assoc.* **2008**, *44*, 654–662. [CrossRef]
35. Weinberger, G.; Livshitz, Y.; Givati, A.; Zilberbrand, M.; Tal, A.; Weiss, M.; Zurieli, A. *The Natural Water Resources Between the Mediterranean Sea and the Jordan River*; Hydro Report 11/1; Israel Hydrological Service: Jerusalem, Israel, 2011; p. 63.
36. Eriksson, E.; Khunakasem, V. Chloride concentration in groundwater, recharge rate and rate of deposition of chloride in the Israel Coastal Plain. *J. Hydrol.* **1969**, *7*, 178–197. [CrossRef]
37. Kass, A.; Gavrieli, I.; Yechieli, Y.; Vengosh, A.; Starinsky, A. The impact of freshwater and wastewater irrigation on the chemistry of shallow groundwater: A case study from the Israeli Coastal Aquifer. *J. Hydrol.* **2005**, *300*, 314–331. [CrossRef]
38. Nativ, R.; Weisbrod, N. Hydraulic Connections among Subaquifers of the Coastal-Plain Aquifer, Israel. *Ground Water* **1994**, *32*, 997–1007. [CrossRef]
39. Yosef, B. *The Hydrogeology of the Coastal Plain of Western Galilee (in Hebrew)*; TAHAL: Tel Aviv, Israel, 1967.
40. Wexler, A. *Hydrogeological Plan for the Water Development, Production, and Utilization in the Kavri Basin*; GSI/36/2001; Water Commission & Planning Department: Tel Aviv, Israel, 2001.
41. Dvory, N. *Hydrographic Survey in the Production Field of Nahariya Municipality Drilling Etgar Engineering LTD Report*; Etgar: Tel Aviv, Israel, 2009.
42. Yakirevich, A.; Kuznetsov, M.; Adar, E. *Modeling Contaminant Migration in Coastal Aquifer (Givon Region)— Internal Report*; Ben-Gurion University of the Negev: Beersheba, Israel, 2015.

Article

The Impact of Capillary Trapping of Air on Satiated Hydraulic Conductivity of Sands Interpreted by X-ray Microtomography

Tomas Princ [1,*], Helena Maria Reis Fideles [1,2], Johannes Koestel [3] and Michal Snehota [1]

[1] Faculty of Civil Engineering, Czech Technical University in Prague, 166 29 Prague, Czech Republic;
    helenamrf@poli.ufrj.br (H.M.R.F.); michal.snehota@fsv.cvut.cz (M.S.)
[2] Laboratório de Métodos de Modelagem e Geofísica Computacional, Universidade Federal do Rio de Janeiro,
    Rio de Janeiro 21941-596, Brazil
[3] Department of Soil and Environment, Swedish University of Agricultural Sciences, 750 07 Uppsala, Sweden;
    john.koestel@slu.se
*   Correspondence: tomas.princ@fsv.cvut.cz; Tel.: +420-22435-3725

Received: 17 December 2019; Accepted: 4 February 2020; Published: 7 February 2020

Abstract: The relationship between entrapped air content and the corresponding hydraulic conductivity was investigated experimentally for two coarse sands. Two packed samples of 5 cm height were prepared for each sand. Air entrapment was created by repeated infiltration and drainage cycles. The value of K was determined using repetitive falling-head infiltration experiments, which were evaluated using Darcy's law. The entrapped air content was determined gravimetrically after each infiltration run. The amount and distribution of air bubbles were quantified by micro-computed X-ray tomography (CT) for selected runs. The obtained relationship between entrapped air content and satiated hydraulic conductivity agreed well with Faybishenko's (1995) formula. CT imaging revealed that entrapped air contents and bubbles sizes were increasing with the height of the sample. It was found that the size of the air bubbles and clusters increased with each experimental cycle. The relationship between initial and residual gas saturation was successfully fitted with a linear model. The combination of X-ray computed tomography and infiltration experiments has a large potential to explore the effects of entrapped air on water flow.

Keywords: residual saturation; porous media; permeability; entrapped air; two-phase flow

1. Introduction

Interaction between water and gas needs to be considered for water fluxes in near-saturated and saturated soil because the entrapped air may block pores and thereby affect the hydraulic conductivity [1–3]. Entrapped air is defined as a discontinuous non-wetting phase that is not connected to the atmosphere [1]. Entrapped air may appear in the form of single pore bubbles, clusters or ganglia [4]. With growing size, such ganglia become more mobile [5] and may escape from the soil to the atmosphere. The pressure in the discontinuous air phase usually differs from the atmospheric pressure [6] and can be measured from the curvature of the water–air interface.

We distinguish between three states of saturation. If the pressure in the soil water is lower than the air entry value, the soil is denoted as unsaturated soil. In this state, it contains a continuous air phase with connection to and in equilibrium with the atmosphere. Quasi-saturated [1] or satiated [7] soil contains air in the form of entrapped air bubbles or air ganglia with water pressures higher than the air entry value [6]. Fully saturated soil, which is rarely found under natural conditions, does not contain any air bubbles or ganglia. The amount of entrapped air in porous media is commonly expressed as a residual gas saturation, $S_{g,r}$ (–), or an entrapped volumetric air content [1], ω (–). The residual gas saturation is the ratio of entrapped air volume to the volume of pores. The entrapped air content is the ratio of entrapped air volume to the bulk volume of soil.

Chatzis et al. [8] define two main mechanisms of non-wetting phase trapping in porous media, termed (i) distinguished-bypassing and (ii) snap-off trapping. Distinguished-bypassing results from the fact that smaller throats are filled with water before larger ones are, due to capillary forces. If a larger pore containing gas is surrounded by water-filled throats, the gas is entrapped. According to Almajid and Kovscek [9], snap-off trapping is a hydraulic process where the wetting phase forms blockages in the pore throat that snap-off gas bubbles as they move through the pore throat. The snap-off mechanism is present predominantly in pore networks with a large aspect ratio, i.e., ratios between pore–body and pore–throat sizes, and seems to be the dominant mechanism in the formation of a discontinuous gas phase within such porous media [10].

The amount of the entrapped non-wetting phase in porous media depends predominantly on the pore structure [8], but also on the history of wetting and drying [11,12] as well as the contact angle [5,13]. Higher values of the contact angle lead to a suppression of the snap-off mechanism, resulting in a smaller amount of entrapped air. Furthermore, it has been shown that the $S_{g,r}$ value is positively correlated with the initial gas saturation, $S_{g,init}$, when porous media is imbibed [14]. Distribution of the air bubbles may be affected by temperature [15].

Generally, it has been assumed that the $S_{g,r}$ value is constant [16]. However, Hilfer [17] showed that the $S_{g,r}$ value is a function of the time and position. He builds a formulation of two-phase immiscible displacement based on the general balance laws. Used assumptions were that the porous medium is macroscopically homogeneous, the fluid is incompressible, the flow has low Reynolds number and body forces are given by gravity and capillary forces.

The presence of entrapped air in soil introduces ambiguity for the definition and measurements of hydraulic conductivity close to saturation. Given the states of saturation, the hydraulic conductivity may be specified as saturated, K_S (cm·s^{-1}), satiated, $K(\omega)$ (cm·s^{-1}), or unsaturated, $K(\theta)$ (cm·s^{-1}). A widely used model that calculates the unsaturated hydraulic conductivity function from the soil water content and the saturated hydraulic conductivity was introduced by Mualem [18]. It represents an estimate of the relative hydraulic conductivity function $K_r(\theta) = K(\theta)/K_s$ (–) based on the capillary model theory [19]. The water retention curves are modeled by the approaches of either Brooks and Corey [20] or van Genuchten [21]. Another model for the satiated hydraulic conductivity was presented by Faybishenko [1].

Laboratory experiments have shown that hydraulic conductivity depends significantly on the entrapped air content ω [22–24]. In a laboratory study on small soil samples, Sakaguchi et al. [25] showed that the measured satiated hydraulic conductivity, $K(\omega)$, was smaller than the unsaturated hydraulic conductivity, $K(\theta)$, although the same amount of air was entrapped. The authors explained this by entrapped air clogging of the largest pores. The effect of entrapped air on infiltration in the field was also confirmed by Alagna et al. [26].

For a long term infiltration experiment carried out on large soil monoliths, Faybishenko [1] observed three stages in which the hydraulic conductivity changed characteristically. During the first stage, the value of the hydraulic conductivity decreased. Faybishenko presumed that at the beginning, the air was entrapped in small pores. But due to the capillary forces, water started to infiltrate into the smaller pores of the matrix. The air was thus continuously displaced into larger pores, forming entrapped air bubbles that then blocked preferential flow pathways. In the second stage, the hydraulic conductivity value started to increase again due to the dissolution of gas bubbles. In the third stage of the experiment, the hydraulic conductivity decreased again, this time due to the buildup of biofilms created by microbial activity.

A fundamental improvement in the description of processes connected with the flow of liquids in porous media was achieved with the possibility to use nondestructive visualization techniques. Noninvasive imaging can be used to obtain information about the distribution of the wetting and non-wetting phases [27].

Snehota et al. [3] illustrated and quantified air redistributions during infiltration experiments in a repacked sand sample using neutron imaging. The sand sample contained blocks of fine sand surrounded by coarse sand. Snehota et al. [3] found that entrapped air was transferred from the fine into the coarse sand while the flow rate decreased from 0.400 cm·min^{-1} to 0.296 cm·min^{-1}. Similar results were obtained in an experiment with a similarly designed sand sample [2] on which repeated infiltration runs were performed. Also here, the redistribution of entrapped air from the finer to the coarser sand caused the infiltration rate to slow down. The flow rate decreased from 3.7 cm·h^{-1} to 1.0 cm·h^{-1} during the first infiltration run and during the second infiltration run from 3.6 to 1.6 cm·h^{-1}. Thereafter, an infiltration run with degassed water was performed during which the flow rate increased to a maximum value of 10.5 cm·h^{-1}.

A frequently used imaging technique in porous media research is synchrotron-based X-ray tomography. Taking advantage of its high image resolution, it was often applied to obtain pore geometry properties [28–30], aggregate microstructures [31–33] and multi-phase flow observations [34,35]. Industrial X-ray scanners are also suited to collect detailed information on soil structure development [36,37] or gas bubbles distribution [38–42]. A disadvantage of industrial X-ray scanners is that they operate with polychromatic X-rays, which may cause beam hardening artifacts [43,44].

Hydraulic conductivity is a crucial parameter in all components of water cycle modeling, rainfall excess determination, water and solute transport in vadose zone as well as in recharge of groundwater [26,45–47]. The impact of entrapped air on measured values of satiated hydraulic conductivity means that instead of a well-defined constant parameter, its value may vary in dependence on many factors, some of them described in this paper.

This study aimed to unravel the relationship between the entrapped air content and the dependence of the air trapping on the initial air saturation in two coarse sands. Recent experimental studies [3,48] on air entrapment involved the same sands as components of more complex, heterogeneous samples. To enable numerical modeling of these experiments in future, the $K(\omega)$ relationship of coarse sands is needed. An additional goal of the current study was to investigate variations of the entrapped air bubbles in space and time during repeated infiltration drainage cycles. Experiments were conducted on packed samples of coarse sand representing homogeneous media with narrow pore size distribution to rule out the effects of heterogeneity. The aim was to use X-ray computed micro-tomography to characterize the sizes, shapes and abundance of entrapped air bubbles.

2. Materials and Methods

2.1. Material Characterization

We prepared four packed samples, each two with different sands that we denote with abbreviations ST and FH, respectively. The ST sand is a commercially available technical quartz sand—with a SiO_2 content 99.40% and Fe_2O_3 content 0.04% (technical designation ST 03/08, Sklopisek Strelec, Czech Republic). The FH sand is a commercially produced quartz sand from with a SiO_2 content of larger than 99% (technical designation FH31, Quarzwerke Frechen, Germany). We evaluated the grain-size distribution of the two different sands by dry sieving. Here we employed mesh-sizes of 0.20, 0.25, 0.40, 0.50, 0.63, 0.80, and 1.00 mm. The particle densities were measured using a pycnometer and turned out to be identical 2.62 g·cm^{-3} for all samples. The dry bulk densities of the samples were calculated from the mass and volume of dry sand samples. The samples' porosities were determined from the particle densities and the dry bulk densities. The grain size distributions of all samples are shown in Figure 1. Table 1 contains the respective values of dry bulk density, porosity as well as packing height and sample volume. The FH sand is finer and has a lower porosity than the ST sand. The value of the uniformity coefficient (D_{60}/D_{10}, where D_x represents the size for which x% particles are smaller) for ST sand was 2.2 and for FH sand the value was 1.7. Both sands are considered to be poorly graded, but the FH sand is more uniform. The effective size (D_{10}) of the ST sand is 0.29 mm and of the FH sand is 0.22 mm.

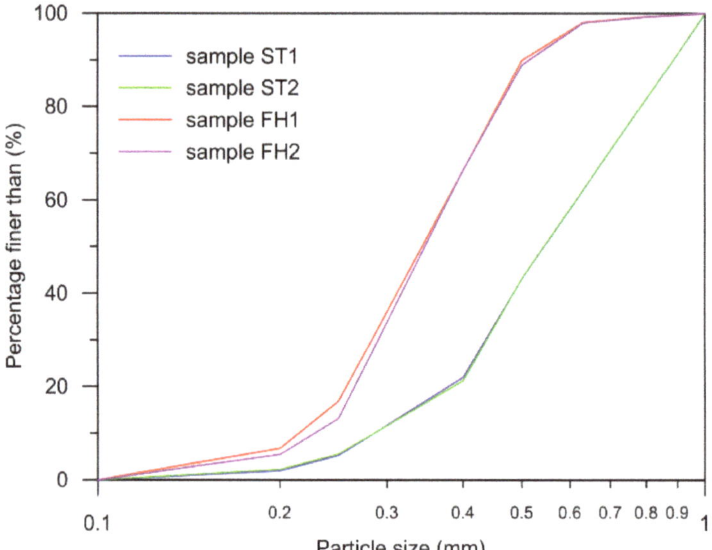

Figure 1. Grain-size distribution of the sands under study.

Table 1. Characteristics of sand samples.

Sand Sample	Dry Bulk Density (g·cm^{-3})	Porosity (−)	Height (cm)	Volume (cm^3)
ST1	1.66	0.372	5.1	207.6
ST2	1.68	0.364	5.1	207.6
FH1	1.71	0.357	5.2	211.7
FH2	1.78	0.327	5.5	223.9

2.2. Experimental Setup

The saturated and satiated hydraulic conductivity of sand were measured with repetitive falling-head infiltration experiments. The respective setup is presented in Figure 2. The sample container consisted of an acrylic glass (PMMA) cylinder (7.2 cm inner diameter, 7.0 cm height) and a PMMA ring with a filter-mesh (size 99 μm) attached to its bottom. An o-ring made of nitrile rubber was used as a hydraulic seal, ensuring no water could escape between the ring and cylinder. The sample container was filled with the sand up to the height of approximately 5.0 cm. The container was placed into a larger, outer PMMA vessel (inner diameter 12.3 cm), where it was mounted onto five PMMA retainers. An overflow was inserted into the wall of the outer vessel (WADI-PZ, M12×1.5, 3.0–6.5 mm, Hugro Plastic Cable Gland, Waldkirch, Germany) through which a constant water level could be kept. To achieve a sufficiently large flow rate through the overflow, we attached a tube connected to a peristaltic pump (see Figure 2). The vessel was placed on a balance with 0.1 g precision to monitor the amount of water and gas in the outer vessel and the sample.

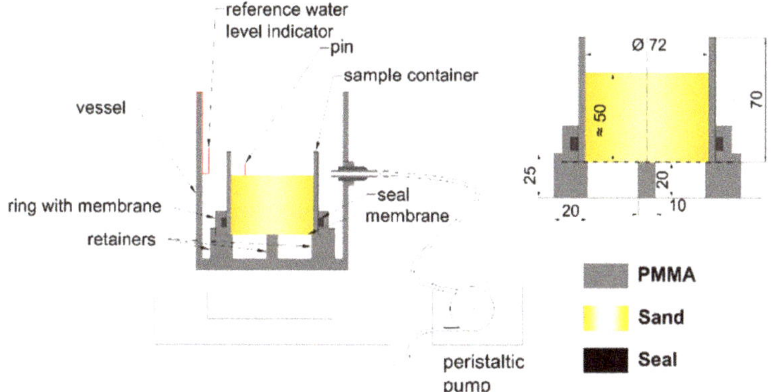

Figure 2. Schema of the experimental setup. Left: The sample container filled with sand and inserted into the outer vessel. The peristaltic pump is connected to the overflow. Right: The sample container and the retainers in detail and with dimensions.

2.3. Sample Preparation

Two samples (ST1 and ST2) were packed with ST sand, while another two samples (FH1 and FH2) were prepared with FH sand. The samples were packed to a height of approximately 5.0 cm (the exact heights are given in Table 1) into cylindrical containers with an inner diameter of 7.2 cm. The sand packing was performed in degassed, deionized water, to minimize the introduction of entrapped air bubbles. The sand was added in 20 steps. After each addition step, the sand in the column was compacted by a small manual compactor, consisting of a rubber stopper and rod. The saturation by vacuum was used to ensure full saturation of the sand in the column [49]. The vessel with a submerged sample container was inserted into a vacuum chamber. The vacuum chamber was then evacuated for approximately 20 min at −95 kPa (relative to the atmospheric pressure). To minimize changes in pore geometry due to the resulting air bubble ebullition, a perforated plate loaded with 1.24 kg weight was placed on the sample's surface during this process.

2.4. Determination of Saturated Hydraulic Conductivity

To determine the value of saturated hydraulic conductivity, the first infiltration experiment with the repetitive falling-head was performed on the fully saturated sample, directly after it had been removed from the vacuum chamber. We assumed that the entrapped air content ω was zero at this point. The water used in infiltration experiments was kept in an open bottle one day before the experiment in order to equilibrate the temperature and dissolved air content to the laboratory conditions.

The workflow of the infiltration experiments is shown in Figure 3. At the first stage (Figure 3a), the sample container was submerged in degassed water. Then a simple repetitive falling-head infiltration experiment was carried out to determine the volumetric flux through the sample at known changes of hydraulic gradient. Here, a pin inserted in the sand (Figure 3b) was used to indicate a reference water level. A batch of water (40 cm^3) was manually added on top of the column each time the water level had dropped to the tip of the pin. The water was allowed to flow freely through the filter-mesh at the bottom of the sample. The saturated hydraulic conductivity, K_S (cm·s^{-1}), was determined in each interval by Darcy's law adapted for falling-head method [50]:

$$K_S = \frac{L}{t} \ln \frac{\Delta H + \frac{V}{A}}{\Delta H} \tag{1}$$

where ΔH (cm) is a hydraulic head difference at the moment when the water level touches the pin, V is batch volume of added water and A is cross-section area of sample ($V/A \approx 1$ cm), t is time of infiltration of the batch volume and L (cm) is the sample height. A range of ΔH was from 1.8 cm to 2.4 cm. After the infiltration experiment, the outer vessel with the sample was flooded precisely to the reference level indicated by the tip of the bent wire (Figure 3c). It was then weighed to obtain the reference mass of the fully saturated sample.

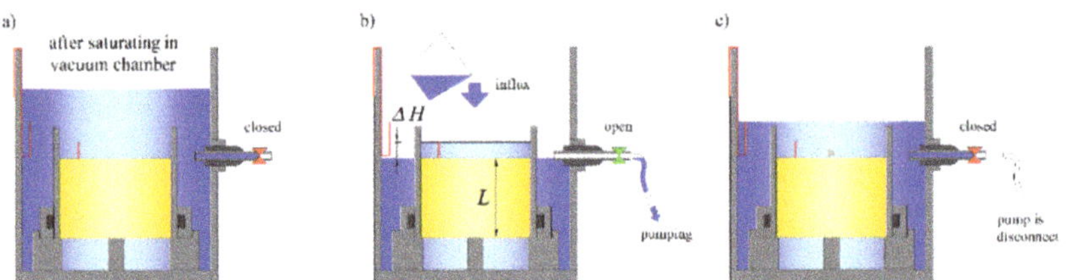

Figure 3. Illustration of the infiltration experiment for measurements of K_S: (**a**) The saturation of the sand sample in the vacuum chamber. The valve is closed; (**b**) the ponding infiltration experiment: L is the height of the sample and ΔH is the difference of hydraulic heads. The valve is opened, and the peristaltic pump keeps the lower water level in the outer vessel constant. When the water level on the sand sample drops to the tip of the pin, another batch of 40 cm^3 of water is poured. This is repeated at least 15 times; (**c**) the sample is prepared for weighing.

2.5. Determination of Satiated Hydraulic Conductivity and Entrapped Air

To determine the satiated hydraulic conductivity for varying contents of entrapped air, several cycles of ponded infiltration runs followed by drainage runs were performed. After each infiltration run, the sample was placed onto a standard laboratory sand box of the design similar to [49]. The box contained saturated fine sand with a high air entry value. The water within sand box was connected to a burette by a flexible hose. Tension was produced by the height difference between the sand surface in the sand box and the water level in the burette. The sample was gradually drained to a specific pressure head. Ten drainage-satiation runs were conducted, each providing one satiated hydraulic conductivity. The chosen pressure heads, as well as the corresponding and drainage durations, are given in Table 2.

Table 2. Pressure heads set on sand box and duration of drainage runs for samples ST and FH.

Run No.	1	2	3	4	5	6	7	8	9	10
Pressure Head (cm)	−4	−10	−20	−30	−35	−40	−50	−50	−50	−50
Duration (h)	21	23	22	21	20	20	17	20	24	24

A perforated plate loaded with a weight of 0.62 kg was placed onto the surface of the sand sample during the drainage process to minimize possible pore geometry changes within the sample. The sample mass, $M_{samp,init}$, was determined after each drainage run to calculate the gas saturation of the drained sample, $S_{g,init}$. The calculation is described in Appendix A. Then, a ponding infiltration experiment was performed in the same way as described in Section 2.4, this time to obtain the satiated hydraulic conductivity, $K(\omega)$. A range of ΔH was from 2.0 cm to 2.5 cm.

The entrapped air content ω was calculated from the difference between the mass of the outer vessel with a satiated sample and the mass of the outer vessel with a degassed, fully saturated sample.

The obtained satiated hydraulic conductivity values were fitted with (i) the relationship published in Faybishenko [1]:

$$K(\omega) = K_0 + (K_S - K_0)\left(1 - \frac{\omega}{\omega_{max}}\right)^n \tag{2}$$

and (ii) with the van Genuchten–Mualem model (VGM) for unsaturated hydraulic conductivity [18,21]:

$$K(S) = K_S S^l \left[1 - \left(1 - S^{1/m}\right)^m\right]^2 \tag{3}$$

where K_S (cm·s^{-1}) is the saturated hydraulic conductivity, n (–), m(–) and l (–) are exponents, the S (–) is the water saturation ($S = 1 - S_{g,r}$) and K_0 (cm·s^{-1}) is the hydraulic conductivity obtained for a particular ω_{max} (–). The parameter ω_{max} is the maximum possible entrapped air content, which depends on the method chosen to saturate the sample. According to Faybishenko [1], it is not a soil property but rather a state variable.

Faybishenko's equation is based on a power-law relationship from Averyanov [51], used for two-phase flow, where the residual saturation was replaced by satiated saturation. The equations were verified on high columns (0.5–5 m in height) of soil during long-term infiltration.

Mualem [18] derived a general equation of relative hydraulic conductivity by the model of pore size distribution. Mualem then used Brooks and Corey's equation of retention curve [20] and the measurements of 45 soils from literature to estimate exponent l, which can be positive or negative. He obtained the value of $l = 0.5$, which is commonly used. van Genuchten [21] combined Mualem's model of pore distributions with an equation for the retention curve, and verified the equation on five samples with good agreement. The equation is known as the Mualem–van Genuchten equation.

We moreover investigated how residual gas saturations during each infiltration run were related to the initial gas saturation. Here we evaluated relationships proposed by Land [14] and by Aissaoui [52]. Land [14] uses the simple term to describe the relationship between initial gas saturation $S_{g,init}$ (–) and residual gas saturation $S_{g,r}$ (–):

$$\frac{1}{S_{g,r}} - \frac{1}{S_{g,init}} = \frac{1}{S_{g,r,max}} - 1 \tag{4}$$

where $S_{g,r,max}$ (–) is a maximum value of the $S_{g,r}$ obtained for $S_{g,init} = 1$. Since the $S_{g,init}$ value was in most cases less than 1, we propose a modified version of Land's term:

$$\frac{1}{S_{g,r}} - \frac{1}{S_{g,init}} = \frac{1}{S_{g,r,max}} - \frac{1}{S_{g,init,max}} \tag{5}$$

where $S_{g,r,max}$ is maximum value of the $S_{g,r}$ and the $S_{g,init,max}$ is appropriate value of the $S_{g,init}$. The term proposed by Aissaoui reads:

$$S_{g,r} = \frac{S_{g,r,max}}{S_{g,C}} S_{g,init} \quad when \ S_{g,init} < S_{g,C} \tag{6a}$$

$$S_{g,r} = S_{g,r,max} \quad when \ S_{g,init} \geq S_{g,C} \tag{6b}$$

where $S_{g,C}$ (–) is the critical gas saturation corresponding to the point where the maximum trapped saturation, $S_{g,r,max}$, is reached.

Land derived the relationship on the published experimental data on six sandstone samples [14]. He presented figures of six plots with good agreement between fitted functions and data. Aissaoui [52] proposed the linear relationship between $S_{g,r}$ and $S_{g,init}$ where after the reaching of the critical gas saturation $S_{g,C}$ the $S_{g,r}$ stays constant. According Pentland [53], Aissaoui's equation gives a better prediction of entrapped air saturation for unconsolidated soil than the Land's equation.

2.6. X-Ray Computed Tomography of Samples

In addition to the ten drainage and infiltration cycles specified in Table 2, two drainage and infiltration experiments were conducted in combination with X-ray computed micro-tomography (CT) imaging. In this fashion, we obtained information on the spatial distribution of entrapped air bubbles. The imaging was conducted using a GE Phoenix v|tome|x m industrial X-ray scanner with a 240 kV X-ray tube and a tungsten target, located at the Swedish University of Agricultural Sciences in Uppsala (Sweden). The result is a stack of 16-bit horizontal slices with pixel size 47 μm and 47 μm slice thickness.

The samples were first fully saturated in a vacuum chamber, after which they were X-rayed while submerged inside the water-filled outer vessel. The samples were then drained at a tension of −30 hPa in the case of material ST and −35 hPa in the case of material FH. The drained ST samples were scanned in air-filled outer vessels. Then an infiltration experiment was applied to all samples followed by immediate imaging. The samples were still submerged inside the water-filled outer vessel during imaging. The second drainage cycle was carried out under a pressure of −50 hPa before a second infiltration run and more X-ray scans were conducted. Details of the experiment schedule are summarized in Table 3 and Figure 4.

Table 3. Pressure heads set on a sand box in the experiment with computed micro-tomography (CT) imaging of the sample.

Parameters of Drainage	ST1	ST2	FH1	FH2
Pressure Head (cm)	−30	−30	−35	−35
Duration (h)	14	15	4.5	7.5
Pressure Head (cm)	−50	−50	−50	−50
Duration (h)	14	13	6.5	6.0

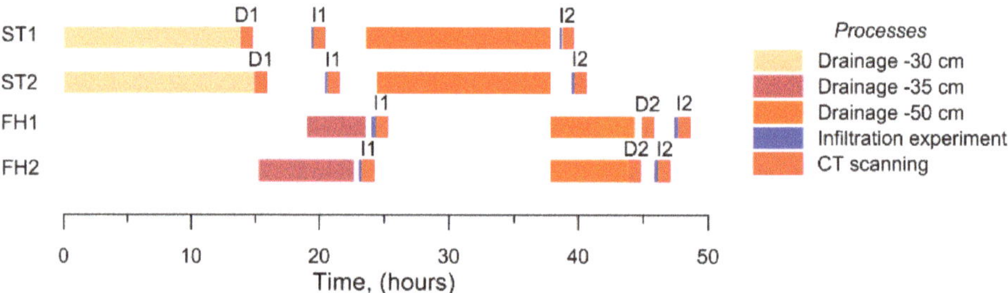

Figure 4. Schedule of the experiment with X-ray micro-tomography imaging. CT imaging (shown in red color) was conducted after the first infiltration (I1) and after the second infiltration (I2). CT imaging was also performed after drainage (D1 and D2) experiment and character D indicates CT scanning after drainage. The duration of scan was 1 hour and duration of infiltration experiment was from 10 to 20 minutes.

2.7. Image Preprocessing

The CT images were preprocessed and analyzed using the software Fiji (ImageJ) [54]. Preprocessing includes the elimination of the noise, edge enhancement and artifact removal [55].

First, all horizontal image layers in the region of interest were selected. The location of the topmost layer depended on the shape of the surface because after the infiltration experiment the surface was not exactly horizontal. The bottommost selected layer was defined by the location of the sample retainer. Then a 3D-median filter with radius 2 voxels and an unsharp mask was used.

The tomograms were further adjusted to minimize the effect of beam hardening. Beam hardening refers to the non-linear filtering of soft (or low-frequency) X-rays during the beam's passage through a sample, leaving only harder (or-high-frequency) X-rays in the beam [43]. For soil samples with a cylindrical shape, beam hardening manifests itself as a bright ring artifact along the column's perimeter [44].

More specifically, the beam-hardening artifacts in this study were detected in two forms: First, along with the entire sample, a general darkening of the image towards the horizontal center of the sand sample. Second, below the PMMA ring, the gray values were attenuated due to the extra X-ray-filtering effect of the ring.

To remove the artifact caused by the PMMA ring, the intensity of each slide was multiplied by a correction coefficient α_i. The average image intensity of a 60 pixels wide region of interest located within the wall of the sample cylinder (Figure 5a) was measured for each slice (I_i). The value of correction coefficient α_i for the slice i was then:

$$\alpha_i = \left(\frac{I_{avg}}{I_i} - 1\right)\beta + 1 \tag{7}$$

where I_{avg} is the average of all the values of I_i and β is a correction factor. The effect of the parameter β on profiles of average image intensity in the central part of the sample (shown in red in Figure 5c) is depicted in Figure 5d–f. A factor of $\beta = 0.5$ led to the best correction results.

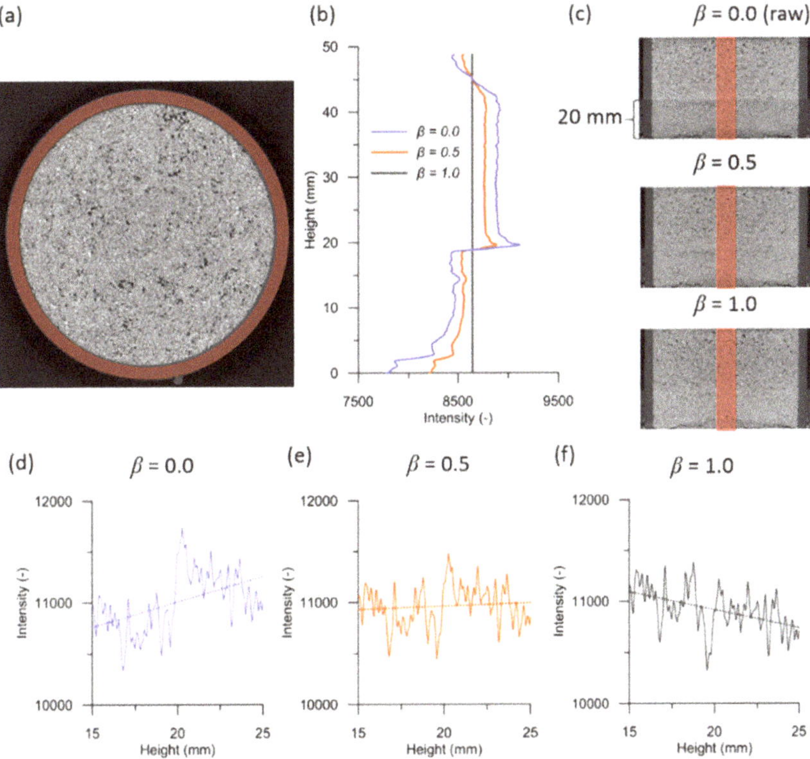

Figure 5. Adjustment of the vertical beam hardening artifact: (**a**) Slice with marked ROI corresponding to sample wall; (**b**) Profile of the I_i (average image intensity inside the wall in particular slice) for different values of β; (**c**) visual comparison for adjustments with different values of β. Note, that lower 20 mm are affected by the presence of the ring; (**d**–**f**) details of the image intensity inside the sand for the region most affected by the artefact.

The next adjustment was related to the second artifact, i.e., the darkening towards the horizontal center of the images. This was done by creating vertical cross-sections through the 3D images. Then, a 9.4 mm (200 pixels) wide strip along the diameter was selected and a profile of intensity was evaluated (Figure 6a,b). This profile was fitted with a center-axis parabola. The ratio between the value on the parabola and the edge values were given by factor ρ. Then each of the images was modified by dividing the intensity by ρ, given the distance from the center of the sample.

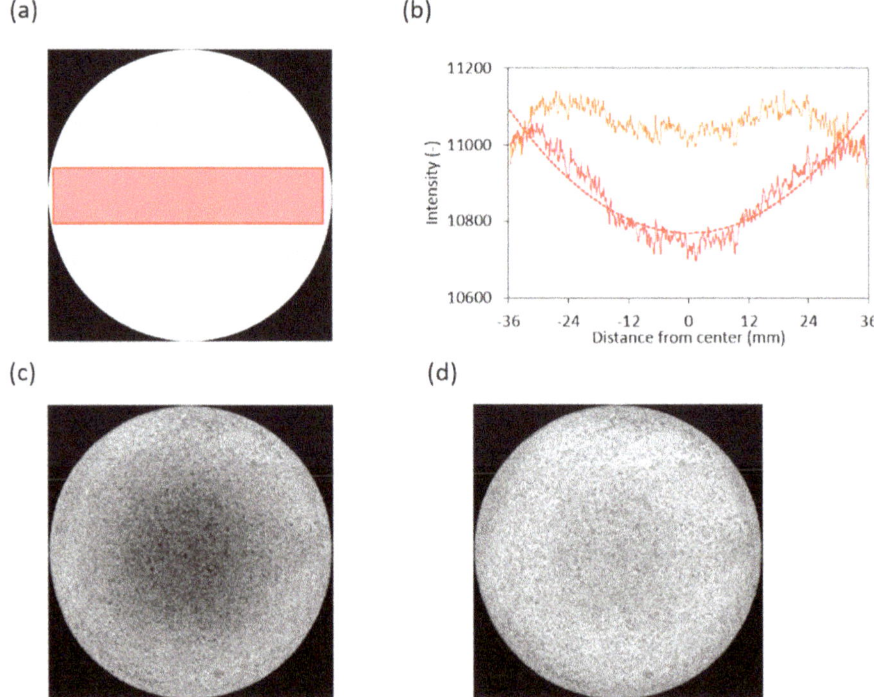

Figure 6. Adjustment of the cupping artifact: (**a**) Marked ROI; (**b**) course of the intensity in ROI from Figure 6a before and after adjustment; (**c**) image before the adjustment; (**d**) image after the adjustment.

2.8. Image Analysis

The images were binarized into two classes—one class depicting the gas phase and the other all other phases, namely the water and solid phases. An automatic binarization was not successful in this case due to poorly recognizable peaks on the intensity histogram. The threshold values were obtained by visual inspection for each sample. Then a morphological opening with a kernel radius of 1 voxel was carried out using the ImageJ plugins 3D Erode and 3D Dilate. We then removed 'holes' in the segmented air bubbles, which represented segmentation artifacts, using the 3D Fill Holes plugin [56].

We determined the imaged gas volume content as a function of the height of the sand samples as the ratio between the area of gas and the column cross-section for each image slice. These results were also used to determine the air volume content of the whole sample. The plugin BoneJ [57] was used to evaluate the thickness of the air bubbles, defined as the diameter of the largest sphere that fits within individual bubbles.

3. Results and Discussion

3.1. Initial Gas Saturation

The relationship between initial and residual gas saturation as fitted using Aissaoui's and Land's equations is shown in Figure 7 together with the results of Pentland et al. We observed a positive correlation between $S_{g,init}$ and the residual saturation $S_{g,r}$. The maximum of the initial gas saturation ranged between 80% to 90%. The evaluated parameters for Land's and Aissaoui's models are shown in Table 4. In our case the growth of $S_{g,r}$ value did not converge to an upper limit as it was supposed by Aissaoui. On the basis of experiments with octane used as a non-wetting phase, Pentland et al. [53]

assumed that the upper limit of $S_{g,r}$ corresponds to the occupancy of all large pores by the non-wetting phase. They also assumed that small pores continue to fill causing further increase of $S_{g,init}$, while $S_{g,r}$ does not increase, because gas is displaced from the small pores. In our study the upper limit of $S_{g,r}$ was not detected, probably due to the apparently unstable pore geometry in the upper portion of the sample, where pores enlarged with increasing $S_{g,init}$. This effect will be discussed in detail in the next paragraph. Yet, Aissaoui's equation results in a better fit of the experimental values than the Land's equation, which is in agreement with Pentland's study.

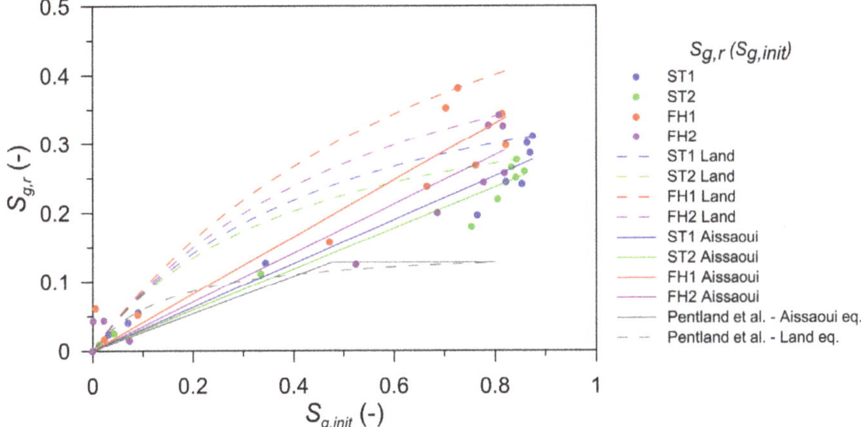

Figure 7. Initial gas saturation vs. residual gas saturation, for experiment without CT. The modified Land's (5) and Aissaoui's (6) equations were used for fitting. Pentland et al. [53] used octane in a sandpack.

Table 4. Values of the parameters and coefficients of determination for Land's equation (5) and Aissaoui's equation (6).

Sand Sample	Land			Aissaoui	
	$S_{g,r,max}$	$S_{g,init,max}$	R^2	$S_{g,r,max}$	R^2
ST1	0.311	0.875	0.9183	0.311	0.9531
ST2	0.276	0.843	0.9149	0.276	0.9682
FH1	0.382	0.727	0.8570	0.382	0.9002
FH2	0.342	0.808	0.8101	0.342	0.9019

3.2. Air Content and Air Distribution

Table 5 lists the volumetric air contents ω of the whole samples, obtained gravimetrically and by X-ray image analysis. The amount of entrapped air, after infiltrations I1 and I2, was higher for FH sand. The values of ω obtained gravimetrically were higher than the values obtained by X-ray imaging. This was probably caused by the fact that bubbles under certain sizes were not detected being below the image resolution. The field of view which determines the image resolution is given by the size of the sample. The size of the sample was designed to be larger than the expected representative elementary volume of sand, which led to an image resolution that neglected the smaller bubbles. Table 5 shows that the gravimetrically measured ω values were significantly larger after drainage than the ω values measured after each infiltration experiment. However, this effect is not reproduced in the X-ray imaging results. On the contrary, we observed the opposite. This could be due to the presence of predominantly small bubbles after drainage that remained too small to be detected in the image data, as can be seen in Figure 8, where vertical cross-sections of the binarized images of the sample ST2 are shown. After drainage (D1) there was a large number of small bubbles, mostly located in the upper half of the sample. We can assume that in this case an additional large number of small bubbles existed, which remained undetected because they were below the resolution of the X-ray images. After the first infiltration experiment (I1), larger pores were created by displacing nearby grains of sand (Figure 8b) and the air had aggregated into larger bubbles, probably due to the effect of Ostwald ripening [58]. After the second infiltration experiment (I2), the bubbles became even larger (Figure 8c). It seems that the higher amount of the visible air bubbles in the sample occurred due to the enlargement of pores.

Table 5. Comparison of values of the air volume content obtained by gravimetric method and image analysis.

Property	Experiment Stage	ST1	ST2	FH1	FH2
Porosity (–)	-	0.372	0.364	0.357	0.327
ω gravimetric (–)	D1	0.244	0.140	0.175	0.197
	I1	0.040	0.080	0.051	0.103
	D2	0.300	0.300	0.283	0.252
	I2	0.045	0.101	0.088	0.108
ω image analysis (–)	D1	0.035	0.040	-	-
	I1	0.026	0.044	0.046	0.045
	D2	-	-	0.055	0.056
	I2	0.039	0.059	0.063	0.065

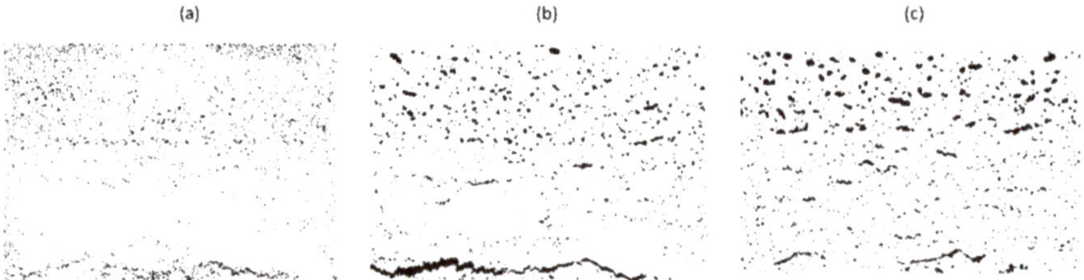

Figure 8. Differences between gas bubbles distribution and bubbles sizes in sample ST2 in different stages: (**a**) stage D1; (**b**) stage I1; (**c**) stage I2. In the figure are the vertical cross-sections through the center.

The air saturation profile along the height of the sample is shown in Figure 9. It should be noted that only relatively large bubbles were detected by x-ray CT. It is evident that the air is predominantly distributed in the upper part of the sample, from 25 mm upwards, especially after infiltration experiments (I1, I2). The accumulation of air in large bubbles at the upper half of the sample was probably caused by the buoyancy force in combination with Ostwald ripening.

The amount of air in the upper part is approximately doubled between I1 and I2. The air in the lower 5 mm is predominantly present in newly created cracks.

The frequency of the thickness of the bubbles can be seen in Figure 10. To eliminate the effect of the air-filled fractures, only parts of the image 2 cm above the bottom were analyzed. In all the cases, the air bubble thickness increased between stages I1 and I2. More specifically, the thickness of the smallest 50% of the bubbles changed from 300 to 405 µm for ST1, from 370 to 470 µm for ST2, from 310 to 405 µm for FH1, and from 370 to 470 µm for FH2. The maximum thickness of all detected air bubbles was approximately 2000 µm.

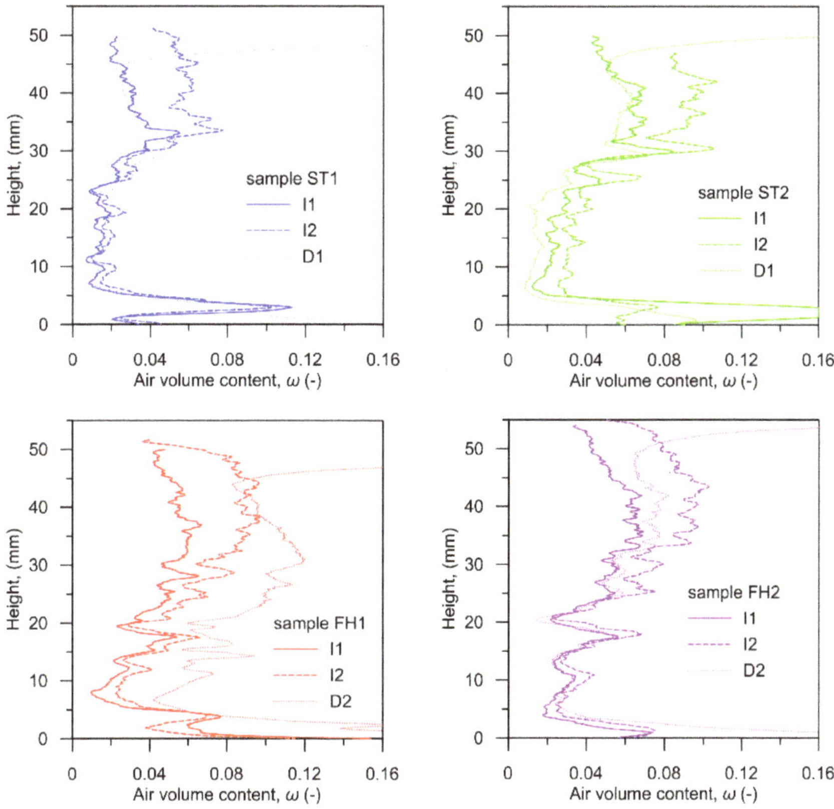

Figure 9. Air volume content of the sand as a function of the height.

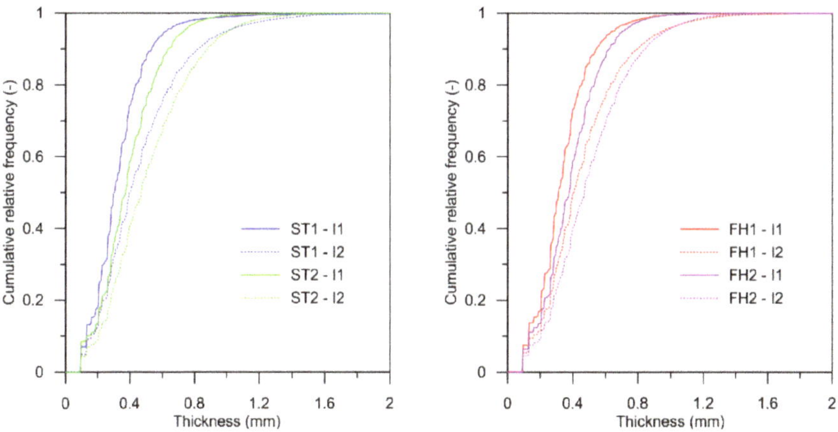

Figure 10. Bubble thickness for the stages after the infiltration experiment (I1 and I2).

3.3. Hydraulic Conductivity

Figure 11 shows the relationship between the satiated hydraulic conductivity and the air content in the sand. Also depicted are fits with Faybishenko's (2) and VGM functions (3). The respective model parameters n and m were obtained by the least squares method. The corresponding values are shown in Table 6. It can be seen that the results of the infiltration run on the ST samples conducted next to the X-ray scanner followed the same trend as observed for the other 10 experiments. This was not the case for the FH samples, especially not for FH2. One possible explanation could be the smaller number of measured points and the experiment FH2 can be considered as an outlier. The coefficients of determination for both models are given in Table 6. The coefficients of determination for Faybishenko's model for batch with CT scanning are high due to fitting three points only. K_S of the coarser sand ST is almost twice as high as the K_S of the finer sand FH. The mean K_S of FH sand (0.035 cm·s^{-1}) was similar to the value reported for the same sand FH31 by Sacha et al. (0.040 cm·s^{-1}) and by Haber-Pohlmeier et al. (0.045 cm·s^{-1}) but higher than value given by Sucre et al. (0.011 cm·s^{-1}). Differences between K_S values might have been caused by different packing density.

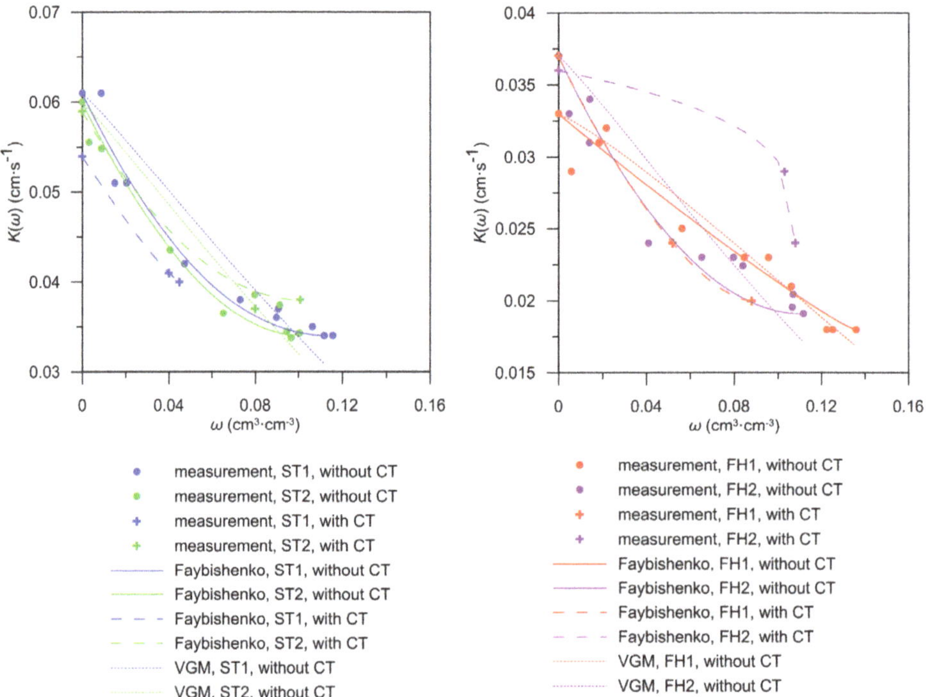

Figure 11. Satiated hydraulic conductivity of sands as a function of entrapped air content, fitted by Faybishenko's and Mualem's (VGM) models (lines). The ST sand is on left and FH sand is on right.

Table 6. Parameters and coefficients of determination of Faybishenko's and VGM model.

Parameters of Models	ST1	ST2	FH1	FH2
Without CT—Faybishenko				
n $(-)$	2.13	1.90	1.14	2.00
K_0 (cm·s^{-1})	0.034	0.034	0.018	0.019
K_S (cm·s^{-1})	0.061	0.060	0.033	0.037
ω_{max} $(-)$	0.116	0.107	0.142	0.117
R^2	0.9694	0.9665	0.9448	0.9327
Without CT—VGM				
m $(-)$	1.15	1.13	1.36	1.17
R^2	0.8701	0.9055	0.9382	0.8178
With CT—Faybishenko				
n $(-)$	1.20	1.97	1.62	0.28
K_0 (cm·s^{-1})	0.040	0.038	0.020	0.024
K_S (cm·s^{-1})	0.054	0.059	0.037	0.036
ω_{max} $(-)$	0.045	0.101	0.088	0.108
R^2	1.0000	0.9877	1.0000	1.0000

Faybishenko's model led to similar fits of the $K(\omega)$ as the VGM equation. An exception was sample FH1, where the overall shape of the $K(\omega)$ was abnormal. In comparison, Marinas et al. [24] determined relatively similar values of coefficients of determination for Faybishenko's and the VGM equations, with one exception, where Faybishenko's equation fitted better. The exception was the Unimin-2010 sand, which has a porosity (0.38) and bulk-density (1.63 g·cm^{-3}) similar to ST sand. The R^2 values for Faybishenko's equation were 0.952 and 0.927 and R^2 values for the VGM equation were 0.890 and 0.841. However, in the VGM equation, Marinas et al. [24] used a fitted exponent l, with values 0.347 and 0.618. In our case, Faybishenko's parameters n and ω_{max} are similar to the values published for Masa sandy loam [25].

4. Conclusions

For the purpose of simulation of two-phase flow, the relationship between entrapped air content and satiated hydraulic conductivity $K(\omega)$ of two coarse sands was studied. Simultaneously with falling head infiltration and drainage experiments, the X-ray computed tomography imaging was performed to reveal temporal changes of the entrapped air distribution and the bubble shapes within the sample.

The positive correlation between initial air saturation $S_{g,init}$ and residual air saturation $S_{g,r}$ was found; this result is in agreement with Land [14] or Mansoori et al. [59]. The maximum value of $S_{g,r}$ was 0.31 for ST sand and 0.38 for FH sand. A good fit of the $S_{g,r}$ ($S_{g,init}$) relationship was obtained using Aissaoui's equation.

As expected, the saturated hydraulic conductivity of coarser sand ST was consistently higher than the hydraulic conductivity of finer sand FH. The hydraulic conductivity of samples at full saturation was approximately twice as high as the hydraulic conductivity of samples with maximal entrapped air content. The relationship between air volume content and satiated hydraulic conductivity was successfully fitted with Faybishenko's equation. Faybishenko's model seems to be appropriate to express the $K(\omega)$ relationship.

X-ray computed tomography revealed that the size of air bubbles and clusters increased significantly with each infiltration-drainage cycle. The final shapes of observed bubbles appeared to be elongated in the horizontal direction. The increase in bubble size was not spatially uniform within an individual sample. Instead, air bubbles grew larger close to the surface of the sample. Bubble ebullition probably also led to the formation of the larger voids in the sand packing.

The obtained $K(\omega)$ relationship can be used as a characteristic in newly developed numerical models of two-phase flow, such as the model by Fucik and Mikyska [60]. It is evident that the entrapped air significantly influences measured values of hydraulic conductivity even in samples of simple material and geometry. The presented methodology can be used to determine the $K(\omega)$ in other natural and artificial porous materials.

Author Contributions: T.P. and H.M.R.F. conducted the experiments and wrote an original draft. T.P. did the development of methodology and formal analysis of results. J.K. provisioned CT imaging of sand samples, contributed to the image analysis and reviewed and edited the manuscript. M.S. did conceptualization, funding acquisition, project administration and supervision, development of methodology and review and editing of the manuscript. All authors have read and agreed to the published version of the manuscript.

Funding: This research was funded by the Czech Science Foundation under project number GA17-06759S and by the Grant Agency of the Czech Technical University in Prague, grant No. SGS18/122/OHK1/2T/11.

Acknowledgments: We thank Milena Cislerova for valuable comments that greatly improved the manuscript.

Conflicts of Interest: The authors declare no conflict of interest.

Appendix A

The initial gas saturation ($S_{g,init}$) value determines the amount of gas that is in the sample after drainage in a sand box. The mass of the sample container with sand after drainage (when the sample was not yet submerged in water) is $M_{samp,init}$. To evaluate $S_{g,init}$, also the value of the mass of the sample container with sand after the full saturation, $M_{samp,sat}$, must be acquired. The difference between $M_{samp,sat}$ and $M_{samp,init}$ determines the mass of water that has been replaced by gas and thus the volume of entrapped gas within the sand sample (assuming $\rho_w =1$ g·cm^{-3}). $S_{g,init}$ value is then:

$$S_{g,init} = \frac{\left(M_{samp,sat} - M_{samp,init}\right)}{\rho_w p V_{sand}} \tag{A1}$$

where ρ_w is water density and p is the porosity of the sand sample. The mass of the fully saturated sample (see Figure A1) $M_{samp,sat}$ is:

$$M_{samp,sat} = M_0 - \left(M_{fill,w} - (V_{sand} + V_{cont})\rho_w\right) \tag{A2}$$

where M_0 is the mass of the outer vessel with a fully saturated sample (see Figure 3c), $M_{fill,w}$ is the mass of the vessel without the sample, when the water level is exactly at the reference level. V_{sand} is the volume of the sand sample, V_{cont} is the volume of the container wall and ring.

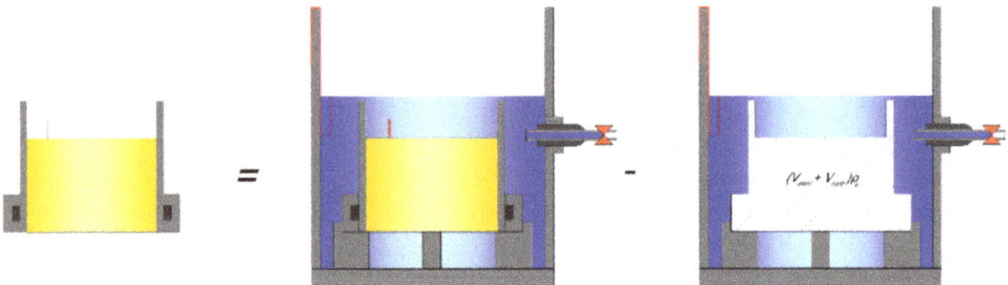

Figure A1. Mass of the saturated sand sample and sample container ($M_{samp,sat}$) is equal to the difference between the mass of the vessel with fully saturated sample flooded to the reference level (M_0) and the mass of the vessel and water surrounding sample ($M_{fill,w} - (V_{sand}+V_{cont})\rho_w$).

References

1. Faybishenko, B. Hydraulic behavior of quasi-saturated soils in the presence of entrapped air - laboratory experiments. *Water Resour. Res.* **1995**, *31*, 2421–2435. [CrossRef]

2. Sacha, J.; Jelinkova, V.; Snehota, M.; Vontobel, P.; Hovind, J.; Cislerova, M. Water and air redistribution within a dual permeability porous system investigated using neutron imaging. *Proc. 10th World Conf. Neutron Radiogr. (Wcnr-10)* **2015**, *69*, 530–536. [CrossRef]

3. Snehota, M.; Jelinkova, V.; Sobotkova, M.; Sacha, J.; Vontobel, P.; Hovind, J. Water and entrapped air redistribution in heterogeneous sand sample: Quantitative neutron imaging of the process. *Water Resour. Res.* **2015**, *51*, 1359–1371. [CrossRef]

4. Roy, J.W.; Smith, J.E. Multiphase flow and transport caused by spontaneous gas phase growth in the presence of dense non-aqueous phase liquid. *J. Contam. Hydrol.* **2007**, *89*, 251–269. [CrossRef] [PubMed]

5. Tzimas, G.C.; Matsuura, T.; Avraam, D.G.; VanderBrugghen, W.; Constantinides, G.N.; Payatakes, A.C. The combined effect of the viscosity ratio and the wettability during forced imbibition through nonplanar porous media. *J. Colloid Interface Sci.* **1997**, *189*, 27–36. [CrossRef]

6. Ausweger, G.M.; Schweiger, H.F. Numerical study on the influence of entrapped air bubbles on the time-dependent pore pressure distribution in soils due to external changes in water level. In Proceedings of the 3rd European Conference on Unsaturated Soils (E-UNSAT), Paris, France, 12–14 September 2016.

7. Boulding, J.R.; Ginn, S.J. *Practical Handbook of Soil, Vadose Zone, and Ground-Water Contamination: Assessment, Prevention, and Remediation, Second Edition*; CRC Press: Boca Raton, FL, USA, 2003.

8. Chatzis, I.; Morrow, N.R.; Lim, H.T. Magnitude and detailed structure of residual oil saturation. *Soc. Pet. Eng. J.* **1983**, *23*, 311–326. [CrossRef]

9. Almajid, M.M.; Kovscek, A.R. Pore-level mechanics of foam generation and coalescence in the presence of oil. *Adv. Colloid Interface Sci.* **2016**, *233*, 65–82. [CrossRef]

10. Kovscek, A.R.; Tang, G.Q.; Radke, C.J. Verification of Roof snap off as a foam-generation mechanism in porous media at steady state. *Colloids Surf. A-Phys. Eng. Asp.* **2007**, *302*, 251–260. [CrossRef]

11. Cislerova, M.; Simunek, J.; Vogel, T. Changes of steady-state infiltration rates in recurrent ponding infiltration experiments. *J. Hydrol.* **1988**, *104*, 1–16. [CrossRef]

12. Li, Y.; Flores, G.; Xu, J.; Yue, W.Z.; Wang, Y.X.; Luan, T.G.; Gu, Q.B. Residual air saturation changes during consecutive drainage-imbibition cycles in an air-water fine sandy medium. *J. Hydrol.* **2013**, *503*, 77–88. [CrossRef]

13. Nguyen, V.H.; Sheppard, A.P.; Knackstedt, M.A.; Pinczewski, W.V. The effect of displacement rate on imbibition relative permeability and residual saturation. *J. Pet. Sci. Eng.* **2006**, *52*, 54–70. [CrossRef]

14. Land, C.S. Calculation of imbibition relative permeability for two and three-phase flow from rock properties. *Soc. Pet. Eng. J.* **1968**, *8*, 149. [CrossRef]

15. Krol, M.M.; Mumford, K.G.; Johnson, R.L.; Sleep, B.E. Modeling discrete gas bubble formation and mobilization during subsurface heating of contaminated zones. *Adv. Water Resour.* **2011**, *34*, 537–549. [CrossRef]

16. Bear, J. *Dynamics of Fluids in Porous Media*; Elsevier: New York, NY, USA, 1972.

17. Hilfer, R. Capillary pressure, hysteresis and residual saturation in porous media. *Phys. A-Stat. Mech. Its Appl.* **2006**, *359*, 119–128. [CrossRef]

18. Mualem, Y. New model for predicting hydraulic conductivity of unsaturated porous-media. *Water Resour. Res.* **1976**, *12*, 513–522. [CrossRef]

19. Childs, E.C.; Collisgeorge, N. The permeability of porous materials. *Proc. R. Soc. Lond. Ser. A-Math. Phys. Sci.* **1950**, *201*, 392–405. [CrossRef]

20. Brooks, R.H.; Corey, A.T. *Hydraulic Properties of Porous Media*, 3rd ed.; Colorado State University: Fort Collins, CO, USA, 1964.

21. Van Genuchten, M.T. A closed-form equation for predicting the hydraulic conductivity of unsaturated soils. *Soil Sci. Soc. Am. J.* **1980**, *44*, 892–898. [CrossRef]

22. Fry, V.A.; Selker, J.S.; Gorelick, S.M. Experimental investigations for trapping oxygen gas in saturated porous media for in situ bioremediation. *Water Resour. Res.* **1997**, *33*, 2687–2696. [CrossRef]

23. Ryan, M.C.; MacQuarrie, K.T.B.; Harman, J.; McLellan, J. Field and modeling evidence for a "stagnant flow" zone in the upper meter of sandy phreatic aquifers. *J. Hydrol.* **2000**, *233*, 223–240. [CrossRef]

24. Marinas, M.; Roy, J.W.; Smith, J.E. Changes in Entrapped Gas Content and Hydraulic Conductivity with Pressure. *Ground Water* **2013**, *51*, 41–50. [CrossRef]

25. Sakaguchi, A.; Nishimura, T.; Kato, M. The effect of entrapped air on the quasi-saturated soil hydraulic conductivity and comparison with the unsaturated hydraulic conductivity. *Vadose Zone J.* **2005**, *4*, 139–144. [CrossRef]

26. Loizeau, S.; Rossier, Y.; Gaudet, J.P.; Refloch, A.; Besnard, K.; Angulo-Jaramillo, R.; Lassabatere, L. Water infiltration in an aquifer recharge basin affected by temperature and air entrapment. *J. Hydrol. Hydromech.* **2017**, *65*, 222–233. [CrossRef]

27. Pohlmeier, A.; Garre, S.; Roose, T. Noninvasive Imaging of Processes in Natural Porous Media: From Pore to Field Scale. *Vadose Zone J.* **2018**, *17*. [CrossRef]

28. Yu, X.; Hong, C.; Peng, G.; Lu, S. Response of pore structures to long-term fertilization by a combination of synchrotron radiation X-ray microcomputed tomography and a pore network model. *Eur. J. Soil Sci.* **2018**, *69*, 290–302. [CrossRef]

29. Udawatta, R.P.; Gantzer, C.J.; Anderson, S.H.; Assouline, S. Synchrotron microtomographic quantification of geometrical soil pore characteristics affected by compaction. *Soil* **2016**, *2*, 211–220. [CrossRef]

30. Zong, Y.T.; Yu, X.L.; Zhu, M.X.; Lu, S.G. Characterizing soil pore structure using nitrogen adsorption, mercury intrusion porosimetry, and synchrotron-radiation-based X-ray computed microtomography techniques. *J. Soils Sediments* **2015**, *15*, 302–312. [CrossRef]

31. Peth, S.; Horn, R.; Beckmann, F.; Donath, T.; Fischer, J.; Smucker, A.J.M. Three-dimensional quantification of intra-aggregate pore-space features using synchrotron-radiation-based microtomography. *Soil Sci. Soc. Am. J.* **2008**, *72*, 897–907. [CrossRef]

32. Zhou, H.; Peng, X.; Peth, S.; Xiao, T.Q. Effects of vegetation restoration on soil aggregate microstructure quantified with synchrotron-based micro-computed tomography. *Soil Tillage Res.* **2012**, *124*, 17–23. [CrossRef]

33. Ma, R.M.; Cai, C.F.; Li, Z.X.; Wang, J.G.; Xiao, T.Q.; Peng, G.Y.; Yang, W. Evaluation of soil aggregate microstructure and stability under wetting and drying cycles in two Ultisols using synchrotron-based X-ray micro-computed tomography. *Soil Tillage Res.* **2015**, *149*, 1–11. [CrossRef]

34. Wildenschild, D.; Hopmans, J.W.; Rivers, M.L.; Kent, A.J.R. Quantitative analysis of flow processes in a sand using synchrotron-based X-ray microtomography. *Vadose Zone J.* **2005**, *4*, 112–126. [CrossRef]

35. Schnaar, G.; Brusseau, M.L. Characterizing pore-scale configuration of organic immiscible liquid in multiphase systems with synchrotron X-ray microtomography. *Vadose Zone J.* **2006**, *5*, 641–648. [CrossRef]

36. Koestel, J.; Schluter, S. Quantification of the structure evolution in a garden soil over the course of two years. *Geoderma* **2019**, *338*, 597–609. [CrossRef]

37. Sandin, M.; Koestel, J.; Jarvis, N.; Larsbo, M. Post-tillage evolution of structural pore space and saturated and near-saturated hydraulic conductivity in a clay loam soil. *Soil Tillage Res.* **2017**, *165*, 161–168. [CrossRef]

38. Suekane, T.; Thanh, N.H.; Matsumoto, T.; Matsuda, M.; Kiyota, M.; Ousaka, A. Direct measurement of trapped gas bubbles by capillarity on the pore scale. *Greenh. Gas Control Technol. 9* **2009**, *1*, 3189–3196. [CrossRef]

39. Setiawan, A.; Nomura, H.; Suekane, T. Microtomography of Imbibition Phenomena and Trapping Mechanism. *Transp. Porous Media* **2012**, *92*, 243–257. [CrossRef]

40. Geistlinger, H.; Mohammadian, S.; Schlueter, S.; Vogel, H.J. Quantification of capillary trapping of gas clusters using X-ray microtomography. *Water Resour. Res.* **2014**, *50*, 4514–4529. [CrossRef]

41. Geistlinger, H.; Mohammadian, S. Capillary trapping mechanism in strongly water wet systems: Comparison between experiment and percolation theory. *Adv. Water Resour.* **2015**, *79*, 35–50. [CrossRef]

42. Herring, A.L.; Gilby, F.J.; Li, Z.; McClure, J.E.; Turner, M.; Veldkamp, J.P.; Beeching, L.; Sheppard, A.P. Observations of nonwetting phase snap-off during drainage. *Adv. Water Resour.* **2018**, *121*, 32–43. [CrossRef]

43. Barrett, J.F.; Keat, N. Artifacts in CT: Recognition and avoidance. *Radiographics* **2004**, *24*, 1679–1691. [CrossRef]

44. Weller, U.; Leuther, F.; Schluter, S.; Vogel, H.J. Quantitative Analysis of Water Infiltration in Soil Cores Using X-Ray. *Vadose Zone J.* **2018**, *17*. [CrossRef]

45. Izadi, B.; Podmore, T.H.; Seymour, R.M. Consideration of air entrapment in surface irrigation - a computer-simulation study. *Agric. Water Manag.* **1995**, *28*, 245–252. [CrossRef]

46. Goncalves, R.D.; Teramoto, E.H.; Engelbrecht, B.Z.; Alfaro Soto, M.A.; Chang, H.K.; van Genuchten, M.T. Quasi-Saturated Layer: Implications for Estimating Recharge and Groundwater Modeling. *Groundwater* **2019**. [CrossRef]

47. Mizrahi, G.; Furman, A.; Weisbrod, N. Infiltration under Confined Air Conditions: Impact of Inclined Soil Surface. *Vadose Zone J.* **2016**, *15*. [CrossRef]

48. Sacha, J.; Snehota, M.; Trtik, P.; Hovind, J. Impact of Infiltration Rate on Residual Air Distribution and Hydraulic Conductivity. *Vadose Zone J.* **2019**, *18*. [CrossRef]

49. Klute, A. Water retention: Laboratory methods. In *Methods of Soil Analysis—Part 1—Physical and Mineralogical Methods*; Klute, A., Ed.; American Society of Agronomy: Madison, WI, USA, 1986; pp. 635–662.

50. Klute, A.; Dirksen, C. Hydraulic conductivity and diffusivity: Laboratory methods. In *Methods of Soil Analysis—Part 1—Physical and Mineralogical Methods*; Klute, A., Ed.; American Society of Agronomy: Madison, WI, USA, 1986; pp. 687–734.

51. Averyanov, S.F. Zavisimost vodopronitsaemosti pochvo-gruntov ot soderzhaniya v nikh vozdukha. *Dokl. Akad. Nauk Sssr* **1949**, *69*, 141–144.

52. Aissaoui, A. Etude théorique et expérimentale de l'hystérésis despressions capillaries et des perméabilitiés relatives en vue du stockagesouterrain de gaz. Ph.D. Thesis, École des Mines de Paris, Paris, France, 1983.

53. Pentland, C.H.; Itsekiri, E.; Al Mansoori, S.K.; Iglauer, S.; Bijeljic, B.; Blunt, M.J. Measurement of Nonwetting-Phase Trapping in Sandpacks. *Spe J.* **2010**, *15*, 274–281. [CrossRef]

54. Schindelin, J.; Arganda-Carreras, I.; Frise, E.; Kaynig, V.; Longair, M.; Pietzsch, T.; Preibisch, S.; Rueden, C.; Saalfeld, S.; Schmid, B.; et al. Fiji: An open-source platform for biological-image analysis. *Nat. Methods* **2012**, *9*, 676–682. [CrossRef] [PubMed]

55. Schluter, S.; Sheppard, A.; Brown, K.; Wildenschild, D. Image processing of multiphase images obtained via X- ray microtomography: A review. *Water Resour. Res.* **2014**, *50*, 3615–3639. [CrossRef]

56. Ollion, J.; Cochennec, J.; Loll, F.; Escude, C.; Boudier, T. TANGO: A generic tool for high-throughput 3D image analysis for studying nuclear organization. *Bioinformatics* **2013**, *29*, 1840–1841. [CrossRef]

57. Doube, M.; Klosowski, M.M.; Arganda-Carreras, I.; Cordelieres, F.P.; Dougherty, R.P.; Jackson, J.S.; Schmid, B.; Hutchinson, J.R.; Shefelbine, S.J. BoneJ Free and extensible bone image analysis in ImageJ. *Bone* **2010**, *47*, 1076–1079. [CrossRef]

58. Garing, C.; de Chalendar, J.A.; Voltolini, M.; Ajo-Franklin, J.B.; Benson, S.M. Pore-scale capillary pressure analysis using multi-scale X-ray micromotography. *Adv. Water Resour.* **2017**, *104*, 223–241. [CrossRef]

59. Al Mansoori, S.; Iglauer, S.; Pentland, C.H.; Bijeljic, B.; Blunt, M.J. Measurements of Non-Wetting Phase Trapping Applied to Carbon Dioxide Storage. *Greenh. Gas Control Technol. 9* **2009**, *1*, 3173–3180. [CrossRef]

60. Fucik, R.; Mikyska, J. Mixed-hybrid finite element method for modelling two-phase flow in porous media. *J. Math. Ind.* **2011**, *3*, 9–19.

 *water*

Article

Numerical Investigation of Techno-Economic Multiobjective Optimization of Geothermal Water Reservoir Development: A Case Study of China

Luyi Zhang [1], Ruifei Wang [2], Hongqing Song [1,*], Hui Xie [1,*], Huifang Fan [1], Pengguang Sun [3] and Li Du [3]

[1] School of Civil and Resource Engineering, University of Science and Technology Beijing, 30 Xueyuan Rd, Beijing 100083, China; 18813050023@163.com (L.Z.); fanhuifang@ustb.edu.cn (H.F.)

[2] College of Petroleum Engineering, Xi'an Shiyou University, Xi'an 710065, China; sirwrf2003@163.com

[3] Sinopec Star Petroleum Co., Ltd., 263 N 4th Ring Road Middle, Beijing 100083, China; sunpengguang.xxsy@sinopec.com (P.S.); duli.xxsy@sinopec.com (L.D.)

* Correspondence: songhongqing@ustb.edu.cn (H.S.); xiehui20000@ustb.edu.cn (H.X.)

Received: 20 September 2019; Accepted: 4 November 2019; Published: 6 November 2019

Abstract: Renewable geothermal utilization is a significant approach for residential heating with two principal modes, direct geothermal district heating systems (DGDHSs) and indirect geothermal district heating systems (IGDHSs). The key principle of geothermal development design is to prevent premature thermal breakthrough, which could result in low efficiency of geothermal heating systems. In this paper, a new approach considering building heating demand, geothermal water resource protection, and optimal economic benefits is presented systematically. The results simulated by OGS software show that well spacing, reinjection temperature, and production rate are the most significant parameters affecting thermal breakthrough in geothermal reservoirs. In addition, production rate and reinjection temperature have a huge effect on the payback period of investment. Comparing IGDHS to DGDHS, the investment in construction of geothermal wells and the annual water consumption decrease by up to 10% and 50%, respectively. Additionally, electricity costs increase by 5% to 30%. The indirect geothermal district heating system with a well spacing of 300 m, a production rate of 100 m^3/h, and a reinjection temperature of 301.15 K is much better for this case, both technically and economically. The systematic calculation approach can be reasonably applied to other regions with geothermal energy utilization.

Keywords: geothermal water reservoir; well spacing; direct geothermal district heating system; indirect geothermal district heating system; multiobjective optimization; technical and economic evaluation

---

1. Introduction

With the development of the economy and improvement of people's living standards, primary energy consumption has maintained a rapid growth. In northern China, energy consumption of building heating and space cooling account for 20% and 14%, respectively [1]. One measure for energy conservation and emission reduction is to improve building insulation levels and the air tightness of building envelopes. Another measure is to use geothermal resources. Geothermal energy is a stable resource from the earth's interior [2,3]. Theoretically, it can be utilized continuously [4]. As long as the upper limit of geothermal utilization is properly controlled, the service life of a geothermal reservoir can be extended [5,6]. Besides, geothermal energy is generally unaffected by geography, climate, and season, unlike solar, water, wind, biomass, and ocean energy [7,8].

Geothermal heating technology is an effective way to fully develop and utilize geothermal energy [9,10]. It has been applied in some countries because of its high operational efficiency and significant environmental benefits [11]. However, the high initial investment and long payback period have hindered the popularization of this technology [12,13]. Geothermal well infrastructure investment can exceed 50% of system construction costs for all types of geothermal utilization systems [14,15]. Statistically, the expense of drilling a geothermal well is about \$300 per meter [16]. The cost increases with deepening of the wells [17]. In addition, all extracted groundwater needs to be reinjected to maintain the pressure of the geothermal reservoir. The reinjection rate is far lower than the production rate, resulting in a waste of ground water [18]. Cold reinjection water also leads premature thermal breakthrough, which affects the system's service behavior and operating efficiency.

Researchers have studied the energy production of geothermal temperature fields. Zhang et al. developed a numerical model to evaluate geothermal reservoirs and optimize the production performance of geothermal systems [19]. Held et al. derived a reservoir mathematical model that considered fluid–rock interaction in the utilization of geothermal resources [20]. Nam carried out 3D numerical simulations and verified the reliability by comparing results with experimental results [21]. Salimzadeh et al. set up a fully coupled thermal–hydraulic–mechanical (THM) finite element model [22]. Pandey et al. found that thermochemical and thermomechanical effects were the most important consequences of heterogeneity [23]. Nikhil et al. simplified the crack model by a statistical linear congruent generator method, and carried out sensitivity analysis of fracture aperture size and fluid flow [24].

The previous efforts offered many insights into modeling heat extraction from geothermal reservoirs. Rock deformation should be emphasized, which may lead to unwanted sand components in the water flow [25]. However, previous studies ignored the economic benefits of geothermal projects. The disadvantages of geothermal projects without economic benefit analysis were huge. It not only increased the cost, but also caused wasted geothermal resources. High investment and operating cost hindered the development of geothermal heating technology.

In this study, based on a coupled mathematical and multiobjective optimization model, numerical investigation is carried out to evaluate the heat extraction and economic efficiency of a geothermal project in Xinji County. First, coupling of the thermal–hydraulic–mechanical model and multiobjective function of the geothermal heating system is established. Second, the main factors causing thermal breakthrough are analyzed based on data of a geothermal reservoir in Xinji City. Third, production parameters of direct and indirect geothermal district heating systems are optimized based on the ground heating load and system operation strategies. Finally, the initial investment and operation cost of direct and indirect geothermal district heating systems are analyzed, and the optimal method and production parameters are thus obtained. This optimization and economic evaluation method can be applied to geothermal heating systems in other regions, which is beneficial to the further development of geothermal heating technology.

2. Target Project

2.1. Background

The area studied in this paper is located in Xinji City, Hebei, China. The heat-providing area is 250,100 m^2, which includes residential and commercial areas. The designed heating load is 9040 kW. After this project is put into operation, the annual heating consumption will be 65,600 GJ.

The sedimentary Miocene Guantao Formation is the most favorable for geothermal heating systems [26,27]. According to a geological investigation conducted by China Petroleum and Chemical Corporation, the geothermal resources of Xinji is regarded as consisting of homogeneous porous media with high permeability. The reservoir thickness is 512 m and the average porosity of this geothermal reservoir is 30%. The depth of the bottom boundary varies from 1600 to 2100 m. The single-well water rate is between 78 and 120 m^3/h, and the production temperature is in the range of 333.15–3337.15 K. Laboratory tests show that the reservoir rock permeability is around 160 mD. The average reservoir temperature at target layers around 1300 m deep is about 333.15 K. The ground water of this geothermal reservoir can be directly used for building heating due to the low relative temperature [26].

To ensure the stable operation of a geothermal system, the geological conditions and well test data need to be considered in the design of the heating system [27]. Reinjection operation data of Xinji oil production well no. 5 are shown in Figure 1. Except during the filter element replacement period, the reinjection rate is mostly maintained at 120 m³/h, which ensures production water being totally reinjected underground.

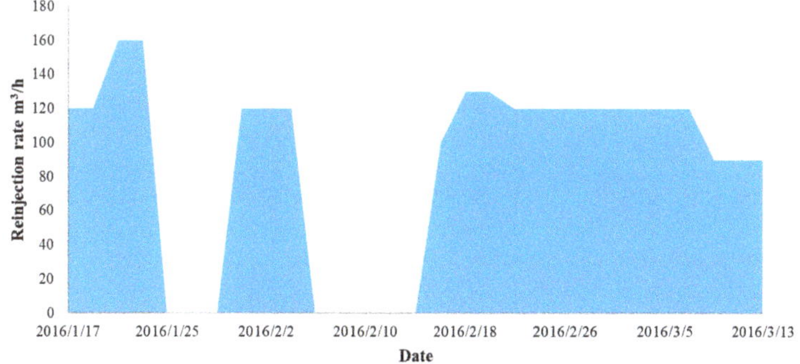

Figure 1. Reinjection rate of Xinji oil production well no. 5.

2.2. Geothermal Heating Strategy

There are two kinds of geothermal heating modes designed for Xinji. One is a direct geothermal district heating system (DGDHS). As shown in Figure 2a, ground water extracted from the geothermal production well is transported to the heat exchanger. The heating circulation water absorbs heat from the geothermal water. Then, the heat is piped to residential areas. The other one is an indirect geothermal district heating system (IGDHS), shown in Figure 2b. The heat of geothermal water is partially transferred to the heating circulation water at the primary heat exchanger with higher returning water temperature, and the heat is piped to some residential areas. After the initial heat transfer, the geothermal water is transported to the sequential secondary heat exchanger, where the heat is transferred from the geothermal water to the intermediate circulation water. Then, the heat pump extracts the heat from the intermediate circulation water and discharges it into the heating circulation water. Finally, the heat is transported to the remaining residential areas.

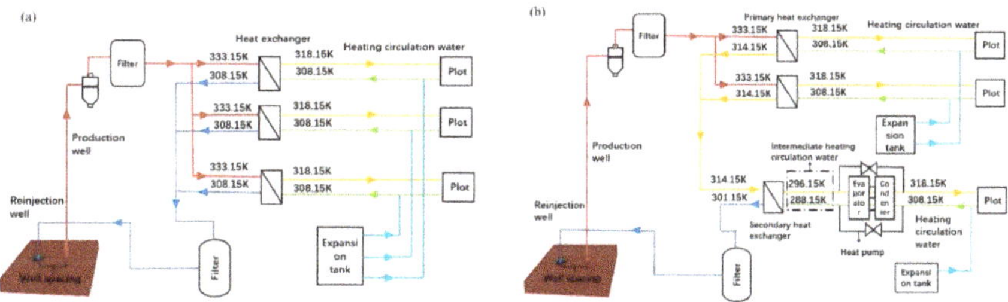

Figure 2. Operation strategies of geothermal heating systems: (**a**) direct geothermal district heating system (DGDHS) and (**b**) indirect geothermal district heating system (IGDHS).

3. Mathematical Model and Optimization Methodology

3.1. Thermal–Hydraulic–Mechanical Model

In this paper, the problems of water flow, heat transport, and rock deformation in the Xinji geothermal reservoir are studied jointly. The process of groundwater extraction is considered to be a flowing process of liquid in porous media consisting of a liquid phase and a solid phase [28]. The process of modeling and theoretical analysis consists of the following assumptions [28–33]:

(1) As the microstructures are well connected, thus, the hydraulic and transport characteristics of the rock matrix can be described by averaged quantities.
(2) Local thermodynamic equilibrium is assumed between liquid phase and solid phase.
(3) There is diffusion, convective and conductive heat transfer in porous media. Radiation heat transfer, which has little effect on geothermal water, is ignored.
(4) As total fluid balance (not transported species) and yields are considered in the model, there is no water loss in the geothermal reservoir.

The mass balance equation, energy balance equation, and rock deformation in porous media are shown in Equations (1)–(3) [31,32]:

$$\frac{\partial(\phi\rho_f)}{\partial t} + \nabla \cdot (\phi\rho_f u_{fs}) = Q_f \tag{1}$$

$$\rho c_p \frac{\partial(T)}{\partial t} + \nabla \cdot (-\lambda\nabla T) + \rho_f c_f u_{fs}\nabla T = 0 \tag{2}$$

$$\nabla \cdot \sigma = 0 \tag{3}$$

where ϕ is rock porosity; ρ_f is water density (kg/m^3); t is time (s); Q_f is the mass flow of sink/source item (kg/(m^3·s)); ρc_p is the heat capacity of porous media (J/(m^3·K)); T is temperature (K); λ is the thermal conductivity of porous media (W/(m·K)); σ is the typical Cauchy stress tensor; and u_{fs} is fluid flow velocity between rock and water flow (m/s). Both the heat capacity and thermal conductivity of porous media can be estimated as an arithmetic mean of each phase property weighted by its volume fraction.

Then, the flow velocity u is divided into the rock velocity u_s and the relative velocity of fluid u_{fs}. Huo et al. and Salimzadeh et al. proved that u_s is associated with the expansion coefficient of rock and fluid $\frac{\partial\varepsilon}{\partial t}$, and ε is calculated from the displacement solved by Equation (3) [32,33]. The term u_s is used to couple the water flow with rock deformation. Equation (1) is rewritten into Equation (4):

$$c_t\phi\rho\frac{\partial p}{\partial t} + \nabla \cdot (\phi\rho_f(u_{fs} + u_s)) = Q_f \tag{4}$$

where p is fluid pressure (Pa); c_t is total compressibility (Pa^{-1}), $c_t = c_f + c_s$; c_f is water compressibility (Pa^{-1}); and c_s is rock compressibility (Pa^{-1}).

In Equation (4), the effect of rock deformation caused by geothermal water extraction is considered. Thus, the flow velocity $u_{fs} + u_s$ depicts the movement of water in porous media. This is a very important step. Although u_s is usually a very small value, it cannot be neglected, as it accounts for the rock deformation mechanism.

As shown in Equation (5), u_{fs} further expressed by Darcy's law that shows the heat–fluid coupling process:

$$u_{fs} = -\frac{k}{\phi\mu}(\nabla p + \rho_f g\nabla z) \tag{5}$$

where g is gravitational acceleration (m/s^2); z is the depth of the geothermal reservoir (m); and k is the intrinsic permeability tensor (m^2).

In order to specify the solution for the above equations, one needs to prescribe appropriate boundary conditions. The details are shown in [29,30].

(1)   Prescribed pressure (Dirichlet condition)

$$p = \bar{p} \text{ on } \Gamma_p \tag{6}$$

for imposing hydrostatic conditions at domain boundaries or pressure at wellbores.

(2)   Prescribed fluid flux (Neumann condition)

$$q_n = q \cdot n \text{ on } \Gamma_q \tag{7}$$

with the normal vector n for imposing inflow or outflow rate at wellbores.

(3)   Prescribed temperature (Dirichlet condition)

$$T = \bar{T} \text{ on } \Gamma_T \tag{8}$$

for imposing natural thermal gradients at model boundaries and the temperature of the injected fluid.

The finite element method is used in Open-GeoSys (OGS) to provide numerical solutions to the aforementioned coupled formulation. More information can obtained from [29–33].

3.2. Calculation of Well Number based on Heating Load

Based on the heating load of residential buildings, the number of geothermal production wells is calculated, which is used to calculate construction costs. The building heat load is estimated based on local weather conditions, the function of the building, and characteristics of the protected exterior construction [34]. The calculation formula is shown in Equation (9):

$$Q_l = W_b \cdot S_b + W_c \cdot S_c \tag{9}$$

where W_c is heat load per unit area of residential building (W/m^2); W_b is heat load per unit area of commercial building (W/m^2); S_c is residential heat-providing area (m^2); S_b is commercial heat-providing area (m^2); and Q_l is total heat load (W).

The number of geothermal production wells is determined by the production rate and heating mode, shown in Equations (10) and (11):

$$n_1 = \frac{(1 + \alpha)Q_l}{c_w \cdot q_p \cdot \rho_f \cdot (t_g - t_h)} \tag{10}$$

where n_1 is the number of production wells for DGDHS (rounded up); α is the heat loss coefficient of the pipe network (5%–10%); c_w is the specific heat of water (J/(kg·K)); t_g is production temperature (K); t_h is reinjection temperature (K); and q_p is production rate (m^3/h); and

$$n_2 = \frac{(1 + \alpha)Q_l}{c_w \cdot q_p \cdot \rho_f \cdot \left[(t_g - t_{h1}) + \frac{COP}{COP-1}(t_{h1} - t_{h2})\right]} \tag{11}$$

where n_2 is the number of production wells for IGDHS (rounded up); t_{h1} is water temperature for the first return (K); t_{h2} is water temperature for the second return (K); and COP is the coefficient of performance.

3.3. Multiobjective Optimization Model for Heating Systems

In this part, a multiobjective optimization model for geothermal heating systems is established. The project's profitability is analyzed by coupling the initial investment and operating cost of the system. The initial investment, annual operating cost, and revenue are shown in Figure 3 [35–37].

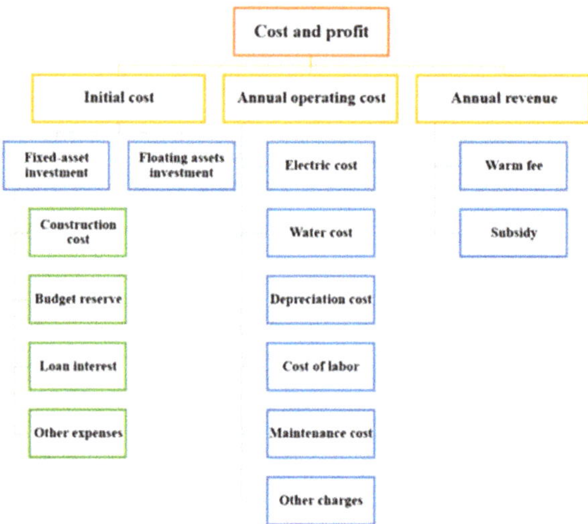

Figure 3. Initial investment, annual operating cost, and revenue of geothermal heating systems.

Geothermal well construction costs account for the main part of all construction costs which are based on well number, type, and construction depth. Due to limited drilling area, a vertical well is adopted for geothermal well construction, and the rest are directional wells, shown in Figure 4. The construction depth of a directional well is related to well spacing and directional angle, as in Equation (12), and the construction cost of the project can be estimated according to Equation (13):

$$h_2 = L + (h_1 - L) / \cos \theta \tag{12}$$

where h_1 is the depth of the vertical well (m); h_2 is the depth of the directional well (m); L is the depth of the first section (m); and θ is directional angle (°); and

$$C_z = (h_1 \cdot P_1 + C_f) + (h_2 \cdot P_2 + C_f) \cdot (n - 1) \tag{13}$$

where C_z is construction cost of a geothermal well (k USD); P_1 is construction cost per meter of vertical well (k USD); P_2 is construction cost per meter of directional well (k USD); C_f is cost of attached ground facility (k USD); and n is total number of geothermal heating wells.

The payback period is the time required for the return on investment to "repay" the total investment of the geothermal heating system [38], as expressed in Equation (14), and is calculated from the beginning year of construction [39,40]:

$$\sum_{t=0}^{Pt} (CI - CO)_t (1 + i_e)^{-t} = 0 \tag{14}$$

where Pt is static payback time (year); CI is cash inflow; CO is cash outflow; and $(CI - CO)_t$ is net cash flow for the year, t.

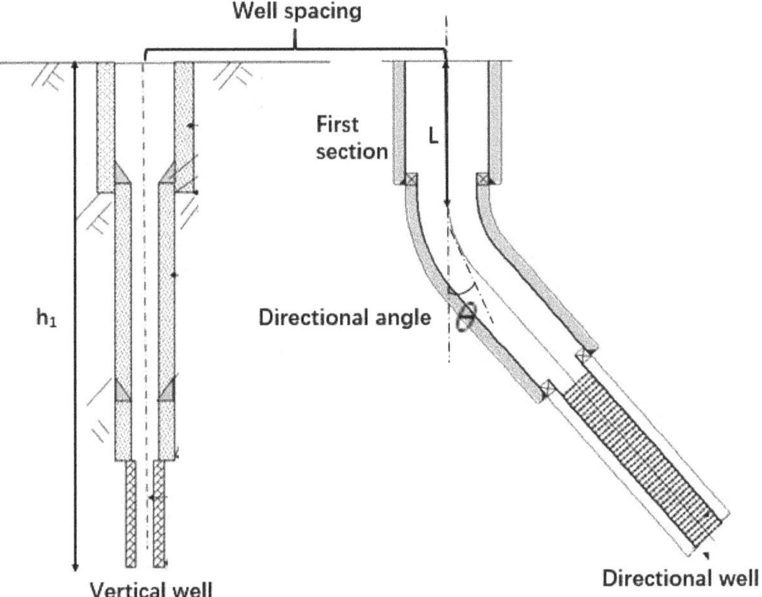

Figure 4. Schematic diagram of vertical well and directional well.

There are two main objectives of geothermal heating system optimization: one is to have the shortest payback period, and the other is to have minimum well spacing. In order to improve the profitability of the system, its production parameters must be optimized. However, three factors restrict the optimization of a geothermal heating system. The first is the area of equipment, which must be smaller than the project area. This factor is becoming more significant due to the limited space of urban areas. Second, a geothermal system is affected by groundwater resources, which is related to the hydrogeological conditions in the specific region. Finally, the system must meet the people's requirements for thermal comfort in the built environment. The multiobjective optimization functions are shown in Equation (15).

Optimization function:

$$\min Pt(x_1, x_2, x_3 \ldots x_n) \tag{15}$$

$$\min l(x_1, x_2, x_3 \ldots x_n)$$

Constraint conditions: Scale constraint:

$$S_g < S$$

Geological constraint:

$$q_x \leq q_{\mathrm{pmax}}(x_1, x_2, x_3 \ldots x_n)$$

Load constraint:

$$Q_x \geq Q_l(x_1, x_2, x_3 \ldots x_n)$$

where l is well spacing (m); S_g is the area of equipment (m^2); S is project area (m^2); q_x is the design value of production rate (m^3/h); q_{pmax} is the maximum of production rate (m^3/h); Q_x is the design value of system load (W); and $x_1, x_2, x_3 \ldots x_n$ represent different working conditions.

If hydrogeological conditions in a specific region allow the use of geothermal heating systems, these multiobjective optimization functions can be used for assessment.

3.4. Simulation Procedure

According to the established thermal–hydraulic–mechanical and multiobjective optimization models, using simulation software, the calculation process is shown in Figure 5. A geothermal heating system can be optimized through the entire calculation process. First, formation parameters are input into the model. The movement of the cold water thermal front is calculated. By comparing the temperature of the production well (T_{P0} is the initial temperature of the production well, and T_{P50} is the temperature at the 50th year), the optimal spacing of wells is selected. Finally, the initial investment and operating cost are calculated. Unprofitable geothermal heating systems are simply discarded. Though the multiobjective optimization model, the optimal production parameters can be obtained.

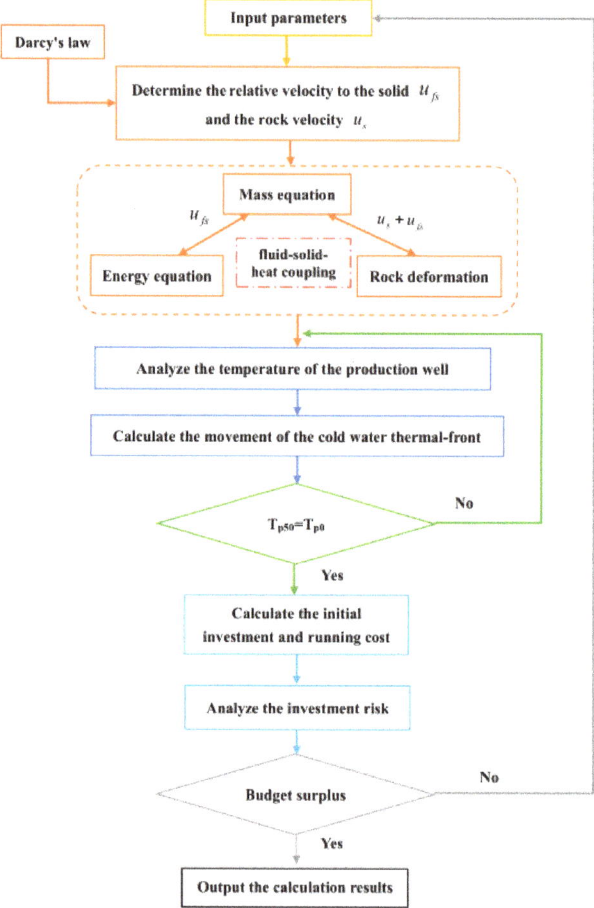

Figure 5. Flowchart of simulation procedure.

4. Result and Discussion

4.1. Sensitivity Analysis

This part studies the influence mechanism of various factors on the formation of thermal breakthrough. The basic conditions of the Xinji geothermal reservoir provided by China Petroleum and Chemical Corporation are listed in Table 1. These parameters, such as well spacing, production rate, and reinjection temperature, affect the formation of thermal breakthrough. They also determine the heat extraction efficiency of a geothermal heating system. In addition, periodic heating is considered in the modeling: a geothermal heating system works for 5 months each winter. Geothermal wells are closed during the remainder of the year. Results are obtained at the final time step at the end of the 50th year.

Table 1. Thermal–hydraulic–mechanical parameters for the base case of the Xinji geothermal reservoir.

Property	Value	Property	Value	Property	Value
Geothermal scale	266.6 km^2	Specific heat capacity of water	4185.4 J/kg·K	Thermal conductivity of water	0.6 W/(m·K)
Thickness of reservoir	512 m	Porosity	30%	Thermal conductivity of rock	3 W/(m·K)
Storage temperature	333.15 K	Rock density	2600 kg/m^3	Poisson's ratio of rock	0.1
Storage pressure	1300 Pa	Rock heat	878 J/kg·K	Young's modulus of rock	10 Gpa
Water density	967.4 kg/m^3	Water compressibility	4.5×10^{-10} Pa^{-1}	Permeability	160 mD
Water viscosity	4.15×10^{-6} m^2/s	Rock compressibility	4.3×10^{-10} Pa^{-1}	Well spacing	200 m

4.1.1. Effect of Well Spacing

Well spacing is the distance between a heat production well and a reinjection well, which determines the placement of geothermal wells. In this section, different well spacings (200, 400, and 600 m) are set into the mathematical model. The remaining parameters are as follows: production rate is 100 m^3/h and reinjection temperature is 311.15 K. Temporal evolutions of the water temperature and pressure at the production well are plotted in Figure 6.

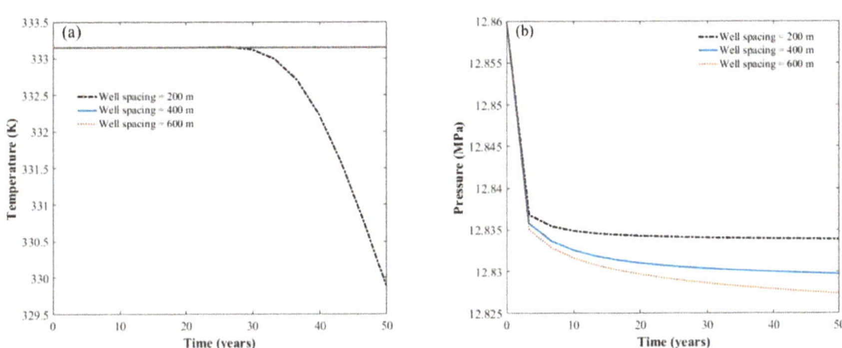

Figure 6. Time evolution of production well in well spacing studies: (**a**) temperature at production well; (**b**) pressure at production well.

In general, large well spacing can slow down the formation of thermal breakthrough. It can be observed from Figure 6a that when the well spacing is 200 m, the temperature of the production well begins to decline at the 25th year. The reason is that the movement of cold water thermal front reaches the production wells. This temperature change would reduce the heat extraction efficiency of a geothermal heating system. Eventually, ground heating demands are not met. For the other well spacings, the temperature of production wells is constant at 333.15 K. Note that there is no thermal breakthrough.

As shown in Figure 6b, the pressure of the production well decreases with increased well spacing. This is because the reinjected water is better able to repressurize the depleted area around the production well with smaller spacing. It means that large well spacing can effectively avoid premature thermal breakthrough. The movement of the cold water thermal front takes a long time to arrive at the production well. However, it loses the significance of reinjection to maintain water pressure. As commercial, land use, economic, and policy factors should be considered in the selection of geothermal well spacing, optimizing the well spacing is a challenging task.

4.1.2. Effect of Reinjection Temperature

The results of first part show that there is an apparent decline in production temperature of a geothermal heating system with well spacing of 200 m and a production rate of 100 m^3/h. These parameters are helpful to visualize the influence of reinjection temperature on thermal breakthrough. In this case, four values of the parameters (288.15, 298.15, 308.15, and 318.15 K) are considered. Temporal evolution of the water temperature and pressure at a geothermal production well is plotted in Figure 7.

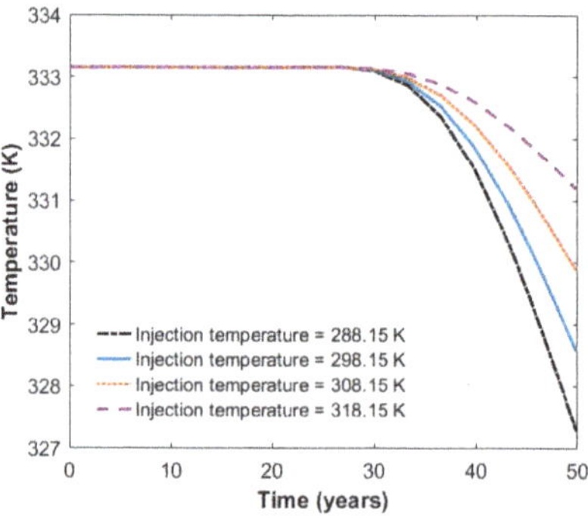

Figure 7. Time evolution of production well in reinjection temperature studies.

The temperature of reinjection water has a great influence on the formation of thermal breakthrough and the temperature of the production well. As is observed in Figure 7, thermal breakthrough is formed first when the reinjection temperature is 288.15 K. The lower the reinjection temperature is, the earlier the thermal breakthrough is formed, and the more the production water temperature drops. This is because the temperature decrease caused by the reinjected water results in the movement of the cold water thermal front originating from the reinjection well.

Although lower reinjection temperature leads to premature thermal breakthrough, it also points to a high utilization ratio of geothermal resources. Therefore, reinjection temperature must be controlled reasonably in geothermal heating projects.

4.1.3. Effect of Production Rate

The production rate is the water production rate at the production well. All geothermal water must be reinjected underground, so the reinjection rate is always equal to the production rate. According to the previous analysis of well spacing and reinjection temperature, when the well spacing is 200 m and the reinjection temperature is 308.15 K, the temperature of the production well changes markedly. Therefore, those parameters help to observe the sensitivity of temperature to production rates. Different production rates (80, 100, and 120 m^3/h) are input into the model. The results are shown in Figure 8.

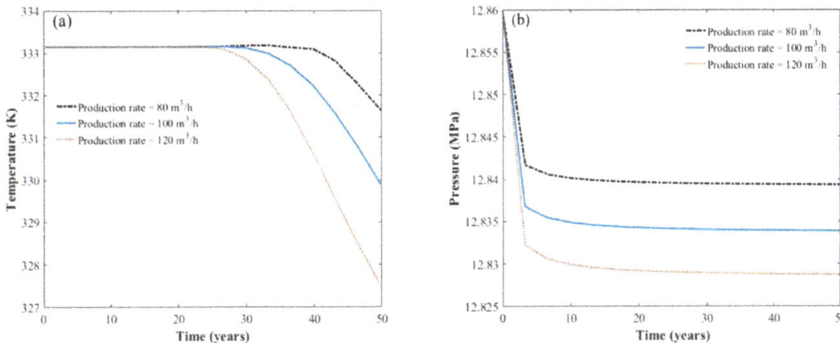

Figure 8. Time evolution of production well in production rate studies: (**a**) temperature at production well; (**b**) pressure at production well.

A large production rate promotes the movement of the cold water thermal front toward the production well. As observed Figure 8a, with decreased production rate, the temperature decrease at the production well is delayed, and the drop in production temperature is not obvious. This is because the decreased well rate slows down the spread of the temperature decrease initiated by the movement of the cold water thermal front.

According to Figure 8b, the greater the production rate, the lower the pressure. This is because a lower well rate results in a smaller pressure gradient between the reinjection well and the production well.

A small production rate delays the temperature drop of production water. However, it also results in poor heating extraction of the geothermal heating system. Therefore, a reasonable range of production rates should be controlled on the basis of local test data.

4.2. Numerical Optimization

4.2.1. Parameter Optimization

Taking IGDHS with a production rate of 100 m^3/h and reinjection temperature of 301.15 K as an example, results are shown in Figure 9. Figure 9a shows that production temperature does not change when well spacing is over 300 m. As can be seen from Figure 9b, with increased well spacing, the pressure of the production well keeps dropping. This is similar to the first part of the simulation results, proving that the results are correct.

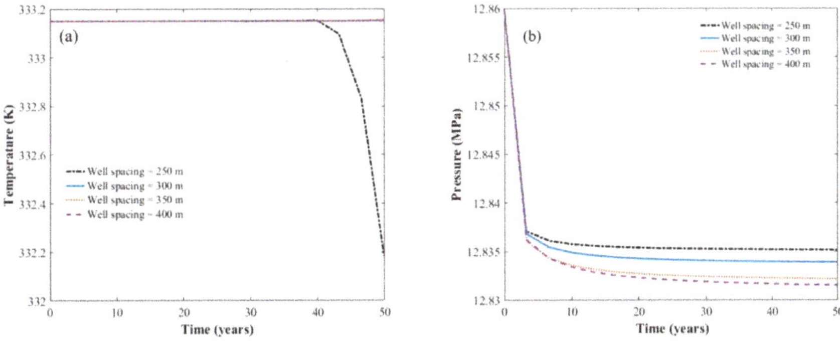

Figure 9. Time evolution of a production well for indirect geothermal district heating systems (IGDHSs) with production rate of 80 m^3/h: (**a**) temperature at production well; (**b**) pressure at production well.

Temperature distribution nephograms of geothermal reservoirs under different situations are shown in Figures 10–12. The results show that the radius of the cold water thermal front is directly proportional to the production rate, and the radius of IGDHS is larger than that of DGDHS. In addition, the radius of the cold water thermal front increases with decreased reinjection temperature. As shown in Figure 11b, the radius of the cold water thermal front is about 260 m. Considering that the uneven distribution of formation and the instability of groundwater accelerate the flow of low-temperature groundwater, it is recommended that well spacing should be over 300 m in this case. The production parameters of each geothermal heating system are shown in Table 2.

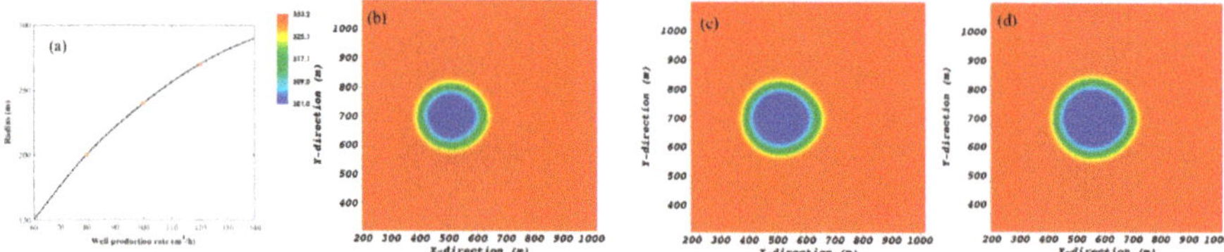

Figure 10. Temperature distribution nephograms of direct geothermal district heating systems (DGDHSs) with different production rates in the 50th year: (**a**) radius of cold water thermal front; (**b**) production rate of 80 m^3/h; (**c**) production rate of 100 m^3/h; (**d**) production rate of 120 m^3/h.

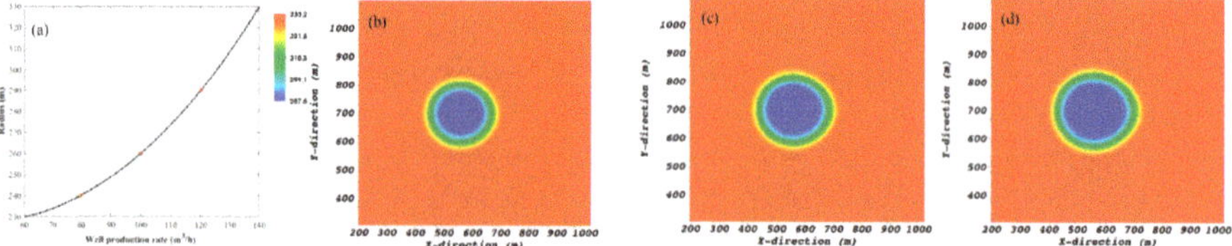

Figure 11. Temperature distribution nephograms of an IGDHS with a reinjection temperature of 301.15 K in the 50th year: (**a**) radius of cold water thermal front; (**b**) production rate of 80 m^3/h; (**c**) production rate of 100 m^3/h; (**d**) production rate of 120 m^3/h.

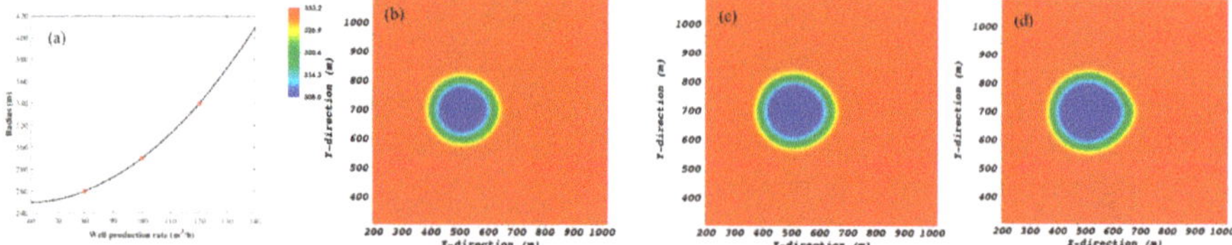

Figure 12. Temperature distribution nephograms of a IGDHS with reinjection temperature of 288.15 K in the 50th year: (**a**) radius of cold water thermal front; (**b**) production rate of 80 m^3/h; (**c**) production rate of 100 m^3/h; (**d**) production rate of 120 m^3/h.

Table 2. Optimal well spacing for different geothermal heating systems.

Type	Reinjection Temperature	Production Rate	Minimum Well Spacing
DGDHS	308.15 K	80 m^3/h	250 m
		100 m^3/h	300 m
		120 m^3/h	350 m
IGDHS	301.15 K	80 m^3/h	300 m
		100 m^3/h	300 m
		120 m^3/h	400 m
IGDHS	288.15 k	80 m^3/h	350 m
		100 m^3/h	400 m
		120 m^3/h	450 m

Projects using groundwater for heating can be optimized with the above-mentioned method. The input parameters are adjusted by the geological characteristics of each area and heating modes. The optimal well spacing of the geothermal heating system can be obtained. This methodology provides a good reference for cities developing geothermal heating technology.

4.2.2. Economic Optimization of Geothermal System

In this paper, a static economic evaluation method is used to calculate the initial investment and operation cost of DGDHS and IGDHS. The basic conditions of geothermal systems provided by China Petroleum and Chemical Corporation are listed in Table 3 (1 USD = 7 RMB). Through the multiobjective optimization model, a geothermal system with high investment return and short investment payback period is selected for the case in Xinji.

Table 3. Economic calculation parameters for the base case.

Construction Cost	Value	Operating Cost	Value	Revenue	Value
Vertical well materials expenses	199 USD/m	Electricity	0.078 USD/kWh	Residential warm fee	2.8 USD/m^2
Directional well materials expenses	208 USD/m	Water	0.65 USD/t	Commercial warm fee	3.9 USD/m^2
Surface ancillary works	32.3 k USD	Maintenance	50 K USD/year		
Geothermal well construction cost	13 USD/m	Salaries	900 USD/month/person		
Other expenses	2.99 USD/m^2				

Figure 13a shows the relationship between well price and production rate. The well price increases with increased production rate, and the same result can be achieved by lowering the reinjection temperature. The reason is that large well spacing is required for a geothermal heating system with large production rates and low reinjection temperatures to prevent thermal breakthrough.

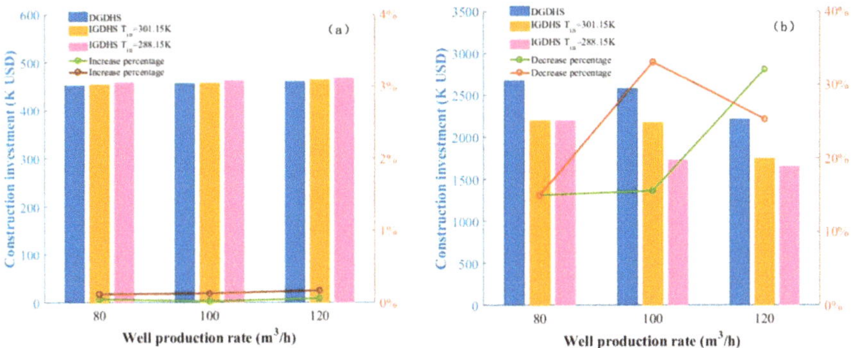

Figure 13. Well price and initial construction investment for geothermal wells of different heating modes: (**a**) construction price of directional well; (**b**) construction investment of heating system.

It can be clearly seen from Figure 13b that the initial construction cost of DGDHS is the highest. Although the construction cost of DGDHS decreases with increased production rate, it is still higher than that of IGDHS. For IGDHS, increasing the production rate or reducing the reinjection temperature can reduce the initial construction cost, and the construction investment of geothermal wells will decrease by up to 30%.

There is a great difference in electricity charges and initial investments of various geothermal heating systems. As shown in Figure 14a, the electric energy consumed by IGDHS is nearly 1.5 times that of DGDHS. The power consumption of IGDHS presents an irregular change with increased well production rate. The reason is that the decreased reinjection temperature leads to increased secondary heat transfer and heat load of the heat pump, which causes increased power consumption. However, the trend in water consumption is reversed, shown in Figure 14b. Water consumption of IGDHS is much less than that of DGDHS, reduced by 10%–50%. This is because IGDHS generates heat by electricity and has better environmental adaptability than DGDHS. Even in extreme weather, IGDHS meets residential heating needs.

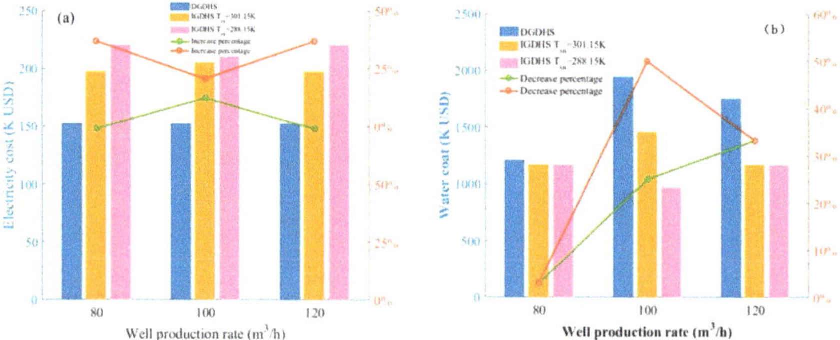

Figure 14. Annual power consumption of different geothermal heating systems: (**a**) electricity cost; (**b**) water cost.

As shown in Figure 15, DGDHS has a longer payback period than IGDHS. Actually, annual power consumption of DGDHS is lower, because there are no heat pumps, but the high cost of construction investment and equipment depreciation causes a longer payback period. On the contrary, IGDHS consumes large amounts of electricity during operation, but the initial investment for a geothermal heating system is lower, which greatly reduces the payback period. As reinjection temperature is lower, the payback period becomes longer. That is because the annual power consumption of a geothermal heating system with low reinjection temperature is significantly higher than other systems. An indirect geothermal district heating system for building heating in the case of Xinji, China, is the best heating mode. The optimal production rate, reinjection temperature, and well spacing for a 50-year heating period are 100 m^3/h, 301.15 K, and 300 m, respectively.

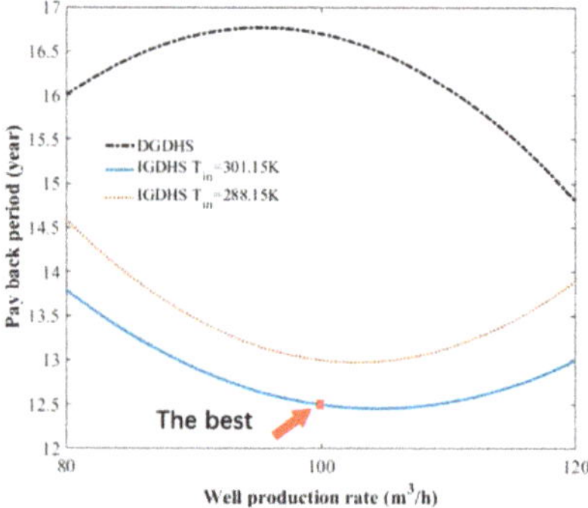

Figure 15. Investment payback period of different geothermal heating systems.

In fact, the income and operating expenses of geothermal heating systems in Xinji fit nicely with economic calculations. It is shown that the calculation results by the static technical economic evaluation are correct. Due to different hydrological conditions and economic levels of cities, all items in this paper cannot be directly used. However, the mathematical and multiobjective optimization models proposed in this paper can be used for assessment, and calculation results can be referenced for the choice of geothermal heating systems.

5. Conclusions

Renewable geothermal utilization is a significant approach for residential heating in geothermal enrichment regions. The development of geothermal reservoirs has to meet building heating demands with optimal economic benefits, and the geothermal water should be protected. In this paper, a new approach with practical engineering requirements and environmental and economic considerations is established systematically.

1. Mathematical models of geothermal development with direct and indirect geothermal district heating systems are established. Furthermore, the optimal well number, well spacing, production rate, and reinjection temperature can be obtained to meet engineering, environmental, and economic criteria.
2. Well spacing, reinjection temperature, and production rate are the most significant parameters affecting thermal breakthrough in geothermal reservoirs. With decreased well spacing and reinjection temperature or increased production rate, premature thermal breakthrough would occur, resulting in low efficiency of geothermal heating systems.
3. For indirect geothermal district heating systems, the construction investment of geothermal wells is reduced by up to 20%, and annual water consumption is reduced by up to 50%, but electricity consumption costs increase by 5% to 30%.
4. Well number and reinjection temperature have a huge effect on the payback period of investment for geothermal development. The payback period of direct geothermal district heating systems is longer than that of indirect systems. Indirect systems have a good environmental adaptability and profitability which is more suitable for building heating.
5. In the case of Xinji, China, an indirect geothermal district heating system is much better than a direct geothermal district heating system, both technically and economically. The optimal production rate, reinjection temperature, and well spacing for 50 years of building heating are 100 m^3/h, 301.15 K, and 300 m, respectively. This optimal production parameters can be the reference for the design of geothermal heating systems in other regions. The systematical calculation approach can be reasonably applied to the selection and optimization of other geothermal systems.

Author Contributions: Conceptualization, L.Z. and H.S.; Data curation, L.Z.; Formal analysis, L.Z., R.W., and H.S.; Funding acquisition, H.S.; Investigation, H.X., P.S., and L.D.; Methodology, R.W.; Project administration, H.S.; Resources, P.S. and L.D.; Software, L.Z.; Supervision, H.S. and H.X.; Validation, H.F.; Visualization, R.W. and H.X.; Writing—original draft, L.Z.; Writing—review & editing, H.X. and H.F.

Funding: This research was funded by the Beijing Nova Program, grant number Z171100001117081.

Acknowledgments: The work was supported by the Beijing Nova Program (grant no. Z171100001117081). The authors also wish to thank Sinopec Star Petroleum Co., Ltd. for the cooperation in this work.

Conflicts of Interest: The authors declare no conflict of interest.

References

1. Shortall, R.; Davidsdottir, B.; Axelsson, G. Geothermal energy for sustainable development: A review of sustainability impacts and assessment frameworks. *Renew. Sustain. Energy Rev.* **2015**, *44*, 391–406. [CrossRef]
2. Bina, S.M.; Jalilinasrabady, S.; Fujii, H. Thermo-economic evaluation of various bottoming ORSs for geothermal power plant, determination of optimum cycle for Sabalan power plant exhaust. *Geothermics* **2017**, *70*, 181–191. [CrossRef]
3. Bina, S.M.; Jalilinasrabady, S.; Fujii, H. Energy, economic and environmental (3E) aspects of internal heat exchanger for ORC geothermal power plants. *Energy* **2017**, *140*, 1096–1106. [CrossRef]

4.   Van Erdeweghe, S.; Van Bael, J.; Laenen, B.; D'haeseleer, W. Feasibility study of a low-temperature geothermal power plant for multiple economic scenarios. *Energy* **2018**, *155*, 1004–1012. [CrossRef]

5.   Aliyu, M.D.; Chen, H.-P. Sensitivity analysis of deep geothermal reservoir: Effect of reservoir parameters on production temperature. *Energy* **2017**, *129*, 101–113. [CrossRef]

6.   Liu, Y.; Hou, J.; Zhao, H.; Liu, X.; Xia, Z. A method to recover natural gas hydrates with geothermal energy conveyed by CO_2. *Energy* **2018**, *144*, 265–278. [CrossRef]

7.   Song, X.; Shi, Y.; Li, G.; Yang, R.; Wang, G.; Zheng, R.; Li, J.; Lyu, Z. Numerical simulation of heat extraction performance in enhanced geothermal system with multilateral wells. *Appl. Energy* **2018**, *218*, 325–337. [CrossRef]

8.   Cui, G.; Ren, S.; Rui, Z.; Ezekiel, J.; Zhang, L.; Wang, H. The influence of complicated fluid-rock interactions on the geothermal exploitation in the CO_2 plume geothermal system. *Appl. Energy* **2018**, *227*, 49–63. [CrossRef]

9.   Omer, A.M. Ground-source heat pumps systems and applications. *Renew. Sustain. Energy Rev.* **2008**, *12*, 344–371. [CrossRef]

10. Kaushal, M. Geothermal cooling/heating using ground heat exchanger for various experimental and analytical studies: Comprehensive review. *Energy Build.* **2017**, *139*, 634–652. [CrossRef]

11. Lund, J.W.; Boyd, T.L. Direct utilization of geothermal energy 2015 worldwide review. *Geothermics* **2016**, *60*, 66–93. [CrossRef]

12. Villarino, J.I.; Villarino, A.; Fernández, F.Á. Experimental and modelling analysis of an office building HVAC system based in a ground-coupled heat pump and radiant floor. *Appl. Energy* **2017**, *190*, 1020–1028. [CrossRef]

13. Self, S.J.; Reddy, B.V.; Rosen, M.A. Geothermal heat pump systems: Status review and comparison with other heating options. *Appl. Energy* **2013**, *101*, 341–348. [CrossRef]

14. Nian, Y.-L.; Cheng, W.-L. Evaluation of geothermal heating from abandoned oil wells. *Energy* **2018**, *142*, 592–607. [CrossRef]

15. Fiaschi, D.; Lifshitz, A.; Manfrida, G.; Tempesti, D. An innovative ORC power plant layout for heat and power generation from medium-to low-temperature geothermal resources. *Energy Convers. Manag.* **2014**, *88*, 883–893. [CrossRef]

16. Stáhl, G.; Pátzay, G.; Weiser, L.; Kálmán, E. Study of calcite scaling and corrosion processes in geothermal systems. *Geothermics* **2000**, *29*, 105–219. [CrossRef]

17. Cheng, W.-L.; Liu, J.; Nian, Y.-L.; Wang, C.-L. Enhancing geothermal power generation from abandoned oil wells with thermal reservoirs. *Energy* **2016**, *109*, 537–545. [CrossRef]

18. Dalampakis, P.; Gelegenis, J.; Ilias, A.; Ladas, A.; Kolios, P. Technical and economic assessment of geothermal soil heating systems in row covered protected crops: A case study from Greece. *Appl. Energy* **2017**, *203*, 201–218. [CrossRef]

19. Zhang, K.; Lee, B.-H.; Ling, L.; Guo, T.-R.; Liu, C.-H.; Ouyang, S. Modeling studies for production potential of Chingshui geothermal reservoir. *Renew. Energy* **2016**, *94*, 568–578. [CrossRef]

20. Held, S.; Schill, E.; Schneider, J.; Nitschke, F.; Morata, D.; Neumann, T.; Kohl, T. Geochemical characterization of the geothermal system at Villarrica volcano, Southern Chile; Part 1, Impacts of lithology on the geothermal reservoir. *Geothermics* **2018**, *74*, 226–239. [CrossRef]

21. Nam, Y.; Ooka, R. Numerical of ground heat and water transfer for groundwater heat pump system based on real-scale experiment. *Energy Build.* **2010**, *42*, 69–75. [CrossRef]

22. Salimzadeh, S.; Paluszny, A.; Nick, H.M.; Zimmerman, R.W. A three-dimensional coupled thermo-hydro-mechanical model for deformable fractured geothermal systems. *Geothermics* **2018**, *71*, 212–224. [CrossRef]

23. Pandey, S.N.; Chaudhuri, A. The effect of heterogeneity on heat extraction and transmissivity evolution in a carbonate reservoir: A thermo-hydro-chemical study. *Geothermics* **2017**, *69*, 45–54. [CrossRef]

24. Bagalkot, N.; Zare, A.; Kumar, G.S. Influence of Fracture Heterogeneity Using Linear Congruential Generator (LCG) on the Thermal Front Propagation in a Single Geothermal Fracture-Rock Matrix System. *Energies* **2018**, *11*, 916. [CrossRef]

25. Papamichos, E.; Furui, K. Analytical models for sand onset under field conditions. *J. Pet. Sci. Eng.* **2019**, *172*, 171–189. [CrossRef]

26. Rubio-Maya, C.; Díaz, V.M.A.; Martíneza, E.P.; Belman-Flores, J.M. Cascade utilization of low and medium enthalpy geothermal resources—A review. *Renew. Sustain. Energy Rev.* **2015**, *52*, 689–716. [CrossRef]

27. Bayer, P.; de Paly, M.; Beck, M. Strategic optimization of borehole heat exchanger field for seasonal geothermal heating and cooling. *Appl. Energy* **2014**, *136*, 445–453. [CrossRef]
28. Pandey, S.N.; Vishal, V.; Chaudhuri, A. Geothermal reservoir modeling in a coupled thermo-hydro-mechanical-chemical approach: A review. *Earth Sci. Rev.* **2018**, *185*, 1157–1169. [CrossRef]
29. Kolditz, O.; Shao, H.; Wang, W.; Bauer, S. *Thermo-Hydro-Mechanical-Chemical Processes in Fractured Porous Media: Modelling and Benchmarking*; Springer: Berlin, Germany, 2016.
30. Watanabe, N.; Blöcher, G.; Cacace, M.; Held, S.; Kohl, T. *Geoenergy Modeling III: Enhanced Geothermal Systems*; Springer: Berlin, Germany, 2017.
31. Kolditz, O.; Bauer, S.; Bilke, L.; Böttcher, N.; Delfs, J.O.; Fischer, T.; Görke, U.J.; Kalbacher, T.; Kosakowski, G.; McDermott, C.I.; et al. OpenGeoSys: An open-source initiative for numerical simulation of thermo-hydro-mechanical/chemical (THM/C) processes in porous media. *Environ. Earth Sci.* **2012**, *67*, 589–599. [CrossRef]
32. Hou, Z.; Gou, Y.; Taron, J.; Gorke, U.J.; Kolditz, O. Thermo-hydro-mechanical modeling of carbon dioxide injection for enhanced gas-recovery (CO2-EGR): A benchmarking study for code comparison. *Environ. Earth Sci.* **2012**, *67*, 549–561. [CrossRef]
33. Guo, X.; Song, H.; Killough, J.; Du, L.; Sun, P. Numerical investigation of the efficiency of emission reduction and heat extraction in a sedimentary geothermal reservoir: A case study of the Daming geothermal field in China. *Environ. Sci. Pollut. Res.* **2018**, *25*, 4690–4706. [CrossRef] [PubMed]
34. Reber, T.J.; Beckers, K.F.; Tester, J.W. The transformative potential of geothermal heating in the U.S. energy market: A regional study of New York and Pennsylvania. *Energy Policy* **2014**, *70*, 30–44. [CrossRef]
35. Von Appen, J.; Braun, M. Interdependencies between self-sufficiency preferences, techno-economic drivers for investment decisions and grid integration of residential PV storage systems. *Appl. Energy* **2018**, *229*, 1140–1151. [CrossRef]
36. Kudełko, J.; Wanielista, K.; Wirth, H. Economic evaluation of mineral extraction projects from fields of exploitation during operational periods. *J. Sustain. Min.* **2013**, *12*, 41–45. [CrossRef]
37. Weinberger, G.; Moshfegh, B. Investigating influential techno-economic factors for combined heat and power production using optimization and metamodeling. *Appl. Energy* **2018**, *232*, 555–571. [CrossRef]
38. Lu, X.; Zhao, Y.; Zhu, J.; Zhang, W. Optimization and applicability of compound power cycles for enhanced geothermal systems. *Appl. Energy* **2018**, *229*, 128–141. [CrossRef]
39. Mi, X.; Liu, R.; Cui, H.; Memon, S.A.; Xing, F.; Lo, Y. Energy and economic analysis of building integrated with PCM in different cities of China. *Appl. Energy* **2016**, *175*, 324–336. [CrossRef]
40. Sajjadian, S.M.; Lewis, J.; Sharples, S. The potential of phase change materials to reduce domestic cooling energy loads for current and future UK climates. *Energy Build.* **2015**, *93*, 83–89. [CrossRef]

Article

Case Study of Heat Transfer during Artificial Ground Freezing with Groundwater Flow

Rui Hu [1,†], **Quan Liu** [2,*,†] and **Yixuan Xing** [2]

[1] School of Earth Science and Engineering, Hohai University, Nanjing 211100, China; rhu@hhu.edu.cn
[2] Geoscience Centre, University of Göttingen, 37077 Göttingen, Germany; yxing@gwdg.de
*   Correspondence: qliu@gwdg.de; Tel.: +49-157-835-53628
†   These authors contributed equally to this work.

Received: 13 August 2018; Accepted: 21 September 2018; Published: 25 September 2018

Abstract: For the artificial ground freezing (AGF) projects in highly permeable formations, the effect of groundwater flow cannot be neglected. Based on the heat transfer and seepage theory in porous media with the finite element method, a fully coupled numerical model was established to simulate the changes of temperature field and groundwater flow field. Firstly, based on the classic analytical solution for the frozen temperature field, the model's ability to solve phase change problems has been validated. In order to analyze the influences of different parameters on the closure time of the freezing wall, we performed the sensitivity analysis for three parameters of this numerical model. The analysis showed that, besides the head difference, the thermal conductivity of soil grain and pipe spacing are also the key factors that control the closure time of the frozen wall. Finally, a strengthening project of a metro tunnel with AGF method in South China was chosen as a field example. With the finite element software COMSOL Multiphysics® (Stockholm, Sweden), a three-dimensional (3D) numerical model was set up to simulate the change of frozen temperature field and groundwater flow field in the project area as well as the freezing process within 50 days. The simulation results show that the freezing wall appears in an asymmetrical shape with horizontal groundwater flow normal to the axial of the tunnel. Along the groundwater flow direction, freezing wall forms slowly and on the upstream side the thickness of the frozen wall is thinner than that on the downstream side. The actual pipe spacing has an important influence on the temperature field and closure time of the frozen wall. The larger the actual pipe spacing is, the slower the closing process will be. Besides this, the calculation for the average temperature of freezing body (not yet in the form of a wall) shows that the average temperature change of the freezing body coincides with that of the main frozen pipes with the same trend.

Keywords: artificial ground freezing method; groundwater flow; temperature field; freezing wall; effective hydraulic conductivity

---

1. Introduction

With its good water barrier and non-pollution characteristics, the artificial ground freezing (AGF) method is widespread in environmental engineering (e.g., excavation of pollutants) [1,2], mining engineering (e.g., mine shafts construction), and especially in underground engineering (e.g., subsea tunnel and urban subway construction) [3–6]. The basic principle of this method is to circulate a fluid refrigerant (ca. −30 °C) or liquid nitrogen (ca. −196 °C) through a pre-buried pipe network in the subsurface in order to form a freezing wall for strengthening construction or blocking groundwater flow.

Since its first application in Germany in 1887, the AGF method has been considerably developed in theory and practice. Especially in the conventional groundwater environment, for example, in freshwater and low flow rate conditions, this technology has been successfully applied in a large number of projects and it tends to be mature [7–10]. However, when this method is implemented under some special groundwater conditions, such as high groundwater flow velocity area, seawater invasion area, or groundwater heavily polluted area, the freezing effect will be affected by groundwater flow velocity, salt content, and different types of pollutants [11–15]. For instance, in the frozen wall project of the Fukushima Power Plant, the "Ice wall" failed to prevent groundwater infiltration. The most likely cause was the complex hydrological environment and the high concentration of radioactive toxic water [1]. This will bring new challenges to the typical application of the AGF method. In this paper, we mainly focus on the effect of flow velocity on the freezing effect from the perspective of groundwater flow.

During the development of freezing walls, groundwater flow and formation temperature distribution will interact with each other [11,16–22]. On the one hand, groundwater flow will take away the heat (coldness) around the freezing columns and slow down the cooling process of the formation. On the other hand, with the lowering of the formation temperature, the groundwater flow process will also be impacted. After the formation temperature reaches freezing point, an ice-water phase transition will occur in the pores, which will result in the formation of ice. Thereby, this will cause changes in the physical properties of the soil, such as permeability, density, thermal conductivity, and strength. The higher the flow velocity is, the stronger that the interaction will be. As the flow velocity increases, the heat transferred by the groundwater flow gradually replaces the heat extracted by the freezing pipes, slowing down the growth of the freezing columns around individual freezing pipes. This is a potential challenge for the AGF method.

In the past, many researchers have carried out research on the effect of groundwater flow on AGF method. First, some scholars have put forward corresponding numerical models from the viewpoint of heat-fluid coupling. Harlan (1973) [23] presented a mathematical model that coupled heat and fluid flow based on analogy and applied to the analysis of heat-fluid phenomena in partially frozen soil. Due to the lack of reliable data, quantitative comparisons of observed and simulated profile responses were not made. Using a finite element scheme, Guymon and Luthin (1974) [24] developed a one-dimensional a coupled model for heat and moisture based on the Richards equation and the heat conduction equation, including convective components, and applied this to analyze moisture movement and storage in Arctic soils. He pointed out that the feasibility of the model depends primarily on the assumption of equations and the computation costs. Hansson et al. (2004) [25] presented a new method that accounts for phase change in a fully implicit numerical model. This model was evaluated by comparing predictions with data from laboratory columns freezing experiments and applied this to simulate hypothetical road problems involving simultaneous heat transport and water flow. Based on some parametric assumptions, McKenzie et al. (2007) [26] simulated groundwater flow with energy transport numerically by modifying the SUTRA computer code and applied this to freezing processes in peat bogs. The model was validated by the analytical solution from Lunardini (1985) without an advection term [27]. Kurylyk et al. (2014) [28] reanalyzed this analytical solution and compared the results between the numerical solution and the Lunardini analytical solution with advection term. In the study of Scheidegger et al. (2017) [29], the development of permafrost was simulated with a two-dimensional (2D) thermo-hydro (TH) model using COMSOL software and this model was validated by this analytical solution. Tan et al. (2011) [30] established a TH coupling model based on the theories of continuum mechanics, thermodynamics, and segregation potential. The model considered the Soret effect, as well as the influence of segregation potential on groundwater flow velocity and water pressure distribution. This model has been verified by comparison with the well-known TH coupling laboratory test that was conducted by Mizoguchi (1990) [31].

Secondly, some scholars have focused on experimental studies of the AGF method combined with groundwater flow. Anton Sers (2009) [32] studied the growth of the freezing body with various groundwater flow velocities through a small-scale sandbox. Pimentel et al. (2012) [11] developed a large-scale physical experimental device for the AGF with high groundwater flow velocities and compared the experimental results of different experimental setups [33–35]. The results showed that, for the conditions of no groundwater flow or single freezing pipe, there were no substantial differences with respect to temperature change. But, with multiple pipes and high groundwater flow velocity, the discrepancies become larger as advection becomes more dominant in relation to heat conduction. Huang et al. (2012) [36] conducted a series of physical experiments and numerical simulations with controlled groundwater flow and boundary conditions. These results have shown that the growth of a freezing body is affected by groundwater flow.

It has been proven that the groundwater flow will affect the temperature distribution around the freezing pipes and further prevent the formation of the freezing wall. However, the quantification of the effect of groundwater flow velocity on the evolution of the freezing wall is still in dispute. Jessberger and Jagow (1996) [37] studied the effect of groundwater flow velocity on the AGF method with brine circulation and pointed to a flow velocity of 2 m/day as the critical limit. That is to say, the frozen body between the pipes will not close when the flow rate is more than 2 m/day. Besides this, Andersland and Ladanyi (2004) [38] claimed that, even with flow velocities between 1 and 2 m/day, no closure of the freezing body will occur. Other scholars have also given some different critical flow velocities [39,40].

In this work, we introduced the parameter of effective hydraulic conductivity and the saturation function of unfrozen water. Based on this parameter and function, we set up a fully coupled three-dimensional (3D) numerical model (COMSOL Multiphysics® model with finite element) with groundwater flow field and frozen temperature field. This model is validated through the comparison of the results from numerical calculation and the analytical solution of the classic frozen problem from Lunardini [27] Its ability to solve the phase change problem is evaluated. With different numerical case studies, sensitivity analysis was performed for three parameters: the pipe spacing, thermal conductivity of soil grain and groundwater flow velocity (derived from hydraulic gradient or head difference). The corresponding influences of each parameter on the closure time of the freezing wall was analyzed. Finally, a strengthening project of a metro tunnel with AGF method in South China was chosen as an example for the field study. A numerical model was set up to simulate the artificial ground freezing process of this project. The regular pattern of frozen wall development was analyzed and the model was validated through the comparison of calculated and in-situ measured temperatures.

2. Methods

During the AGF process, thermal-hydraulic (TH) coupling is a key issue to be solved. As soil temperatures decrease below the freezing point, free water in the pores starts to freeze and the resulting ice crystals clog the original pore channels of groundwater flow. Simultaneously, due to the existence of convection, the change of the groundwater flow field will affect heat transfer in soil in return. In particular, the water-ice phase transition will lead to small changes in the physical properties of soil materials. The typical phase diagram of the freezing process is shown in Figure 1. In a saturated representative volume element (RVE), the pore space ε consists of ice, free water, and unfrozen film water, which are denoted by subscripts i, w, and rs, respectively. The saturation S, which describes the proportion of the volume occupied by the components in the pores, is introduced.

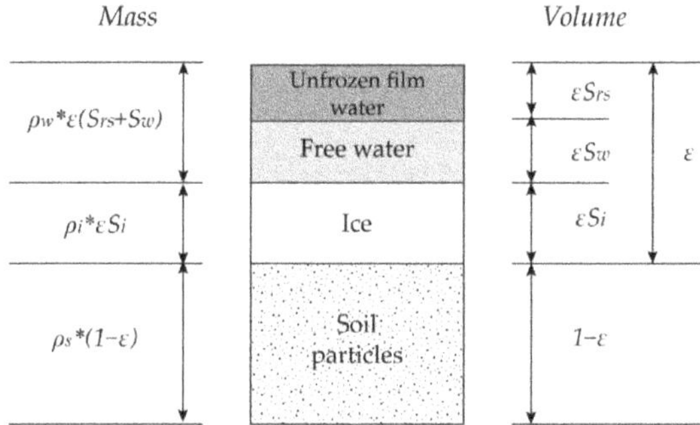

Figure 1. Schematic of a typical phase diagram for the freezing process.

Based on the above coupling mechanism, we construct the mathematical model of this problem by establishing the governing equations of the groundwater flow field and the freezing temperature field, respectively. Through the multi-physics finite element software COMSOL Multiphysics®, this problem can be numerically solved. Assumptions on this issue are required, as followed:

1. the ice is immobile and the medium is non-deformable;
2. the aquifer is fully saturated and its total porosity remains constant; and,
3. the freezing point depression due to solute concentrations is negligible.

2.1. Water Mass Conservation Equation

Due to the barrier effect of ice, water mass transfer is inevitably affected by the crystallization, which is controlled by temperature. Moreover, water-ice phase transition will be accompanied by density change between liquid and solid, resulting in a mass loss. Thus, the total water mass conservation could be governed by Darcy's law, as follows:

$$(1 - \varepsilon S_i) S_{OP} \frac{\partial p}{\partial t} + \nabla [-k_r K \cdot (\nabla p + \rho_w g \nabla D)] = Q_S + Q_T \tag{1}$$

where S_{OP} is the specific pressure storativity; p is the pressure (Pa); ε is the porosity; t is the time (s); k_r is the effective hydraulic conductivity that used to represent the decrease of permeability in the water-ice phase transition zone; K is the hydraulic conductivity (m/s); ρ_w denotes the fluid density (kg/m³); ∇D is the gravitational potential gradient tensor; and, Q_S denotes sources and sinks. Q_T represents temperature-induced mass increments, and in this case, this term can be defined as

$$Q_T = \varepsilon (\rho_w - \rho_i) \frac{\partial S_i}{\partial t} \tag{2}$$

where S_i is the saturation of ice in the pores; and, ρ_i denotes the ice density (kg/m³).

2.2. Energy Conversation Equation

Due to the groundwater flow, heat transfer in the form of convection becomes non-negligible. The higher the groundwater flow velocity is, the more intense its impact on heat transfer will be. Moreover, the heat transfer, in this case, contains the water-ice phase transition process. It causes a dramatic energy shift in a narrow range of phase transition. Therefore, in addition to heat conduction, the energy conversation equation should also include heat convection and phase transition terms,

$$C_{eq} \frac{\partial T}{\partial t} - \nabla \left(\lambda_{eq} \nabla T \right) + C_w \vec{u} \nabla T - \rho_w L \frac{\partial S_w}{\partial T} = Q_G \tag{3}$$

where T is the temperature (K); $\vec{u}$ is seepage velocity vector, as calculated by Darcy's law $\left(\vec{u} = -k_r K\left(\frac{\nabla p}{\rho_w g} + \nabla D\right)\right)$; L is the latent heat released during phase transition (J/kg); C_w is the fluid heat capacity ((J/(kg·K)); S_w is the saturation of water in the pores; and, $\partial S_w / \partial T$ represents the water volume fraction per volume of pore medium that is produced by freezing as T decreases. C is the bulk heat capacity and λ is the bulk thermal conductivity. The subscript *eq* denotes that the physical variables are calculated by using the volumetrically weighted arithmetical mean method, which is,

$$C_{eq} = \varepsilon(S_w \rho_w c_w + S_i \rho_i c_i) + (1 - \varepsilon)\rho_s c_s \tag{4}$$

$$\lambda_{eq} = \varepsilon(S_w \eta_w + S_i \eta_i) + (1 - \varepsilon)\eta_s \tag{5}$$

where c is the specific heat capacity (J/(kg·K)) and η is the thermal conductivity (W/(m·K)). The subscript i, w, and s denote ice, water, and grain particles, respectively.

2.3. TH Coupling Parameters Description

Based on the theory of groundwater flow and heat transfer in porous media, the key coupling factors of TH governing equations are effective hydraulic conductivity and water saturation. They will have a major impact on the distribution of temperature and groundwater flow fields

2.3.1. Effective Hydraulic Conductivity

With the decreasing of temperature, the ice that forms in the pore will block the original groundwater flow channel and it results in lower soil permeability. The effective hydraulic conductivity, defined as a function of temperature, is introduced to describe this permeability reduction. Related research shows that the hydraulic conductivity during freezing depends on temperature and soil type [41–43]. The experimental results from Burt and Williams (1976) [42] show that the hydraulic conductivity will drop down to 10^{-4} times below the original value within a narrow temperature range below freezing point. Some other scholars describe this process by introducing the parameter of relative permeability [26,44]. The common functions that describe the relationship between the physical parameters and changing temperatures with phase change are piecewise linear function [28] and exponential expression [26,29]. In addition to the well-known geotechnical software GeoStudio [45], numerical software using the smoothed step function to describe parameter changes during phase transitions is less common. Due to the release of latent heat during the ice-water phase change, the nonlinearity of the convection-diffusion equation is strengthened. In this paper, we adopt smoothed Heaviside step function to characterize the relationship between permeability and temperature. The effective hydraulic conductivity is defined as

$$k_{ef} = (1 - k_r) flc2hs(T - T_p, \delta T) + k_r \tag{6}$$

where k_r is residual hydraulic conductivity, i.e. the hydraulic conductivity in the frozen body, and it is usually set very small but cannot be equal to zero; and, *flc2hs* represents the smoothed Heaviside function with a continuous second derivative without overshoot, which is a built-in step-function in COMSOL. The function *flc2hs*(x,d) is equal to 0 for $x < -d$, and to 1 for $x > d$. In between $(-d < x < d)$, the function goes from 0 to 1, and at $x = 0$, *flc2hs* = 0.5 (more details refer to [29]); T_p is the freezing point; and, δT is the radius of the phase transition interval.

2.3.2. The Function of Water Saturation and Freezing

The function of water saturation describes the changing water content in the pores during the phase change. Obviously, it is affected by temperature and therefore a function of temperature. The saturated soils in the unfrozen zone are filled with water in their pores. In this case, the water saturation is 1. As the soil temperature drops below the freezing point of pure water, phase change of the pore water will take place, resulting in a gradual decrease in water saturation. Due to the presence of a certain amount of dissolved salt in the groundwater, the purification effect of ice causes a gradual increase of pore water concentration in the local area. In addition, some of the water exists in the form of a thin film. Hence, when more water turns into ice, the capillary and the absorption force will be enlarged, which makes it harder for the water to freeze [46]. As a result, some super-cooled unfrozen water remains in the frozen zone (S_{rw}). Besides the descriptions with linear function, some scholars use exponential functions to describe the residual saturation [15,34]. In order to facilitate the numerical model, the water saturation function is described by the following Heaviside function,

$$S_w = (1 - S_{rs})flc2hs(T - T_p, \delta T) + S_{rw} \tag{7}$$

The freezing function is used to determine the amount of released latent heat in Equation (3) during the phase change, as $-\partial S_w / \partial T$.

2.4. Numerical Implementation

The numerical implementation of the mathematical model is performed through the finite element numerical software COMSOL Multiphysics®. This software can provide an interface between the interconnected variables of different physical fields. In the study of Scheidegger et al. (2017) [29], the TH coupling permafrost model was also established with this software. Compared with the permafrost model in their study, our AGF model faces a more complicated TH coupling process (mainly reflected in the interactional influence of freezing front shape and seepage velocity). In order to characterize this process, the numerical calculation requires much more detailed grids, especially at the vicinity of the freezing front. In our numerical study, we therefore utilized the self-adapting gridding function to initialize the detailed finite elements in order to improve the convergence of the model. Two smoothed step functions are used to describe the change of effective thermal conductivity and unfrozen water saturation with changing temperature, respectively, with the expectation that the non-linearity at the freezing front could be decreased. The feasibility of the new functions will be verified through the subsequent analytical solution and the case study of the engineering project. Previous experience has shown the obvious advantages of this software in solving complex models of coupled multi-physics. Therefore, the numerical solution of the artificial ground freezing method combined with groundwater flow in this paper is implemented in a COMSOL model, which is based on a full coupling of temperature and flow fields by combining the physical interfaces of Darcy's Law and Heat Transfer in Porous Media. The effective hydraulic conductivity (k_r) and the function of water saturation (S_w) and its derivative with respect to T ($\partial S_w / \partial T$) in this model are defined with the smoothed step function. The relationships between them and various temperatures are shown in Figure 2.

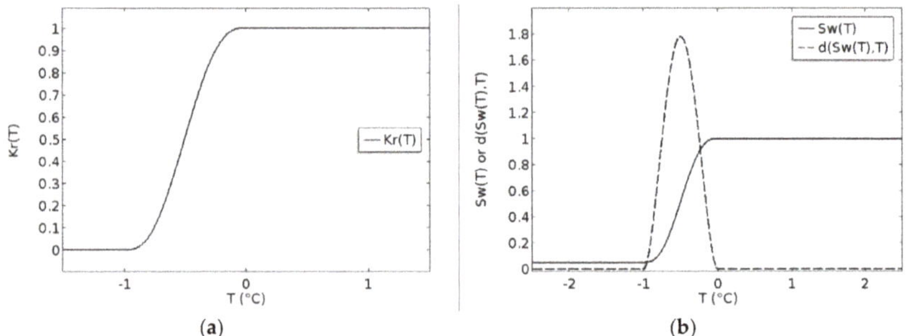

Figure 2. The temperature dependent parameters employed in the numerical implementation: (a) Effective hydraulic conductivity curve with various temperatures; and, (b) The function of water saturation and freezing with various temperatures.

3. Validation and Simulation Studies

3.1. Comparison with Analytical Solution Based Results

The established numerical model for the heat transfer with phase change is verified by comparing the numerically calculated temperature with the results that are based on the classical analytical solution of freezing temperature field proposed by Lunardini (1985) [27] (hereafter referred to as the "Lunardini solution"). The Lunardini solution is firstly developed to solve a one-dimensional, semi-infinite soil thawing problem. It accommodates heat advection via water flow. In the study of Kurylyk et al. (2014) [28], the authors compared their numerical results with this analytical solution. In this study, we also use this analytical solution to validate our numerical model. The employed geometry, initial conditions, and boundary conditions in this study are shown in Figure 3. The detailed physical parameters of materials are given by Kurylyk et al. (2014) [28].

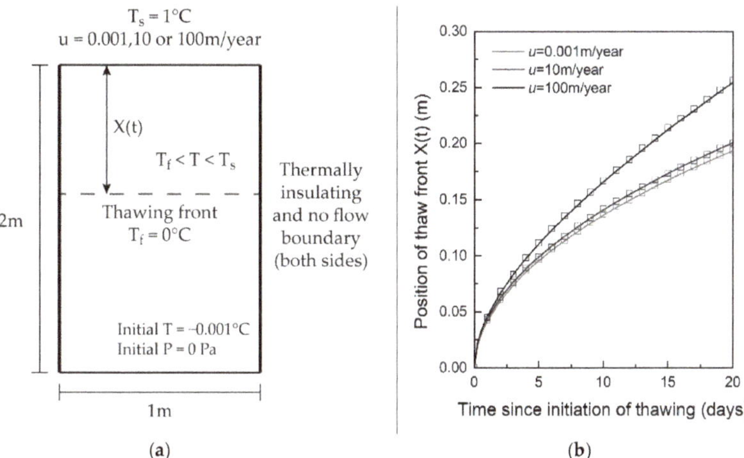

Figure 3. Comparison of the results of analytical and numerical solutions. (**a**) Model geometry, initial conditions and boundary conditions employed in COMSOL; and, (**b**) position of the thawing front versus time at three Darcy velocities: 0.001 m/year (light gray, negligible advection), 10 m/year (medium gray), and 100 m/year (black) calculated by Lunardini's analytical solution (lines) and numerical solution (squares).

To verify the performance of this model at different groundwater flow velocities, we specified three different boundary conditions at the top: 0.001 m/year, 10 m/year, 100 m/year. The comparison of the results of analytical and numerical solutions shows a good consistency in Figure 3, which indicates that this numerical model is reliable for solving the heat transfer problem with phase change and existing groundwater flow. Beyond this, this model will be further verified by the in-situ case study in Section 4.

3.2. Simulation Studies

Groundwater flow will affect the soil temperature distribution in the freezing process. During this process, the groundwater flow rate is one of the key factors. A large amount of research work has been done on the influence of the groundwater flow rate on the closure time of the freezing wall [27,37,47–49]. However, apart from the groundwater flow rate, other parameters, such as pipe spacing, thermodynamic parameters, and so on are also important factors that affect the freezing temperature field [50]. In our study through a series of numerical experiments, we simulate and analyze the effects of pipe spacing, thermal conductivity of soil particles, and groundwater flow velocity on the temperature distribution and the closure time of the freezing wall. The geometric model used in the numerical experiment is shown in Figure 4. The 16 freezing pipes are arranged in a square with the same pipe radius of $r_0 = 0.054$ m. The ambient temperature (T_0) is 15 °C and the initial hydraulic head has a difference of =0.8 m (ΔH) between left and right boundary. The temperature of the freezing pipes (T_p) is -30 °C. For detailed parameter setup, please refer to Table 1.

Based on the numerical model introduced above, three sets of parametric sweep experiments are designed for three factors, which are the head difference (ΔH), thermal conductivity of soil grain (λ_s), and pipe spacing (l): (1) when $l = 1.1$ and $\lambda_s = 2.0$, ΔH values are 0.5, 1.0, 1.5; (2) when $l = 1.1$ and $\Delta H = 1.0$, the λ_s values are 1.5, 2.0, 2.5; and, (3) when $\Delta H = 1.5$ and $\lambda_s = 2.0$, the values of l are 1.0, 1.1, 1.2 (for the individual unit please refer to Table 1). The experimental results of temperature distribution are shown in Figure 5. Parametric sweep results show that as the head difference and pipe spacing increase, the formation of freezing walls becomes more difficult. On the contrary, as the thermal conductivity of soil grain increases, the thickness of the frozen wall gradually increases.

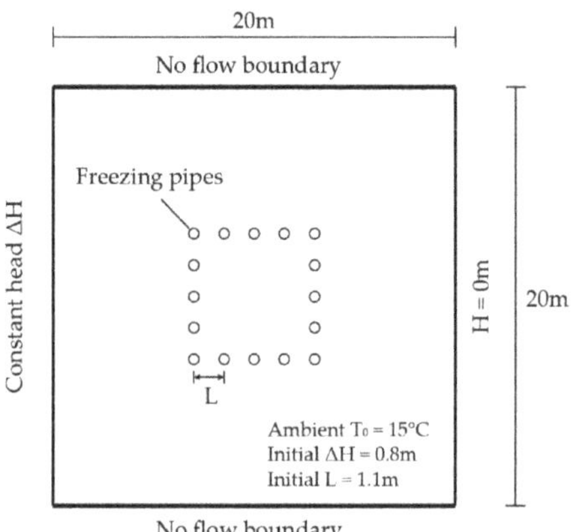

Figure 4. Model geometry, initial and boundary conditions for the simulation of the artificial ground freezing (AGF) project.

Table 1. Parameters used in numerical experiments.

Parameters	Value
Density of Soil ρ_s (kg/m^3)	2600
Density of Water ρ_w (kg/m^3)	1000
Density of Ice ρ_i (kg/m^3)	920
Pipe Spacing l (m)	1.0, 1.1, 1.2 [1]
Head Difference ΔH (m)	0.5, 1.0, 1.5 [1]
Thermal Conductivity of Soil Grain λ_s (W/(m·K))	1.5, 2.0, 2.5 [1]
Thermal Conductivity of Water λ_w (W/(m·K))	0.6
Thermal Conductivity of Ice λ_i (W/(m·K))	2.14
Heat Capacity of Soil c_s (J/(kg·K))	840
Heat Capacity of Water c_w (J/(kg·K))	4200
Heat Capacity of Ice c_i (J/(kg·K))	2100
Porosity n	0.3
Permeability Coefficient k (m/day)	20
Latent Heat of Formation L (J/kg)	334,720

Note: [1] Value for parametric sweep.

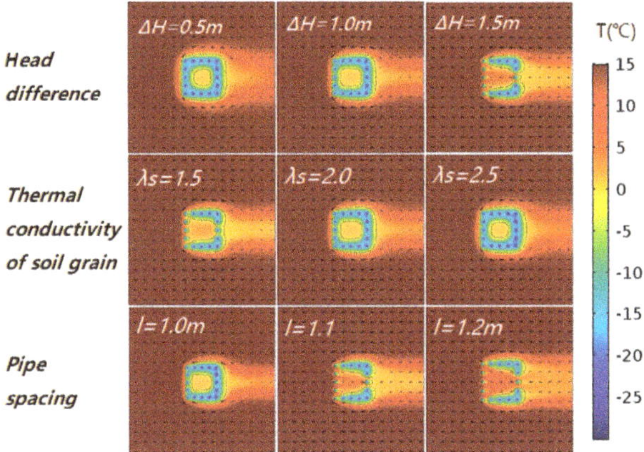

Figure 5. The temperature distribution results of parametric sweep experiments.

Except for the shape of the frozen wall, the closure time of the frozen wall is the key factor in evaluating the ground freezing project. The case studies introduced above show that the parameters of λ_s, l, and ΔH have an important influence on the closure time, among which the roles of λ_s and l are explicit. The third parameter, ΔH, is the controlling factor of the groundwater velocity. The influence of groundwater velocity on the ground freezing process has been discussed for years. In the ground freezing project, if the groundwater flow exceeds a certain speed, which is defined as the critical velocity, the freezing wall will not be able to close. However, this definition is difficult to quantify, as it is essentially an economic problem. As shown in Figure 6, the closure time increases exponentially with increasing groundwater flow velocity, especially after the velocity has exceeded 1.5 m/day.

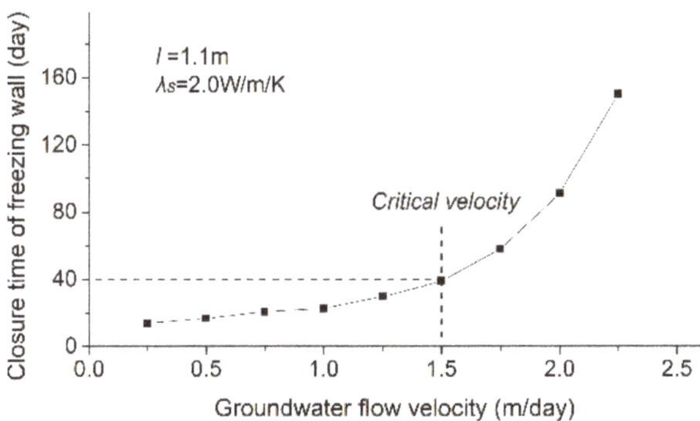

Figure 6. Closure times of freezing wall with various groundwater flow velocities and definition of the critical velocity.

In order to study further the determination and influence of critical velocity, we firstly consider the fact in our case that the freezing wall is planned to be closed within 40 days and define the allowed maximum groundwater flow velocity as the critical velocity. As shown in Figure 6, when $l = 1.1$ and $\lambda_s = 2.0$, the critical velocity is 1.5 m/day. Subsequently, we set these two parameters within certain ranges ($l \in [0.5, 1.5]$, $\lambda_s \in [1.0, 3.0]$) and randomly sample 500 pairs of data to calculate numerically the corresponding critical velocity values. With respect to this critical value, linear regression analyses of pipe spacing and thermal conductivity are performed, and the results are shown in Figure 7. The slope of the regression line indicates that the pipe spacing is negatively correlated to the critical velocity and has more influence on the critical velocity than the thermal conductivity. As shown in Figure 7a, the degree of dispersion with a lower pipe spacing value (smaller than 0.8 m) has an increasing trend, which indicates that the pipe spacing value is no more linearly correlated to the critical velocity. On the contrary, as shown in Figure 7b, the sampling points become more scattered with increasing thermal conductivity. This indicates that, when the thermal conductivity value becomes larger, the critical velocity becomes less sensitive to it. Based on both figures, the critical velocity can be mainly found within the range of 0.25 m/day~1 m/day. If the groundwater velocity does not exceed 1 m/day, in most (about 90.6%) cases the freezing wall should be able to close within 40 days. Once exceeded, extra measures (e.g., lowering the freezing temperature, decreasing the pipe spacing, etc.) should be taken to ensure the closure of the freezing wall.

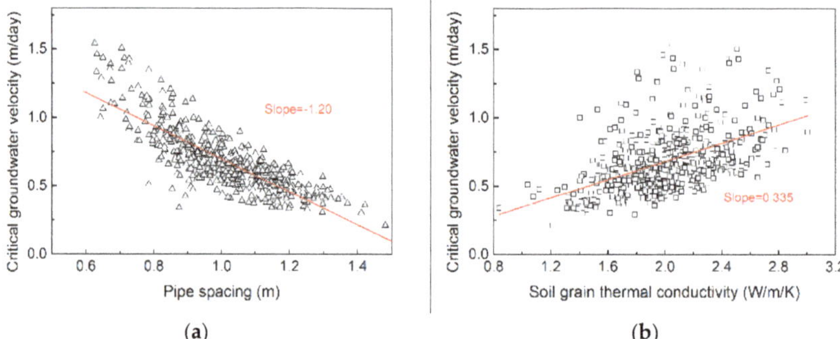

Figure 7. Scatter plot of changes in critical groundwater flow velocity with respect to (**a**) Pipe spacing; and, (**b**) Soil grain thermal conductivity based on 500 random sampling pairs. The red line represents the linear regression analysis.

4. Engineering Application

4.1. Engineering Background and the 3D Model

The application case of this study is an AGF strengthening project for the jointing section of a tunnel and a subway station located in Southern China. According to the geological survey, a saturated aquifer is at the depth of the frozen zone with silty fine sand. According to the in-situ groundwater flow test, the horizontal groundwater flow (the arrows shown in the Figure 8b) is from the left to the right side almost perpendicular to the crosscut of the tunnel. In this AGF project, 35 frozen pipes are arranged with the diameter of 108 mm each, the designed pipe spacing 1.1 m, and the length of ca. 7 m each. According to the in-situ measurements, the arrangement of freezing pipes and monitoring the hole is shown in Figure 8. The 3D model has a geometry of 20 m × 20 m × 15 m, in which the excavated tunnel is 8 m × 7.5 m × 5 m.

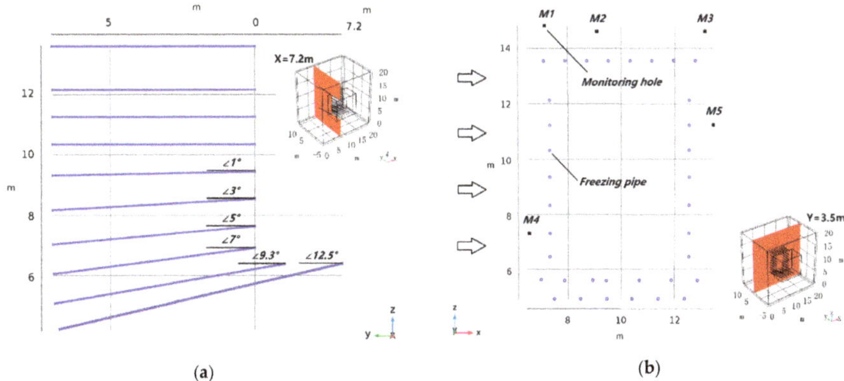

(a) (b)

Figure 8. The arrangement of freezing pipes and monitoring points in three-dimensional (3D) model. (**a**) Side view with blue lines representing the freezing pipes; (**b**) Front view (the arrows represent groundwater flow direction).

4.2. Boundary and Intial Conditions

According to the prior in-situ temperature monitoring data at the desired freezing depth, the initial ground temperature is about 15 °C. The box chart of initial temperatures in different thermo-observation holes is shown in Figure 9.

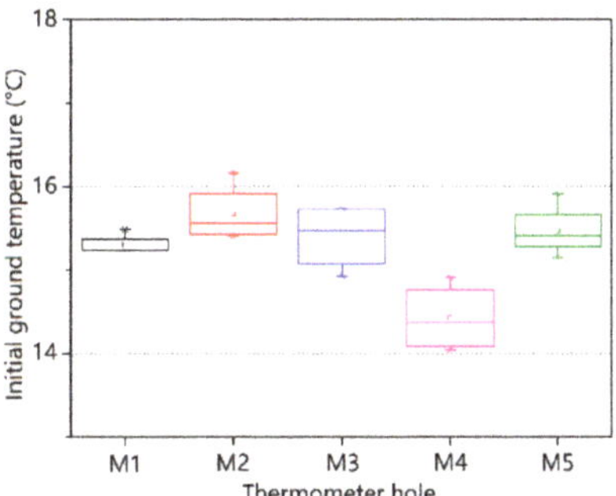

Figure 9. The box chart of initial ground temperatures in different thermo-observation holes.

The lateral wall of the freezing pipe is the cooling source of the freezing system. The change in lateral wall temperature has a great effect on the temperature distribution in the whole system. The temperature monitoring values of the main-pipe, through which the refrigerant can flow to the freezing wall, can be used to represent the lateral wall temperature. According to the measured temperature of the loop main-pipe (T_{out}) during the freezing period (50 days), the lateral wall temperature change curve is shown in Figure 10.

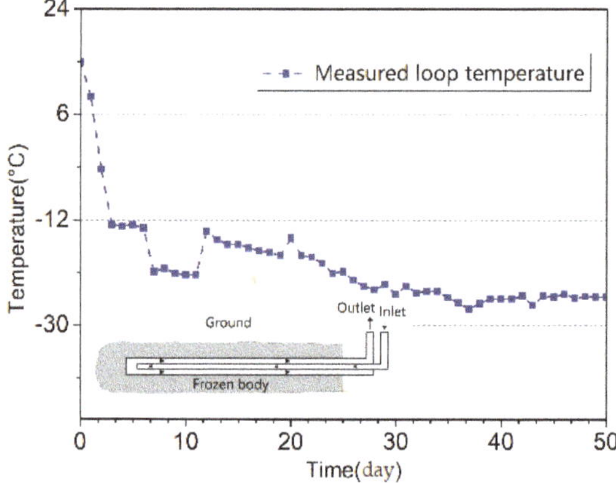

Figure 10. The measured temperature of the loop main-pipe during freezing period.

The groundwater velocity obtained through the in-situ field test is 0.5 m/day. Based on Darcy's law, the head difference between up and downstream boundary is calculated as 0.8 m.

The parameters used for the engineering example in this model is shown in Table 2.

Table 2. Parameters used for engineering example.

Parameters	Value
Density of Soil ρ_s(kg/m^3)	1800
Density of Water ρ_w (kg/m^3)	1000
Density of Ice ρ_i (kg/m^3)	920
Head Difference ΔH (m)	0.8
Thermal Conductivity of Soil Grain λ_s (W/(m·K))	0.73
Thermal Conductivity of Water λ_w (W/(m·K))	0.6
Thermal Conductivity of Ice λ_i (W/(m·K))	2.14
Heat Capacity of Soil c_s (J/(kg·K))	1020
Heat Capacity of Water c_w (J/(kg·K))	4200
Heat Capacity of Ice c_i (J/(kg·K))	2100
Porosity n	0.4
Permeability Coefficient k (m/day)	5
Latent Heat of Formation L (J/kg)	334,720

4.3. Simulation Results, Verification and Discussion

The simulation results of the temperature distribution and freezing body development at various times (10 days, 20 days, 30 days, 40 days, and 50 days) are shown in Figure 11. The freezing body and streamlines development show a clear coupling effect of groundwater flow field and the freezing temperature field. After 50 days, the freezing wall has been completely closed. But, the thickness of the frozen wall at the upstream side (left side) is relatively thin, especially at the junction of this side to the top of the frozen wall (top left corner). The simulated temperature of the freezing wall at the section of Y = 3.5 m shows obvious non-uniform distribution. Due to the arrangement of freezing pipes and the groundwater flow, the temperatures at the top and bottom of the frozen wall are lower than those at both sides of the up- and downstream. The actual pipe spacing has an important influence on the temperature field and the closure time of the freezing wall. For example, at the upper left corner of the freezing circle, the decrease of temperature is slower, resulting in a slower closing process.

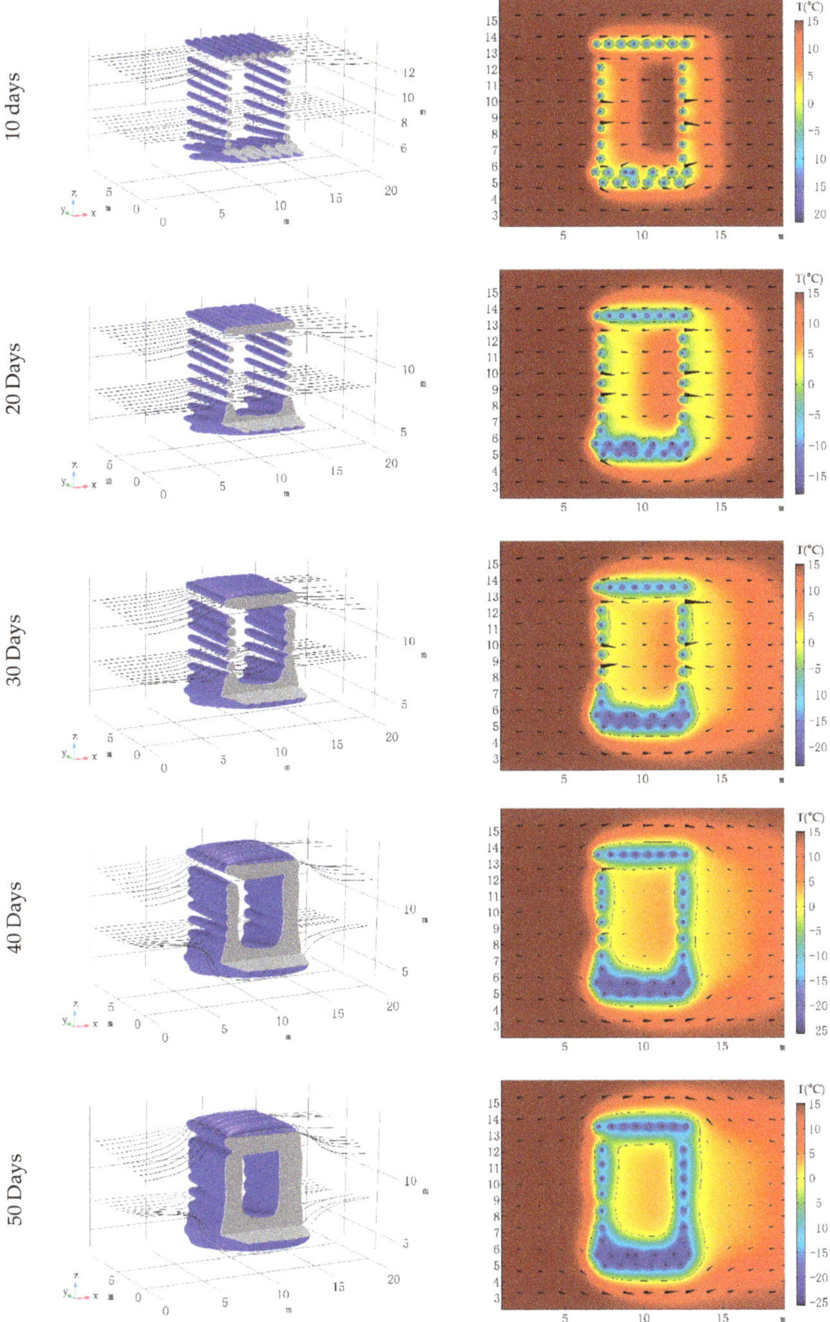

Figure 11. The development of freezing body with streamlines (black lines) (**left**) and temperature distribution with velocity vector (black arrows) at $Y = 3.5$ m (**right**) at various times.

According to the study in Section 3.2, the freezing wall should be able to close within 40 days if the groundwater velocity is 0.5 m/day, the soil thermal conductivity is 2.14 W/(m·K), and the pipe spacing is around 1 m. However, Figure 11 shows a closure time of 50 days. On the 40th day the freezing wall was not yet completely closed. This is due to the fact that the freezing pipes are not uniformly distributed and the pipe spacing values vary. For example, at the upper left corner the pipe spacing is 1.5 m, while at the bottom, the smallest distance between the pipes is only 0.5 m. As the frozen wall was built on the other side, and therefore the groundwater flow was limited for the left side, the unfrozen area went frozen in the next 10 days after the 40th day.

The comparison between the in-situ measured and the numerically calculated temperatures at the five monitoring points (positions shown in Figure 8) is shown in Figure 12. After the 50-day freezing period, the temperature changes of the top three monitoring points (M1, M2, M3) are relatively gentle and among which the middle measuring point M2 has a lower temperature than M1 and M3. Both monitoring points at the upstream side (M4) and the downstream side (M5) have reached below 0 °C after 50 days of freezing. The temperature of the downstream monitoring point decreased more rapidly and reached 0 °C earlier than the upstream monitoring point. Overall, the calculated curves of the temperatures at the monitored points coincide with the ones that are measured in-situ.

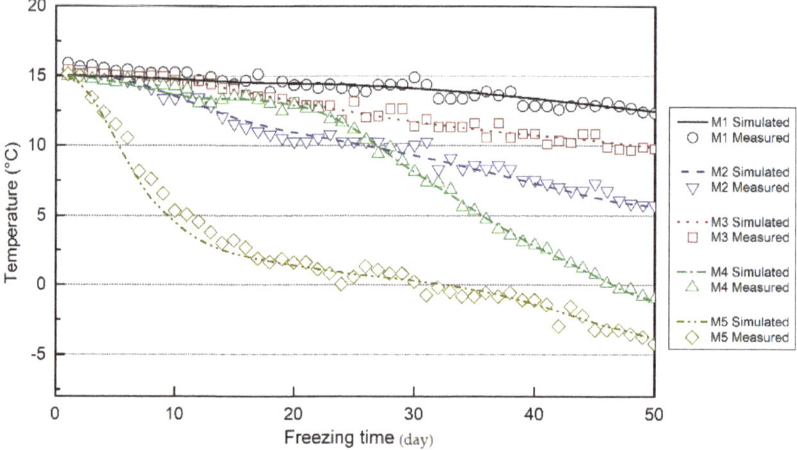

Figure 12. The temperature comparison curve of monitoring points between simulated and measured values.

The average temperature of the frozen body is one of the most important indices to evaluate the strength of the frozen wall. According to the calculated results, the curve of the average temperature of the frozen wall is plotted against time in Figure 13. The results show that the average temperature change of the frozen wall coincides with that of the main frozen pipes (Figure 10) with the same trend. After a 50-day freezing period, the average temperature of the frozen body reached below −10 °C.

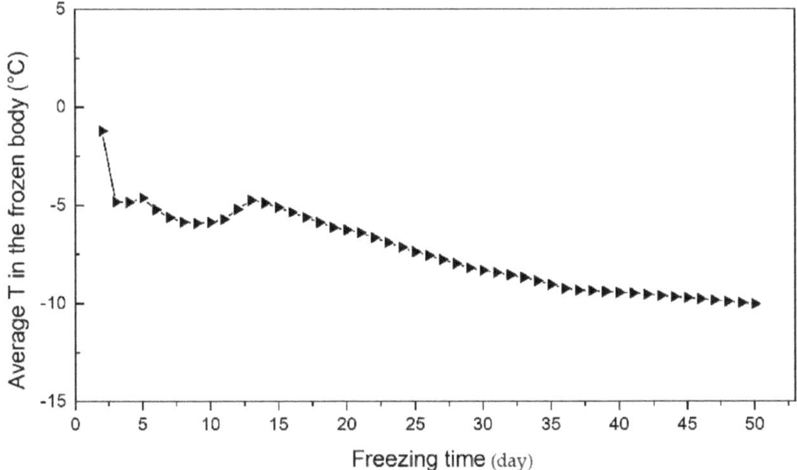

Figure 13. Average temperature curve of the frozen body versus freezing time.

5. Conclusions

In this study, based on the theories of heat transport and groundwater flow in a porous medium, we set up a fully coupled numerical model with finite element software COMSOL Multiphysics®. The coupled process of heat transport and groundwater flow with phase change was simulated. Through the model validation with an analytical solution, the sensitivity analysis, as well as the simulation of a field example with in-situ measurement data, we have drawn the following conclusions.

During the phase change, the effective hydraulic conductivity and the saturation of unfrozen water are functions of the temperature that controls the heat transport process. This function is generally described with a piecewise linear function. In this paper, we decided to use the smoothed step function. The comparison of numerically calculated and analytically solved results, as well as the validation of the field model have shown that it is feasible to use the smoothed step function to describe the change of effective hydraulic conductivity k_r and the saturation of unfrozen water S_w during the phase change.

With numerical case studies, sensitivity analysis was performed for three parameters: head difference ($\cdot H$), thermal conductivity of soil grain (λ_s), and pipe spacing (l). The corresponding influence of these parameters on the critical groundwater flow velocity was analyzed. The results show that when the groundwater flow velocity is within a range of 0.25 m/day~1.0 m/day, in about 90% of cases the freezing wall can be closed within 40 days. Extra measures are recommended to ensure the closure of the freezing wall, if the groundwater flow velocity exceeds this range.

The 3D model for the field example with further validation has shown that, along the groundwater flow direction, the freezing wall forms slowly and on the upstream side the thickness of the freezing wall is thinner than on the downstream side. The actual pipe spacing has an important influence on the temperature field and closure time of the freezing wall. The larger the actual pipe spacing is, the slower the closing process will be. Besides this, the average temperature change of the freezing body coincides with that of the main frozen pipes with the same trend.

Author Contributions: R.H. and Q.L. performed the studies and wrote the paper. Y.X. has contributed to the numerical modelling.

Acknowledgments: The work is supported by the Ministry of Education of the People's Republic of China through the program "Research on Mechanism of Groundwater Exploitation and Seawater Intrusion in Coastal Areas" (Project Code 20165037412) and by "the Fundamental Research Funds for the Central Universities" ("Research on the hydraulic tomographical method for aquifer characterization ", Project Code 2015B29314). It is also supported by Jiangsu Provincial Department of Education, Project Code 2016B1203503.

Conflicts of Interest: The authors declare no conflict of interest.

References

1. Gallardo, A.H.; Marui, A. The aftermath of the Fukushima nuclear accident: Measures to contain groundwater contamination. *Sci. Total Environ.* **2016**, *547*, 261–268. [CrossRef] [PubMed]
2. Dash, J. Ground freezing technology for environmental. *Ice Phys. Nat. Environ.* **2013**, *56*, 241.
3. Wallis, S. Freezing under the sea rescues oslofjord highway tunnel. *Tunnel* **1999**, *8*, 19–26.
4. Hu, X.D.; Zhang, L.Y. *Artificial Ground Freezing for Rehabilitation of Tunneling Shield in Subsea Environment*; Advanced Materials Research; Trans Tech Publications: Zurich, Switzerland, 2013; pp. 517–521.
5. Papakonstantinou, S.; Anagnostou, G.; Pimentel, E. Evaluation of ground freezing data from the Naples subway. *Proc. Inst. Civ. Eng. Geotech. Eng.* **2013**, *166*, 280–298. [CrossRef]
6. Marwan, A.; Zhou, M.-M.; Abdelrehim, M.Z.; Meschke, G. Optimization of artificial ground freezing in tunneling in the presence of seepage flow. *Comput. Geotech.* **2016**, *75*, 112–125. [CrossRef]
7. Russo, G.; Corbo, A.; Cavuoto, F.; Autuori, S. Artificial ground freezing to excavate a tunnel in sandy soil. Measurements and back analysis. *Tunnelling Underground Space Technol.* **2015**, *50*, 226–238. [CrossRef]
8. Haß, H.; Schäfers, P. International Symposium on Geotechnical Aspects of Underground Construction in Soft Ground. In *Application of Ground Freezing for Underground Construction in Soft Ground*; Taylor & Francis London: Oxfordshire, UK, 2005; pp. 405–412.
9. Colombo, G.; Lunardi, P.; Cavagna, B.; Cassani, G.; Manassero, V. The artificial Ground Freezing Technique Application for the Naples Underground. In Proceedings of the Word Tunnel Congress, Agra, India, 22–24 September 2008.
10. Vitel, M.; Rouabhi, A.; Tijani, M.; Guérin, F. Thermo-hydraulic modeling of artificial ground freezing: Application to an underground mine in fractured sandstone. *Comput. Geotech.* **2016**, *75*, 80–92. [CrossRef]
11. Pimentel, E.; Sres, A.; Anagnostou, G. Large-scale laboratory tests on artificial ground freezing under seepage-flow conditions. *Géotechnique* **2012**, *62*, 227. [CrossRef]
12. Arenson, L.U.; Sego, D.C. Freezing Processes for a Coarse sand with Varying Salinities. In Proceedings of the Cold Regions Engineering & Construction Conference, Edmonton, Alta, 16–19 May 2004.
13. Bing, H.; Ma, W. Laboratory investigation of the freezing point of saline soil. *Cold Reg. Sci. Technol.* **2011**, *67*, 79–88. [CrossRef]
14. Frivik, P.; Comini, G. Seepage and heat flow in soil freezing. *J. Heat Transf.* **1982**, *104*, 323–328. [CrossRef]
15. Nishimura, S.; Gens, A.; Olivella, S.; Jardine, R. THM-coupled finite element analysis of frozen soil: Formulation and application. *Géotechnique* **2006**. [CrossRef]
16. Lai, Y.; Pei, W.; Zhang, M.; Zhou, J. Study on theory model of hydro-thermal–mechanical interaction process in saturated freezing silty soil. *Int. J. Heat Mass Transf.* **2014**, *78*, 805–819. [CrossRef]
17. Yang, P.; Ke, J.-M.; Wang, J.; Chow, Y.; Zhu, F.-B. Numerical simulation of frost heave with coupled water freezing, temperature and stress fields in tunnel excavation. *Comput. Geotech.* **2006**, *33*, 330–340. [CrossRef]
18. Wu, D.; Lai, Y.; Zhang, M. Heat and mass transfer effects of ice growth mechanisms in a fully saturated soil. *Int. J. Heat Mass Transf.* **2015**, *86*, 699–709. [CrossRef]
19. Gao, J.; Feng, M.; Yang, W.-H. Research on distribution law of frozen temperature field of fractured rock mass with groundwater seepage. *J. Min. Saf. Eng.* **2013**, *1*, 013.
20. Vitel, M.; Rouabhi, A.; Tijani, M.; Guérin, F. Modeling heat transfer between a freeze pipe and the surrounding ground during artificial ground freezing activities. *Comput. Geotech.* **2015**, *63*, 99–111. [CrossRef]
21. Vitel, M.; Rouabhi, A.; Tijani, M.; Guérin, F. Modeling heat and mass transfer during ground freezing subjected to high seepage velocities. *Comput. Geotech.* **2016**, *73*, 1–15. [CrossRef]
22. Grenier, C.; Anbergen, H.; Bense, V.; Chanzy, Q.; Coon, E.; Collier, N.; Costard, F.; Ferry, M.; Frampton, A.; Frederick, J.; et al. Groundwater flow and heat transport for systems undergoing freeze-thaw: Intercomparison of numerical simulators for 2d test cases. *Adv. Water Resour.* **2018**, *114*, 196–218. [CrossRef]
23. Harlan, R. Analysis of coupled heat-fluid transport in partially frozen soil. *Adv. Water Resour.* **1973**, *9*, 1314–1323. [CrossRef]
24. Guymon, G.L.; Luthin, J.N. A coupled heat and moisture transport model for arctic soils. *Water Resour. Res.* **1974**, *10*, 995–1001. [CrossRef]

25. Hansson, K.; Šimůnek, J.; Mizoguchi, M.; Lundin, L.-C.; Van Genuchten, M.T. Water flow and heat transport in frozen soil. *Vadose Zone J.* **2004**, *3*, 693–704. [CrossRef]

26. McKenzie, J.M.; Voss, C.I.; Siegel, D.I. Groundwater flow with energy transport and water–ice phase change: Numerical simulations, benchmarks, and application to freezing in peat bogs. *Adv. Water Resour.* **2007**, *30*, 966–983. [CrossRef]

27. Lunardini, V. Freezing of soil with phase change occurring over a finite temperature difference. In Proceedings of the 4th International Offshore Mechanics and Arctic Engineering Symposium, Dallas, TX, USA, 17–21 February 1985.

28. Kurylyk, B.L.; McKenzie, J.M.; MacQuarrie, K.T.; Voss, C.I. Analytical solutions for benchmarking cold regions subsurface water flow and energy transport models: One-dimensional soil thaw with conduction and advection. *Adv. Water Resour.* **2014**, *70*, 172–184. [CrossRef]

29. *British Geological Survey Commissioned Report: 2017: Coupled Modelling of Permafrost and Groundwater. A Case Study Approach*; British Geological Survey: Keyworth, Nottingham, UK, 2017; p. 157.

30. Tan, X.; Chen, W.; Tian, H.; Cao, J. Water flow and heat transport including ice/water phase change in porous media: Numerical simulation and application. *Cold Reg. Sci. Technol.* **2011**, *68*, 74–84. [CrossRef]

31. Mizoguchi, M. Water, Heat and Salt Transport in Freezing Soil. Ph.D. Thesis, University of Tokyo, Tokyo, Japan, 1990.

32. Sres, A. *Theoretische und Experimentelle Untersuchungen zur Künstlichen Bodenvereisung im Strömenden Grundwasser*; vdf Hochschulverlag AG: Zurich, Switzerland, 2010; p. 18378.

33. Ständer, W. *Mathematische Ansätze zur Berechnung der Frostausbreitung in Ruhendem Grundwasser im Vergleich zu Modelluntersuchungen für Verschiedene Gefrierrohranordnungen im Schactund Grundbau*; Technische Hochschule Fridericiana, Institut für Bodenmechanik und Felsmechanik: Karlsruhe, Germany, 1967.

34. Victor, H. *Die Frostausbreitung beim Künstlichen Gefrieren von Böden unter dem Einfluss Strömenden Grundwassers*; Univ. Inst. f. Bodenmechanik u. Felsmechanik: Karlsruhe, Germany, 1969; p. 42.

35. Sanger, F.; Sayles, F. Thermal and rheological computations for artificially frozen ground construction. *Eng. Geol.* **1979**, *13*, 311–337. [CrossRef]

36. Huang, R.; Chang, M.; Tsai, Y.; Lu, S.; Wu, P. Influence of seepage flow on temperature field around an artificial frozen soil through model testing and numerical simulations. In Proceedings of the 18th Southeast Asian Geotechnical Conference, Singapore, 29–31 May 2013; Leung; Leung, C.F., Goh, S.H., Shen, R.F., Eds.; Geotechnical Society of Singapore, Research Publishing Services: Singapore, 2013.

37. Jessberger, H.; Jagow-Klaff, R. *Bodenvereisung. Grundbau-Taschenbuch, Teil 2: Geotechnische Verfahren*; Country Ernst & Sohn: Berlin, Germany, 2001; pp. 121–166.

38. Andersland, O.B.; Ladanyi, B. *Frozen Ground Engineering*; John Wiley & Sons: Hoboken, NJ, USA, 2004.

39. Endo, K. *Artificial Soil Freezing Method for Subway Construction*; Japan Society of Civil Engineers: Tokyo, Japan, 1969.

40. Zhou, X.; Wang, M.; Zhang, X. Model test research on the formation of freezing wall in seepage ground. *J. Chin. Coal Soc.* **2005**, *30*, 196–201.

41. Williams, P.; Burt, T. Measurement of hydraulic conductivity of frozen soils. *Can. Geotech. J.* **1974**, *11*, 647–650. [CrossRef]

42. Burt, T.; Williams, P.J. Hydraulic conductivity in frozen soils. *Earth Surf. Process. Landf.* **1976**, *1*, 349–360. [CrossRef]

43. Horiguchi, K.; Miller, R. Experimental studies with frozen soil in an "ice sandwich" permeameter. *Cold Reg. Sci. Technol.* **1980**, *3*, 177–183. [CrossRef]

44. Jame, Y.W.; Norum, D.I. Heat and mass transfer in a freezing unsaturated porous medium. *Water Resour. Res.* **1980**, *16*, 811–819. [CrossRef]

45. *Heat and Mass Transfer Modeling with GeoStudio 2018*, 2nd ed.; GEOSLOPE International Ltd.: Calgary, AB, Canada, 2017.

46. Embleton, C. *The Frozen Earth: Fundamentals of Geocryology*; Cambridge University Press: Cambridge, UK, 1992.

47. Ziegler, M.; Baier, C.; Aulbach, B. Optimization of artificial ground freezing application of tunneling subject to water seepage. In Proceedings of the 17th Internal Conference on Soil Mechanics and Geotechnical Engineering, Alexandria, Egypt, 5–9 October 2009.

48. Zhou, M.; Meschke, G. A three-phase thermo-hydro-mechanical finite element model for freezing soils. *Int. J. Numer. Anal. Methods Geomech.* **2013**, *37*, 3173–3193. [CrossRef]

49. Shen, C.; McKinzie, B.; Arbabi, S. A reservoir simulation model for ground freezing process. In Proceedings of the SPE Annual Technical Conference and Exhibition, Florence, Italy, 19–22 September 2010.

50. Lackner, R.; Amon, A.; Lagger, H. Artificial ground freezing of fully saturated soil: Thermal problem. *J. Eng. Mech.* **2005**, *131*, 211–220. [CrossRef]

The Spatial Distribution of the Microbial Community in a Contaminated Aquitard below an Industrial Zone

Noa Balaban [1], Irina Yankelzon [1], Eilon Adar [1], Faina Gelman [2], Zeev Ronen [1,*] and Anat Bernstein [1]

[1] Department of Environmental Hydrology & Microbiology (EHM), The Zuckerberg Institute for Water Research (ZIWR), The Jacob Blaustein Institutes for Desert Research, Ben-Gurion University of the Negev, Sede Boqer Campus 8499000, Israel; balabann@post.bgu.ac.il (N.B.); irinaya@post.bgu.ac.il (I.Y.); eilon@bgu.ac.il (E.A.); anatbern@bgu.ac.il (A.B.)

[2] Geological Survey of Israel, Yesha'yahu Leibowitz 32, Jerusalem 9692100, Israel; faina@gsi.gov.il

* Correspondence: zeevrone@bgu.ac.il; Tel.: +972-8-6596895

Received: 4 September 2019; Accepted: 9 October 2019; Published: 14 October 2019

Abstract: The industrial complex Neot Hovav, in Israel, is situated above an anaerobic fractured chalk aquitard, which is polluted by a wide variety of hazardous organic compounds. These include volatile and non-volatile, halogenated, organic compounds. In this study, we characterized the indigenous bacterial population in 17 boreholes of the groundwater environment, while observing the spatial variations in the population and structure as a function of distance from the polluting source. In addition, the de-halogenating potential of the microbial groundwater population was tested through a series of lab microcosm experiments, thus exemplifying the potential and limitations for bioremediation of the site. In all samples, the dominant phylum was *Proteobacteria*. In the production plant area, the non-obligatory organo-halide respiring bacteria (OHRB) *Firmicutes* Phylum was also detected in the polluted water, in abundancies of up to 16 %. Non-metric multidimensional scaling (NMDS) analysis of the microbial community structure in the groundwater exhibited clusters of distinct populations following the location in the industrial complex and distance from the polluting source. Dehalogenation of halogenated ethylene was demonstrated in contrast to the persistence of brominated alcohols. Persistence is likely due to the chemical characteristics of brominated alcohols, and not because of the absence of active de-halogenating bacteria.

Keywords: fractured aquitard; groundwater pollution; microbial community's diversity; dehalogenation; tribromoneopentyl alcohol; 1-bromo-1-chloroethane

1. Introduction

Groundwater contamination, by persistent, bioaccumulative and bioactive compounds due to intensive industrialization is a critical environmental issue [1–3]. The wide range of chemical compounds complicates the attempts to protect aquifers from their contamination [4,5]. In addition to pollution from industry, agriculture, and the energy sector, personal care products, and pharmaceutical residues are emerging organic contaminants (EOC) in groundwater [6]. In a comprehensive review by Lapworth et al. (2012), EOCs, such as detergents, antioxidants, fire retardants, plasticizers, as well as antibiotics, anti-inflammatories and barbiturates, were detected in substantial levels between $10–10^4$ ng/L in groundwater globally [7]. One of the largest groups of chemicals that cause environmental contamination are halogenated organic compounds [8]. Due to their high common resistance to degradation, they often accumulate in the environment, and their fate depends highly upon microbial degradation [6–10].

The microbial community structure in the ecosystem is critical for the biodegradation of pollutants to occur. In pristine aquifers, microbial diversity is explained by water chemistry, and therefore changes in geochemical conditions and contaminant content will affect the community structure [11–13]. Bacteria adaptation accelerates the biodegradation of organic chemicals, where shifts in community composition or abundances are one of the adaptive processes. For example, in a pilotscale aquifer, Ma et al. [14] presented clear shifts in the microbial community exposed to a chlorinated ethanol fuel plume. Interestingly, after community shifts led to biodegradation of the contaminant and a geochemical restoration of the aquifer, the microbial community structure was also restored. Long monitoring of a coal tar waste-contaminated groundwater ecosystem revealed that natural attenuation processes are accompanied by temporal changes in the microbial communities [15].

Nonetheless, the existence of a microbial potential in the ecosystem does not always lead to in situ biodegradation [11]. Multiple limitations, such as insufficient biomass, geochemical and hydraulic fluctuations, utilization of a wide range of substrates, competitive inhibition, or catabolite repression, can all inhibit biodegradation [16]. Therefore, integrated approaches, such as geochemical-molecular or isotopic-molecular approaches are used for the actual monitoring of biodegradation in natural environments. For example, Tarnawski et al. [17] followed the reductive dechlorination of chlorinated-ethenes in contaminated aquifers and presented an integrated long-term geochemical-microbial monitoring approach, in order to identify biogeochemical processes limiting or supporting biodegradation. Révész et al. [18], on the other hand, quantified the relevant bacterial groups along with in situ trichloroethylene (TCE) anaerobic biodegradation, while monitoring degradation through the use of stable carbon isotopes.

The Neot-Hovav industrial complex accommodates numerous chemical industries in the pharmaceutical, bromine-based organic compounds and agrochemicals. The complex is situated above the fractured Eocene chalk formation [19]. The fractures and joints in the chalk form an aquitard with preferable pathways for water flow and solute migration, thus transmitting contaminants from the surface to groundwater at relatively high fluxes in a short time [20,21]. For this reason, and due to inadequate treatment of industrial waste, the groundwater underlying the industrial area has been contaminated by numerous organic compounds: Volatile and non-volatile halogenated organic compounds, heavy metals, and halogen-containing salts [4,22,23].

In contrast to porous aquifers, the hydraulic heterogeneity in a fractured system is large, and strongly limits particle transport (like bacteria), solutes including electron donors and acceptors, as well as nutrients along the hydrologic gradients [24]. Thus, the effect of hydrologic discontinuity on microbial community structure and catabolic potential is still an open question. Furthermore, in polluted groundwater like Neot-Hovav, where hundreds of different compounds are present, a unique approach to the aggregated effect of inorganic and organic compounds upon microbial community composition and function should be used. In turn, the diversity and richness of the groundwater microbiota, as well as their metabolic potential, can provide a broader perspective on the "health" of the groundwater and the possibility for natural recovery or the need for active remediation management.

In this study, we examine how the extent and type of contamination above ground determine and affect the microbial community structure in the groundwater. For this cause, we investigated spatial changes in the microbial community of the polluted groundwater that underlies the Neot-Hovav industrial complex in relation to the industrial sources of specific halogenated pollutants. Changes in communal structure as a factor of groundwater quality were examined while providing a more sustainable approach to the understanding of these multi-contaminant, complex environments. Microbial activity of the groundwater bacterial communities was also tested through a series of biodegradation microcosm experiments in the lab with various brominated organic compounds like TBNPA-tribromoneopentyl alcohol, DBNPG-dibromoneopentyl glycol, BCE-bromo-chloro-ethane, EDB ethylene dibromide, and EDC ethylene dichloride. These compounds were chosen, as they are part of the wide contaminant palette that exist in the groundwater, and are produced and used in the industrial area.

2. Materials and Methods

2.1. Material

Tribromoneopentyl alcohol (TBNPA, >98% pure) was obtained from TCI chemicals, Japan; dibromoneopentyl glycol (DBNPG) from Israel Chemicals Ltd. (ICL Israel), 1-bromo-1-chloroethane (BCE), 1,2-Dibromoethane (EDB), and, 1,2-dichloroethane (EDC) (99.8% pure) were obtained from Sigma Aldrich (Rehovot, Israel). Methanol, acetone, di-propanol and acetonitrile were all HPLC grade (Supra-gradient, Bio-Lab, Israel).

2.2. Study Area and Groundwater Sampling

The Neot Hovav industrial site (180,063, 560,798 Israel Transverse Mercator), hosts various chemical industrial plants that overlie an extensively fractured chalk aquitard in which the fractures were shown to play a major role in infiltration, groundwater flow and solute transport [20,21,25]. The groundwater underlying the industrial site is characterized by a slightly acidic to slightly alkaline pH range (6.0–8.5), high electrical conductivity (EC) (14.1–55.1 mS/cm), low dissolved oxygen (DO) concentrations (0.22–0.84 mg/L), and low metal concentrations (<5 µg/L).

In 2015, 17 boreholes were sampled for microbial biodiversity and dissolved organic carbon (DOC), with a spatial distribution from the production plant [20], and up to 3.5 km downstream of the industrial area. In spring 2016 the RH49 pumping well (RH49P) was sampled for establishing microcosms for TBNPA and DBNPG. In February 2019, RH49P was sampled for establishing microcosms BCE, EDB, and EDC degradation.

Based on the land uses in the site, the industrial area was divided into three parts: (1) Adjacent to the brominated organic compounds (BOC's) production plant (groundwater depth: 30–50 m below ground). (2) Along Wadi Hovav—a dry riverbed that drains the erratic runoff in the industrial area (groundwater depth: 2–10 m below ground). (3) Downstream to the production plant, up to ca. 3.5 km from the plant (groundwater depth: 30–40 m below ground) (Figure 1). The area that is downstream to the production plant hosted in the past various wastewater treatment procedures that released along it the year's contaminants to the sub-surface environment. For example, forced evaporation of effluents by sprinklers was practiced over hillslope located next to observation borehole KN4 until the late 1980s [26], and unlined wastewater evaporation ponds were used adjacent to borehole KN201 until 1982 [27]. The locations of the sampled boreholes are summarized in Table 1 [23]. See SI for a detailed description of groundwater sampling procedures.

Table 1. Location of sampled boreholes and their characterization.

Borehole	Location	Comments
RH49, RH49P, RH50, and RH50P	West to the production plant	Slanted borehole drilled through vertical fractures that lie beneath the BOC's production plant
RH26	30 m' east to the production plant	
W1, W2, S2, S3, and S4	Along Wadi Hovav	Drains shallow groundwater within saturated fractures intersected by the Wadi Hovav Trench (an east-west trench)
WT		Drains a north-south trench intersecting with Wadi Hovav
KN4		Located at the downstream edge of the industrial area, on the northern bank of Wadi Hovav
RH33		South to Wadi Hovav
RH10	Downstream of the production plant	Upstream to the Wadi Hovav-Wadi Sekher confluence
KN201		North to Wadi Sekher
W8		In Wadi Sekher, north to the wastewater evaporation ponds. Drains a trench that drains polluted groundwater of Wadi Secker, including leaks from the wastewater ponds
RH22		Located west to W8, on the northern bank of Wadi Sekher

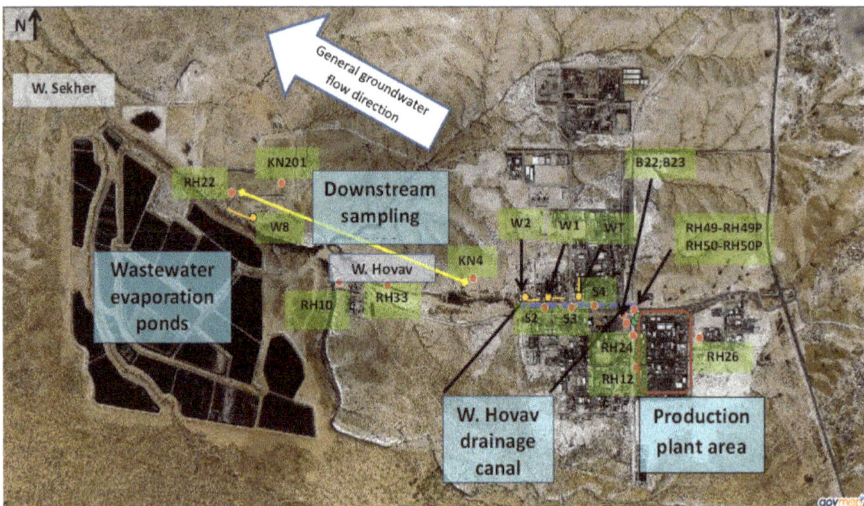

Figure 1. Boreholes sampled in the Neot Hovav industrial site: Adjacent to the production plant; Along the Hovav drainage canal; and downstream to the production plant. Red circles—monitoring boreholes; orange circles-boreholes along drainage canal. Govmap https://www.govmap.gov.il/2016.

2.3. DNA Extraction, PCR Amplification and Sequencing

Seventeen groundwater samples from September 2015 were extracted for total genomic DNA after filtration through 0.22 µm sterilized filters. DNA was extracted by using PowerSoil® kits (MoBio) (Carlsbad, CA, USA), with slight modifications to the manufacturer's instructions: The 0.22 µm sterilized filters were added to the PowerBead Tubes instead of 0.25 g soil. In the elution, 100 µL of solution C6 was added in two steps (twice 50 µL), and centrifuged at $10,000 \times g$ for 30 s in-between. The 16S rRNA gene was amplified by the polymerase chain reaction (PCR) with the primers CS1-341F-CS2-806R for Bacteria and CS1_Arch344F-CS2_Arc806R for Archaea [28]. The thermocycler conditions for DNA amplification were: 5 min at 95 °C, followed by 26 cycles of 95 °C (30 s) → 55 °C (45 s) → 68 °C (30 s) and finalizing the reaction with 68 °C (7 min); a set of PCR products was run and visualized on agarose gel (2%) to verify product specificity. The DNA concentration was determined using a NanoDrop 1000 spectrophotometer (Thermo Scientific, Wilmington, NC, USA).

PCR products were sent to the DNA Services Facility, Research Resources Center, University of Illinois at Chicago (UIC), for sequencing of 16S rRNA using the Illumina MiSeq platform. Operational taxonomic units (OTUs) were assigned according to the company's pipeline: Through the NGS analysis pipeline of the SILVA rRNA gene database project (SILVAngs 1.3) [29]. BLAST (version 2.2.30+) in combination with SILVA SSU or LSU Ref datasets were used for theclassification of reference sequences [30].

2.4. Data Analysis of Sequencing Results

Statistical analysis of the sequencing data was implemented using various R packages (version 3.3.1) [31]. The alpha biodiversity of the microbial community was estimated for each sample using the abundance-based diversity indices Shannon and Chao1, through the plot_richness function of the phyloseq package in R (version 1.14.0) [32]. Significant differences between diversity indices to the sample location were tested by the Kruskal-Wallis Analysis of Variance (ANOVA) by ranks test (STATISTICA software (version 10). Weighted UniFrac distances of the community data was calculated by the UniFracfunction of the phyloseq package, and the Bray-Curtis distance matrix by the bray function of the phyloseq and vegan (version 2.4.0) packages [33]. Nonmetric multidimensional scaling

(NMDS) OTU-based community composition, based on Bray-Curtis and weighted UniFrac distances was conducted by the ordinate function in the phyloseq package, and by the metaMDS function of the vegan package. Significant differences in the microbial community composition between samples were tested by Permutational multivariate ANOVA (PERMANOVA) using the adonis function of the vegan package. Envfit function in the vegan package was applied to illustrate the correlations between environmental parameters and the microbial communities. Microbial phylotypes with their corresponding read counts were imported into METAGENassist [34] to analyze the metabolic features of the microbial communities.

2.5. Microcosm Experiments

The biodegradation potential of groundwater bacteria to degrade TBNPA, DBNPG, BCE, EDB, and EDC was tested in a 100 mL anaerobic groundwater microcosm batch experiment. In all experiments, a RH49P groundwater sample (two liters in a sterile polypropylene bottle) was retrieved from the continuously pumping RH49 borehole (Figure 1, RH49P) [35]. The TBNPA; DBNPG (50 mg/L)/or both TBNPA and DBNPG (25 mg/L each) experiment included three different supplied electron donors: H_2 (1.7–3.8 mL), lactate (0.4 g/L), or acetate (6.8 g/L) to encourage degradation (i.e., each electron donor treatment was given to three bottles that contained 1-TBNBA; 2-DBNPG; and 3-TBNPA+DBNPG), and was done under anaerobic conditions. Three negative controls contained sterile (autoclaved) groundwater with TBNPA (50 mg/L); DBNPG (50 mg/L); or both TBNPA and DBNPG (25 mg/L each). The microcosms were sampled anaerobically at the beginning at a period of weeks, then months, and finally after three years. For BCE, EDB, and EDC (1 mM) microcosms, water was sampled from RH49P in February 2019 and lactate was added to all bottles (0.39 gr/L) (in duplicates). One bottle was autoclaved and served as a control. Samples were taken anaerobically every week. All bottles were held at room temperature and were constantly shaken.

2.6. Analytical Methods

TBNPA and DBNPG concentrations were determined by high performance liquid chromatography (HPLC; Agilent 1100 series, Palo Alto, CA, USA). BCE, EDB and EDC concentrations were determined by Gas Chromatograph-Mass Spectrometer (6890–5975 Agilent Palo Alto, CA, USA. Samples were automatically diluted 100 times, and 5 mL of the solution was purged for 11 min (He flow 40 mL/min). Desorption was performed at 190 °C for 4 min. For GC separation a DB-5 capillary column (30 m × 0.25 mm id × 0.25 μm) was used. The following conditions were applied: Split injection mode 20:1, injector temp. 220 °C; oven heating from 50 °C to 200 °C at a rate of 10 °C/min; He flow was 1 mL/min. The concentration was compared to chloroform as an internal standard. DOC was quantified by a Multi N/C 21005 (Analytic Jena, Jena, Germany).

3. Results

3.1. Molecular Analysis of the Groundwater Indigenous Bacterial Communities

3.1.1. Sequencing and Taxa Identification

The September 2015 16S rRNA amplicon sequencing of groundwater DNA from 17 sampled boreholes yielded 2,101,353 raw reads, with 1,629,795 quality reads (after filtering). The number of reads per sample ranged from 74,591 to 148,679, with a mean of 94,821 ± 18,101. Identification at different taxonomic levels of the normalized data yielded 67 phyla, 209 classes, 412 orders, 647 families, 1052 genera, and 1092 species. Identification at different taxonomic levels of the non-normalized data yielded 72 phyla, 230 classes, 457 orders, 726 families, 1216 genera, and 1271 species (See OTU table in SI). The dominant taxa (above 2.0% abundance) at the Phylum, Class and Family level are presented in Table 2.

Table 2. Taxa identification and abundances above 2.0% abundance at the Phylum, Class and Family level.

Genera	Name	Abundance (%)
phylum	*Proteobacteria*	54.2
	Bacteriodetes	7.3
	OD1	5.8
	Chloroflexi	3.6
	OP3	3.0
	Firmicutes	2.8
	unclassified	0.3
Class	*Gammaproteobacteria*	19.3
	Epsilonproteobacteria	10.4
	Deltaproteobacteriav	8.0
	Alphaproteobacteria	11.5
	unclassified	2.4
Family	*Helicobacteraceae*	7.0
	Pseudomonadaceae	4.6
	Desulfobacteraceae	3.8
	Campylobacteraceae	3.5
	Piscirickettsiaceae	2.8
	Flavobacteriaceae	2.7
	Rhodobacteraceae	2.5
	Rhodospirillaceae	2.5
	unclassified	36.3

3.1.2. Overall Bacterial Community Structure and Relationship with Environmental Variables

Bacterial community structure and compositional differences among the different locations in the industrial area were determined by Nonmetric Multidimensional Scaling (NMDS) analysis with Bray-Curtis and weighted Unifrac distance matrices (Figure 2). Three clusters following spatial location: Production plant, Wadi Hovav, and Downstream, were found. Based on Bray-Curtis distances, which do not incorporate phylogenetic relatedness in community differences, bacterial composition differed significantly ($p < 0.001$, $F_{2,14} = 1.99$) with the three locations. NMDS analysis for Archaea also resulted in clustering (Figure S1).

The microbial communities found in boreholes WT, RH26, and S2 do not cluster with their geographical group (i.e., according to the area). Following the NMDS ordination clustering and the differences in groundwater composition found at both locations, for further analysis, the bacterial community of RH26 was classified as part of the downstream group, and WT as part of the production plant group. The contaminant load (defined as the sum of semi-volatile and volatile compounds that were detected) and DOC (Table 3, Table S1) may explain why WT and RH26 do not cluster with their geographical group (i.e., according to the area).

Table 3. Mean contaminant load and dissolved organic carbon (DOC) of the three locations based on the ordination results. contaminant load-taken from the Neot Hovav annual sampling report for year 2014 [36]; DOC-2015 sampling.

Location	Boreholes in group	Contaminant Load		DOC	
		mean	Range (mg/L)	mean	Range (mg/L)
Production plant	RH49P; RH50P; WT	203.6	103.7–297.4	233.9	25–821
Wadi Hovav crossing the industrial area	W1; W2; S2; S3; S4	8.2	1.5–13.9	41.2	29–49
Downstream to the industrial area	RH26; KN4; RH10; RH22; RH33; W8; KN201	0.2	0.0007–0.4	26.9	4.9–48

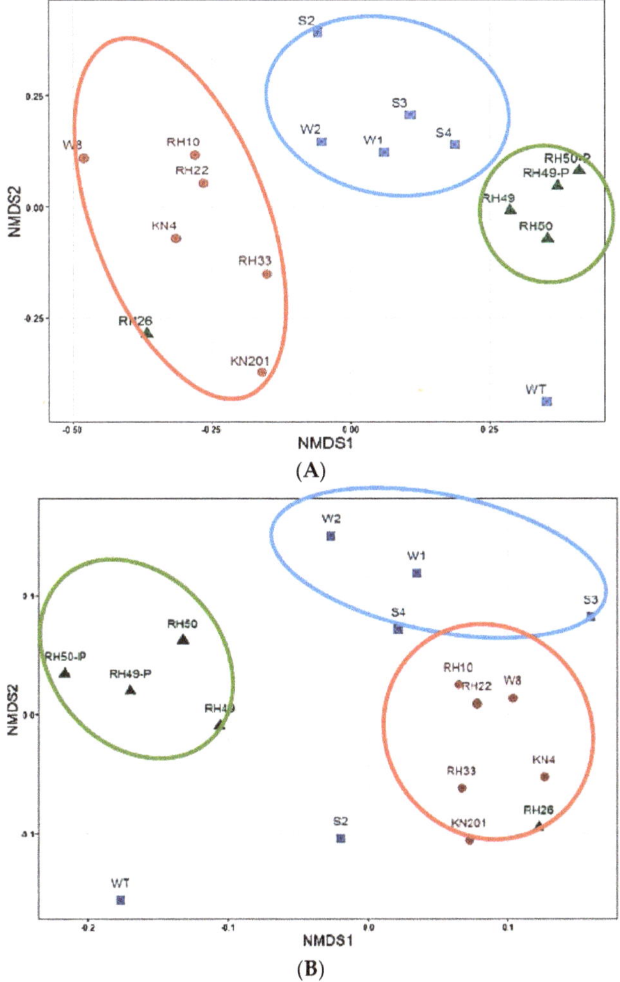

Figure 2. Nonmetric multidimensional scaling (NMDS) ordination for the microbial communities in the examined 17 boreholes using (**A**) the Bray-Curtis distance matrix, and (**B**) the weighted UniFrac distance matrix. Groups are categorized according to borehole location in the industrial area (symbols and colors), green triangles represent the Production plan, blue squares, the Wadi Hovav; red circles indicate Downstream. Circles represent clustered groups according to borehole location in the industrial area. Distance matrix based on 16S rRNA Illumina sequencing. Stress: (**A**) 0.153 (**B**) 0.098.

The effect of various environmental parameters: EC, pH, ion ratios, total nitrogen (TN), DOC, and contaminant load (Tables S2 and S3) on the 2015 groundwater microbial communities was tested, with the assistance of the Envfit function in the vegan package (R). From all tested parameters, only DOC, SO_4^{2-}/Cl^- and contaminant load were found to have a significant influence ($p = 0.05$) on the groundwater microbial community structure (Figure 3). As the ordination used is unconstrained (NMDS), this correlation is not linear. The use of Envfit enabled the identification of environmental parameters with a significant effect on the groundwater microbial communities. This projection shows that in groundwater areas with high contaminant levels, this is the most dominant factor affecting the communal microbial structure. In areas where contamination is low, its effect is not significant, and in those areas the microbial variability is high, making it more challenging to identify the main environmental parameter/s if they exist.

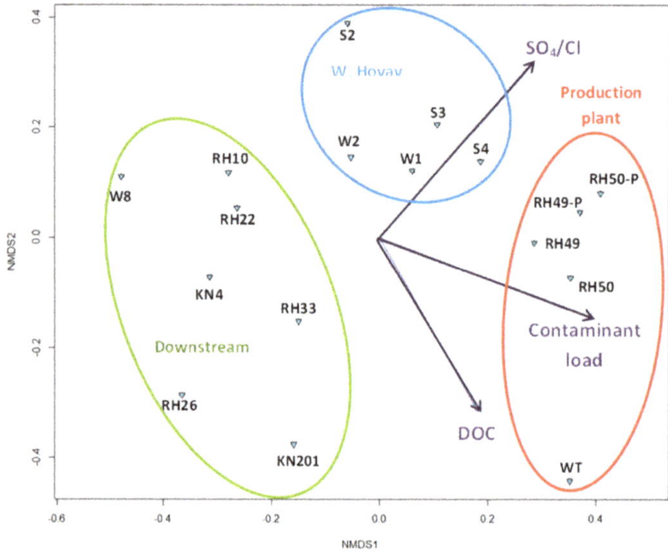

Figure 3. NMDS ordination based on the Bray-Curtis distance matrix, where stress was 0.153, correlated with groundwater parameters, which had a significant correlation ($p = 0.05$) with the microbial communities.

3.1.3. Variations in Microbial Community Diversity

Bacterial α-diversity indices (Chao1, Observed, Shannon) were examined in the context of spatial location in the industrial area (Table S4). No significant correlation (Kruskal-Wallis, all $p > 0.05$) with the location in the industrial area was found (Table S5). The compositional shifts between the microbial communities in the different areas were mainly in terms of relative abundance, and not a microbial taxa turnover. Although not significant, the mean Observed and Chao1 diversities increase from the production plant to the downstream group. As for the Shannon index, the production plant has the highest value (Table 4).

Table 4. Mean values of the Observed, Chao1 and Shannon diversity indices for each area in the industrial site.

Location	Boreholes in Group	N	Observed Mean	Observed Range	Chao1 Mean	Chao1 Range	Shannon Mean	Shannon Range
production plant	RH49; RH49P; RH50; RH50P; WT	5	4247	3317–4678	5505	4016–6228	6.33	6.13–6.51
Wadi Hovav crossing the industrial area	W1; W2; S2; S3; S4	5	4781	2769–6281	6952	3838–8877	6.17	5.38–6.71
Downstream to the industrial area	RH26; KN4; RH10; RH22; RH33; W8; KN201	7	5115	3392–7241	7430	4621–9962	6.27	5.60–7.12

3.1.4. Microbial Community Composition

The production plant area is dominated at the Phyla level by *Proteobacteria* (72.7%), *Bacteriodetes* (8.4%), and *Firmicutes* (6.0%), with all other phyla beneath 2%. The dominance of these three phyla in the most contaminated area can be attributed to their ability to adapt to contaminated and diverse environments [37–39]. With contaminant load decrease, more phyla with abundance greater than 2% appear in Wadi Hovav; *Proteobacteria* were (49.3%), *Bacteriodetes* (8.4%), six phyla had an abundance between 2–8%, and all the rest were <2%. In the downstream area *Proteobacteria* were (43.3%) dominated, eight phyla had an abundance of between 2–8% and the rest are below 2%. In addition, the percent of other phyla that were below 1% increase from the production plant downstream. Combining diversity with relative abundance data reveals that in the production plant area a lower amount of rare taxa exists. With distance from the production plant, as the contaminant load decreases, the amount of rare taxa increases (Figure 4).

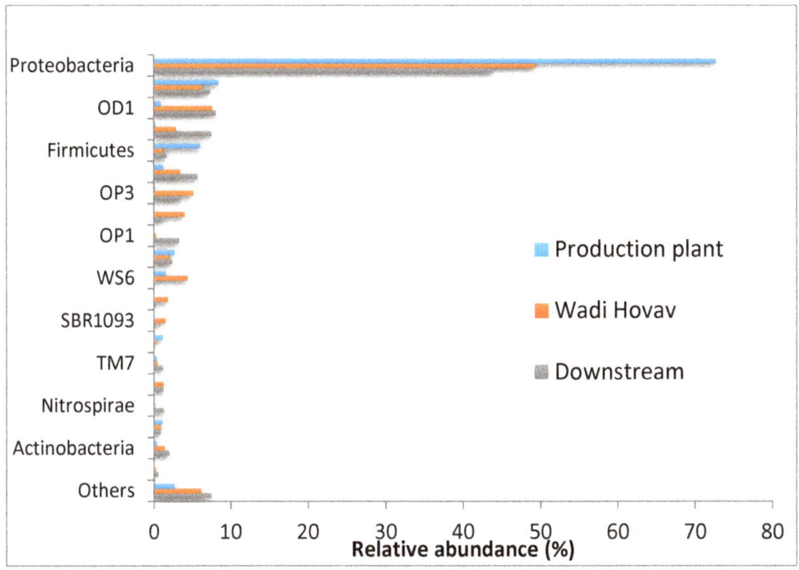

Figure 4. Relative abundance (%) of dominant phyla (>1% abundance) in the three locations of the industrial area-production plant (blue bars); Wadi Hovav (red bars); and downstream (gray bars).

3.2. Dehalogenation Potential of the Groundwater Microbial Community

The biodegradation potential and activity of the groundwater microbial community was tested by microcosm experiments with TBNPA, DBNPG, BCE, EDB, and EDC. Over the course of three years, no decrease in TBNPA and DBNPG was detected. The persistence of TBNPA and DBNPG, despite the presence of the right anaerobic bacteria, raised the question of whether the microbial community has the ability to dehalogenate other contaminates. For this purpose, BCE, EDB, and EDC anaerobic biodegradation was tested with fresh RH49P (February 2019) groundwater. For all three compounds, biodegradation was observed, clearly exhibiting the dehalogenation ability and potential of the groundwater microbial community (Figure 5). Significant loss of BCE was observed in the control bottle. Nevertheless, this is not due to degradation because the isotopic composition of carbon in this bottle remains stable (δ^{13}C of −43.02‰), while in the active one a strong ^{13}C enrichment of BCE was observed (Yankelzon et al. 2019 [40]).

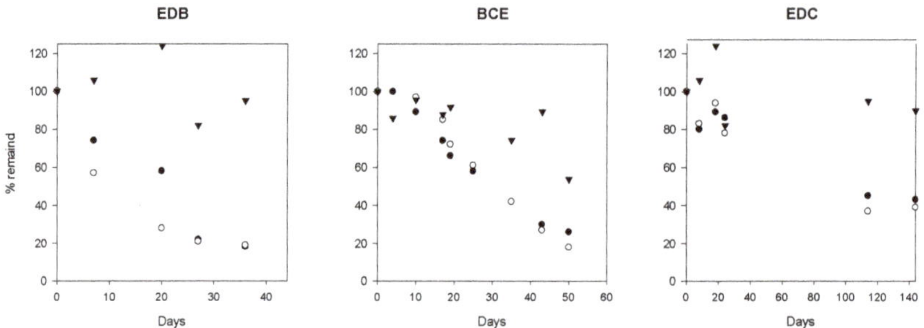

Figure 5. Degradation of Dibromoethane (EDB), 1-bromo-1-chloroethane (BCE) and 1,2-dichloroethane (EDC) by fresh RH49 groundwater microbial community. Open and filled circles represent individual sample bottles and filled triangles represent autoclaved control bottles.

4. Discussion

4.1. Site Hydrology

The studied site is characterized by vertical and horizontal permeable fractures [20,22,25,41,42]. While the directions of the major fracture systems under the industrial complex and downstream Wadi Hovav are the same, their orientation and density are entirely different underneath the downstream ponds [25,42]. It was proposed that active mixing between the different locations over the Neot Hovav aquitard occurs mainly through the vertical multi-layers fractures and less likely through horizontal flows within the bedding planes [4,25]. Therefore, the spatial variability in groundwater quality is controlled mostly by the sources located directly above the fractures. Indeed, as reflected in Table 3, contaminant loads varied from an average of 95.4 mg/L near-production area to 8.2 mg/L in Wadi-Hovav, and down to 0.2 mg/L in the pond/downstream area boreholes. Thus, three discrete zones in the site can disturb the microbial community of the groundwater. Because aquifers are physically and chemically heterogeneous in various extents, the relationship of microbial diversity and activity is expected with a variation of their characteristics of rock mineralogy, water organic content and pollution, hydraulic connectivity, and the physical features of the water flow paths.

4.2. Spatial Distribution of Microbial Community Affected by Environmental Parameters

From testing the correlation between environmental factors and microbial communities (Figure 3), it was seen that DOC, SO_4/Cl, and contaminant load significantly influence the microbial community structure. In this analysis, the environmental values in the boreholes increase in accordance with the arrow direction. They reflect the historical and current input of industrial waste into the groundwater. Each area is characterized by the waste input of different industries, leading to a different contaminant load concentration as well as composition.

Indeed, the results in Figure 2 indicate that groundwater in each area contains a characteristic microbial community, probably due to its typical groundwater chemical (inorganic and organic) composition. Nevertheless, in both distance matrix ordinations, three outliers appear RH26, WT, and S2. These are due to a complex hydrological regime leading to the groundwater chemical composition of certain boreholes, which does not correlate to their geographical location [4]. It has been shown that microbial communities are largely affected by environmental parameters, including contaminant composition and concentration [37–39]. In the Neot Hovav aquitard, numerous organic contaminants are found [43]. Therefore, it is extremely challenging to isolate each component and test its specific effect on the microbial community. For that reason, the parameter "contaminant load" of each borehole was defined. This parameter takes into account all semi-volatile and volatile compounds that were detected at the site during a 2014 monitoring program. The contaminant load calculated from 2014 was compared to the 2015 DOC analysis (Table 3). Both parameters were found to follow a similar trend at the site, e.g., decreasing along the regional groundwater gradient from the production plant, through Wadi Hovav and downstream. Due to their similarity over time and space, both parameters were perceived relevant, and were used for the 2015 analysis, and may explain why WT and RH26 do not cluster with their geographical group (i.e., according to the area). In borehole WT, very high DOC (dissolved organic compound concentrations) was found, and in accordance, the bacterial community at this site does not cluster with any group. Furthermore, the WT water is affected mostly by pharmaceutical contamination that may drive the evolution of a unique community. The groundwater composition of RH26 is more characteristic of the downstream boreholes, and in accordance with the bacterial community of this borehole lies in the downstream cluster. Nevertheless, these parameters do not explain the difference in the S2 community. Following the NMDS ordination clustering and the differences in groundwater composition, during analysis the bacterial community of RH26 was classified as part of the downstream group and WT as part of the production plant group.

4.3. Possible Metabolic Features in the Microbial Communities Based on Spatial Location

Although 16S rRNA gene sequencing does not by itself provide metagenomics data (functionality, abundances, changes), a functional understanding of the microbial communities can be inferred by using the [34] database. Importantly, this analysis is based on DNA sequencing, thus presenting the potential functionality of the community and not their actual activity. The analysis (Figure S2) suggests that functional diversity is larger in the production plant in relation to Wadi Hovav and downstream areas. Bacteria genomes containing dehalogenation function (35.7%) and aromatic hydrocarbon degradation functions (28.4%) are most abundant in the groundwater sampled near the production plant. As for the anaerobic dehalogenating bacteria, the main abiotic parameters that will affect their activity is the availability of suitable electron donors. Competition on electron donors, particularly by sulfate reducers, can render the activity of obligatory reductive de-halogenaters in the groundwater.

These observations are expected due to the high contaminant load in this area, and the high concentrations of halogenated and aromatic compounds. In all areas, bacteria containing ammonia oxidation, nitrate reduction, and sulfate reduction metabolic pathways are abundant. The abundance of these metabolic routes is highest in the production plant area. The presence of sulfate reducers is indeed expected due to the sulfate content in the groundwater and anoxic conditions. Bacteria with an oxidative metabolism are less expected to be active in the site on the one hand, due to the lack of oxygen in the groundwater, and perhaps these are dormant functions. On the other hand, there are oxidizing functions that have been seen to occur under anaerobic conditions. For example, the ANAMMOX bacteria which were found in the groundwater promote ammonia oxidation, or sulfur-oxidizing bacteria may have the ability to reduce ferric iron using elemental sulfur as an electron donor [44], or grow on crude oil [45] under anaerobic conditions.

The quantity of unknown, predicted, metabolic pathways are lowest in the production plant area, and increases in Wadi Hovav and the downstream. This can be the result of database bias, stemming from the fact that most of the existing knowledge is derived from highly contaminated environments. Therefore, the bacteria that accommodate these environments dominate the database, relatively to the moderately-contaminated studied areas.

4.4. Pollution Extent Affects Community Diversity

In all diversity indices, the intergroup variance (range) increases from the production plant downstream, i.e., the high contaminant load in the production plant results in a less diverse community. Nevertheless, no linear correlation can be found between contaminant loads to diversity. These non-significant correlations are in agreement with the subsidy stress effect. From a certain level of anthropogenic disturbance, the whole ecosystem is depressed (i.e., production plant area), whereas in areas with a lower contamination load (i.e., Wadi Hovav and Downstream), its effect on the microbial community is less prominent [46]. In these areas, the dominant factor is contaminant composition and not loads. This is seen in the less-contaminated areas, Wadi Hovav and Downstream;

at times low diversity indices occur in boreholes with low contaminant load, such as RH26, and vice versa, while high diversity indices occur in boreholes with relatively high contaminant load, such as S3. A recent report on the functional diversity of the microbial community in nitrate and uranium-polluted groundwater had suggested that diversity decreases with increasing pollution [47]. However, as in our study, the relationship between contaminates load (uranium) and functional genes richness was insignificant. In another study however, diversity in groundwater polluted with hydrocarbons in China, was lower than the downstream groundwater, likely due to the dominance of degrading bacteria [48].

4.5. Active Dehalogenation Potential

Analysis of the groundwater microbial communities at the site reveals that a large variation of organo-halide-respiring bacteria (OHRB) exists in all three areas. Moreover, the predicted metabolic functions (Figure S1) also point towards the existence of aerobic and anaerobic dehalogenation pathways in the communities. Five microbial genera commonly found in anoxic aquifers contaminated by halogenated compounds were found in the groundwater in all areas [49,50] (Table S6) [51]. Former studies in the Neot Hovav industrial area have also shown the presence of *Desulfovibrio* in stream sediment [52] and industrial wastewater, and of *Dehalococcoides*, *Geobacter* and *Dehalobacter* in groundwater [51,53]. *Dehaloccocoides* have been shown to reductively debrominate brominated diphenyl ethers, Tetrabromobisphenol A, and bromophenol blue [54]. They were also shown to reductively debrominate brominated compounds in saline environments [55], as well as dehalogenate brominated benzene-polluted groundwater as part of a bioaugmentation culture [56]. *Desulfitobacterium* was found in a 2-bromophenol 2-BP-degrading consortium, and additional studies have also exhibited the anaerobic debromination abilities of this genus [57]. *Desulfovibrio*, which was isolated from different environments in the industrial area (stream sediment and wastewater), also exhibited reductive debromination abilities [58]. A tetrabromobisphenol degrading culture was isolated from the sulfate-rich stream sediments in the industrial area and halorespiration of tribromophenol was seen [59].

Moreover, members of the community "supporting" the dehalogenating bacteria, e.g., species found in co-culture, co-metabolic, or other interactions with the organohalide respiring microorganisms, such as *Spirochaeta*, *Pseudomonas*, *Pelobacter*, *Sedimentibacter* and the candidate phyla *OD1* are also found in the groundwater in relatively high abundances (Table S7) [52,60].

The microbial community in the four boreholes in the production plant area (RH49; RH49P; RH50; and RH50P) include the non-obligatory organo-halide respiring bacteria (OHRB) *Firmicutes*. *Firmicutes* are non-obligatory OHRB capable of utilizing a wide range of electron acceptors. Their metabolic versatility results in higher tolerance to high contamination levels, as seen in the production plant area. The abundance of *Firmicutes* in the production wells (RH49P, RH50P), which are constantly pumped to lower groundwater levels to the regular boreholes (RH49, RH50), which are not constantly pumped, was compared. This comparison resulted in a similar phenomenon for both borehole couples, a higher *Firmicutes* abundance in the regular boreholes than in the production well. 15.85% vs. 1.46% for RH49P and RH49 respectively, and 7.72% and 0.38% for RH50P and RH50 respectively.

The *Clostridia* class, harboring the *Dehalobacter* genus with unexplored dehalogenation potential [61], was also identified in these four boreholes. A similar picture rises in this comparison: 28 times more abundant in RH49P vs. RH49, and 25 times more abundant in RH50P vs. RH50. The presence of these classes with a high sulfate-reducing potential [62] is in accordance with the high sulfate concentrations which characterize the Neot Hovav groundwater. Between both wells, the non-production wells, e.g., RH49 and RH50, offer a better representation of the groundwater microbial community.

Over the course of three years of microcosm anaerobic incubation with water from RH49P, a decrease in TBNPA, and DBNPG concentrations were not detected. This is in accordance with the finding in Balaban et al., 2016 [35], that molecular oxygen was necessary for biodegradation to occur.

But, when similar anaerobic microcosms with BCE, EDB and EDC were tested for dehalogenation, this occurred, thus demonstrating the dehalogenation abilities of the microbial groundwater community. In terms of degradation rate, EDB was the fastest with a first-order constant of 0.049 day^{-1}, following by BCE 0.0265 day^{-1}, and the slowest was degradation of EDC that was 0.0068 day^{-1}. These results demonstrate that the groundwater community can degrade halogenated contaminants with a faster degradation of brominated, over-chlorinated compounds. Hence, the stability of TBNPA and DBNPG in the anaerobic microcosm, as well as in the anaerobic site water [35], is due to the chemical characteristics of these compounds and not because of the absence of active de-halogenating bacteria. Moreover, as the groundwater used for these experiments are from the RH49P (production) well, where the abundance of OHRB was found to be lower than in the non-production well (RH49), we can only assume that the presented activity is lower than that of the microbial community in the groundwater.

5. Conclusions

This study exhibits the importance of combining multiple techniques when addressing the natural bioremediation potential of a microbial community. When dealing with a site with multi-contaminants, it is crucial to test various pollutants' fate, to gain a better understanding of this potential. In addition, this study exemplifies how the characteristics of the groundwater matrix govern the pollutant distribution, microbial community structure, and perhaps the natural bioremediation potential.

The complex microbial community that lies in the polluted groundwater within a chalk aquitard that underlies the Neot Hovav industrial site was presented. Hydrologic discontinuity in the aquitard results in spatial heterogeneity of hydro-chemical conditions in the site, leading to variations in microbial community structure. The connection between contaminant loads to bacterial community structure shows that the contamination differences in the distinct groundwater section under separate sections of the industrial site lead to the variation in the microbial community. High contaminant load leads to lower communal diversity and the presence of more tolerant bacteria. When contamination decreases, diversity increases, with pollutant mixture seemingly affecting community structure. Anaerobic de-halogenating activity and potential of the groundwater microbial community from the production plant area were exhibited. While three of the tested compounds were degraded, two were not, thus exemplifying the complication of remediating multi-contaminant locations.

Supplementary Materials: The following are available online at http://www.mdpi.com/2073-4441/11/10/2128/s1, Groundwater sampling procedure, Table S1: Contaminant load of the sampled boreholes. Taken from the Neot Hovav annual sampling report (Adar et al., 2015 [23]), Table S2: Electrical conductivity mS/cm (EC); dissolved oxygen mg/L (DO); pH values measured during the three sampling campaigns (2014–2016), Table S3: Chemical parameters of the sampled boreholes, Neot Hovav annual sampling report (Adar et al., 2015 [23]), Table S4: Diversity indices, Table S5: P values of diversity indices in relation to location in the industrial area. Tested by Kruskal-Wallis ANOVA by ranks (done in Statistica software), Table S6: Relative abundance in the groundwater of reductive dehalogenating microorganisms in the three areas of the industrial site, Table S7: Relative abundance in the groundwater of supporting dehalogenation microorganisms in the three areas of the industrial site, Figure S1: NMDS ordination for the archaeal microbial communities in the examined 17 boreholes using (**a**) the Bray-Curtis distance matrix, and (**b**) a weighted UniFrac distance matrix. Groups are categorized according to borehole location in the industrial area (symbols and colors), Figure S2: Comparison of metabolic groups in the bacterial communities in the three areas, as described using METAGENassist analysis.

Author Contributions: Conceptualization, Z.R., F.G. and A.B.; investigation, N.B. and I.Y.; data curation, N.B.; writing—original draft preparation, N.B.; writing—review and editing, Z.R., F.G., E.A. and A.B.; supervision, Z.R., F.G., E.A. and A.B.

Funding: This research was supported in part by the Israel Science foundation grant number 1482/16.

Acknowledgments: We acknowledge the assistance of Neot Hovav municipality and the help of Almog Gafni, Nadav Knossow in fieldwork and the assistance of Tali Bruner, Chen Adler, and Damiana Diaz Reck in lab work.

Conflicts of Interest: The authors declare no conflict of interest.

References

1. Griebler, C.; Avramov, M. Groundwater ecosystem services: A review Christian. *Freshw. Sci.* **2015**, *34*, 355–367. [CrossRef]
2. Rodríguez-Delgado, M.; Orona-Navar, C.; García-Morales, R.; Hernandez-Luna, C.; Parra, R.; Mahlknecht, J.; Ornelas-Soto, N. Biotransformation kinetics of pharmaceutical and industrial micropollutants in groundwaters by a laccase cocktail from *Pycnoporus sanguineus* CS43 fungi. *Int. Biodeterior. Biodegrad.* **2016**, *108*, 34–41. [CrossRef]
3. An, D.; Xi, B.; Wang, Y.; Xu, D.; Tang, J.; Dong, L.; Ren, J.; Pang, C. A sustainability assessment methodology for prioritizing the technologies of groundwater contamination remediation. *J. Clean. Prod.* **2016**, *112*, 4647–4656. [CrossRef]
4. Ezra, S.; Feinstein, S.; Yakirevich, A.; Adar, E.; Bilkis, I. Retardation of organo-bromides in a fractured chalk aquitard. *J. Contam. Hydrol.* **2006**, *86*, 195–214. [CrossRef]
5. Postigo, C.; Barceló, D. Synthetic organic compounds and their transformation products in groundwater: Occurrence, fate and mitigation. *Sci. Total Environ.* **2015**, *503*, 32–47. [CrossRef]
6. Montes-Grajales, D.; Fennix-Agudelo, M.; Miranda-Castro, W. Occurrence of personal care products as emerging chemicals of concern in water resources: A review. *Sci. Total Environ.* **2017**, *595*, 601–614. [CrossRef]
7. Lapworth, D.J.; Baran, N.; Stuart, M.E.; Ward, R.S. Emerging organic contaminants in groundwater: A review of sources, fate and occurrence. *Environ. Pollut.* **2012**, *163*, 287–303. [CrossRef]
8. Van Liedekerke, M.; Prokop, G.; Rabl-Berger, S.; Kibblewhite, M.; Louwagie, G. *Progress in the Management of Contaminated Sites in Europe*; Joint Research Centre: Ispra, Italy, 2014.
9. Haggblom, M.M.; Bossert, I.D. Halogenated organic compounds—A global perspective. In *Dehalogenation: Microbial Processes and Environmental Applications*; Haggblom, M., Bossert, I., Eds.; Kluwer Academic Publishers: New York, NY, USA, Country, 2003.
10. Birnbaum, L.S.; Staskal, D.F. Brominated flame retardants: Cause for concern? *Environ. Health Perspect.* **2004**, *112*, 9–17. [CrossRef]
11. Alexander, M. *Biodegradation and Bioremediation*; Academic Press: San Diego, CA, USA, 1994.
12. Poursat, B.A.J.; van Spanning, R.J.M.; de Voogt, P.; Parsons, J.R. Implications of microbial adaptation for the assessment of environmental persistence of chemicals. *Crit. Rev. Environ. Sci. Technol.* **2019**, *49*, 2220–2255. [CrossRef]
13. Flynn, T.M.; Sanford, R.A.; Ryu, H.; Bethke, C.M.; Levine, A.D.; Ashbolt, N.J.; Santo Domingo, J.W. Functional microbial diversity explains groundwater chemistry in a pristine aquifer. *BMC Microbiol.* **2013**, *13*, 146. [CrossRef]
14. Dojka, M.; Hugenholtz, P.; Haack, S.; Pace, N. Microbial diversity in a hydrocarbon-and chlorinated-solvent-contaminated aquifer undergoing intrinsic bioremediation. *Appl. Environ. Microbiol.* **1998**, *64*, 3869–3877. [PubMed]
15. Yagi, J.M.; Neuhauser, E.F.; Ripp, J.A.; Mauro, D.M.; Madsen, E.L. Subsurface ecosystem resilience: Long-term attenuation of subsurface contaminants supports a dynamic microbial community. *ISME J.* **2010**, *4*, 131–143. [CrossRef] [PubMed]
16. Meckenstock, R.U.; Elsner, M.; Griebler, C.; Lueders, T.; Stumpp, C.; Aamand, J.; Agathos, S.N.; Albrechtsen, H.J.; Bastiaens, L.; Bjerg, P.L.; et al. Biodegradation: Updating the concepts of control for microbial cleanup in contaminated aquifers. *Environ. Sci. Technol.* **2015**, *49*, 7073–7081. [CrossRef] [PubMed]
17. Tarnawski, S.-E.; Rossi, P.; Brennerova, M.V.; Stavelova, M.; Holliger, C. Validation of an integrative methodology to assess and monitor reductive dechlorination of chlorinated ethenes in contaminated aquifers. *Front. Environ. Sci.* **2016**, *4*, 7. [CrossRef]
18. Révész, K.M.; Lollar, B.S.; Kirshtein, J.D.; Tiedeman, C.R.; Imbrigiotta, T.E.; Goode, D.J.; Shapiro, A.M.; Voytek, M.A.; Lacombe, P.J.; Busenberg, E. Integration of stable carbon isotope, microbial community, dissolved hydrogen gas, and 2HH2O tracer data to assess bioaugmentation for chlorinated ethene degradation in fractured rocks. *J. Contam. Hydrol.* **2014**, *156*, 62–77. [CrossRef] [PubMed]
19. Bahat, D. Early single-layer and late multi-layer joints in the lower Eocene chalk near Beer Sheva, Israel. *Ann. Tect.* **1988**, *2*, 3–11.

20. Dahan, O.; Nativ, R.; Adar, E.M.; Berkowitz, B.; Weisbrod, N. On fracture structure and preferential flow in unsaturated chalk. *Groundwater* **2000**, *3*, 444–451. [CrossRef]

21. Nativ, R.; Adar, E.; Assaf, L.; Nygaard, E. Characterization of the hydraulic properties of fractures in chalk. *Groundwater* **2003**, *41*, 532–543. [CrossRef]

22. Nativ, R.; Adar, E.; Dahan, O.; Nissim, I. Water salinization in arid regions-observations from the Negev desert, Israel. *J. Hydrol.* **1997**, *196*, 271–296. [CrossRef]

23. Adar, E.; Kaplan, A.; Gilerman, L. *Groundwater Quality Monitoring in the Neot Hovav Industrial Complex*; Monitoring report for the years 2012–2013, 2013–2014; Neot Hovav Industrial Council: Neot Hovav, Israel, 2015. (In Hebrew)

24. Maamar, S.B.; Aquilina, L.; Quaiser, A.; Pauwels, H.; Michon-Coudouel, S.; Vergnaud-Ayraud, V.; Labasque, T.; Roques, C.; Abbott, B.W.; Dufresne, A. Groundwater isolation governs chemistry and microbial community structure along hydrologic flowpaths. *Front. Microbiol.* **2015**, *6*, 1–13.

25. Weiss, M.; Rubin, Y.; Adar, E.; Nativ, R. Fracture and bedding plane control on groundwater flow in a chalk aquitard. *Hydrogeol. J.* **2006**, *14*, 1081–1093. [CrossRef]

26. Arnon, S.; Ronen, Z.; Yakirevich, A.; Adar, E. Evaluation of soil flushing potential for clean-up of desert soil contaminated by industrial wastewater. *Chemosphere* **2006**, *62*, 17–25. [CrossRef] [PubMed]

27. Nativ, R.; Nissim, I. Characterization of a desert aquitard-hydrologic and hydrochemical considerations. *Groundwater* **1992**, *30*, 598–606. [CrossRef]

28. Takahashi, S.; Tomita, J.; Nishioka, K.; Hisada, T.; Nishijima, M. Development of a prokaryotic universal primer for simultaneous analysis of Bacteria and Archaea using Next-Generation sequencing. *PLoS ONE* **2014**, *9*, e105592. [CrossRef] [PubMed]

29. Quast, C.; Pruesse, E.; Yilmaz, P.; Gerken, J.; Schweer, T.; Yarza, P.; Peplies, J.; Glockner, F.O. The SILVA ribosomal RNA gene database project: Improved data processing and web-based tools. *Nucleic Acids Res.* **2013**, *41*, 590–596. [CrossRef] [PubMed]

30. Staley, C.; Unno, T.; Gould, T.J.; Jarvis, B.; Phillips, J.; Cotner, J.B.; Sadowsky, M.J. Application of Illumina next-generation sequencing to characterize the bacterial community of the Upper Mississippi River. *J. Appl. Microbiol.* **2013**, *115*, 1147–1158. [CrossRef]

31. Team, R. *RStudio: Integrated Development Environment for R*; RStudio: Boston, MA, USA, 2012; Available online: https://rstudio.com/products/rstudio/ (accessed on 1 July 2016).

32. McMurdie, P.J.; Holmes, S. phyloseq: An R package for reproducible interactive analysis and graphics of microbiome census data. *PLoS ONE* **2013**, *8*, e61217. [CrossRef]

33. Oksanen, J.; Blanchet, F.G.; Kindt, R.; Legendre, P.; Minchin, P.; O'Hara, R.; Simpson, G.; Solymos, P.; Stevens, M.; Wagner, H. *Vegan: Community Ecology Package*, R Package Version 2.0-10; 2013. Available online: https://cran.r-project.org/web/packages/vegan/index.html (accessed on 1 July 2016).

34. Arndt, D.; Xia, J.; Liu, Y.; Zhou, Y.; Guo, A.C.; Cruz, J.A.; Sinelnikov, I.; Budwill, K.; Nesbø, C.L.; Wishart, D.S. METAGENassist: A comprehensive web server for comparative metagenomics. *Nucleic Acids Res.* **2012**, *40*, 88–95. [CrossRef]

35. Balaban, N.; Bernstein, A.; Gelman, F.; Ronen, Z. Microbial degradation of the brominated flame retardant TBNPA by groundwater bacteria: Laboratory and field study. *Chemosphere* **2016**, *156*, 367–373. [CrossRef]

36. Adar, E. *Groundwater Quality Monitoring in the Neot Hovav Industrial Complex*; Neot Hovav Industrial Council: Neot Hovav, Israel, 2014. (In Hebrew)

37. Weber, K.P.; Gehder, M.; Legge, R.L. Assessment of changes in the microbial community of constructed wetland mesocosms in response to acid mine drainage exposure. *Water Res.* **2008**, *42*, 180–188. [CrossRef]

38. Singh, B.K.; Quince, C.; Macdonald, C.A.; Khachane, A.; Thomas, N.; Al-Soud, W.A.; Soren, J.S.; He, Z.; White, D.; Sinclair, A.; et al. Loss of microbial diversity in soils is coincident with reductions in some specialized functions. *Environ. Microbiol.* **2014**, *16*, 2408–2420. [CrossRef] [PubMed]

39. Fan, M.; Lin, Y.; Huo, H.; Liu, Y.; Zhao, L.; Wang, E.; Chen, W.; Wei, G. Microbial communities in riparian soils of a settling pond for mine drainage treatment. *Water Res.* **2016**, *96*, 198–207. [CrossRef] [PubMed]

40. Yankelzon, I.; Ronen, Z.; Gelman, F. Multi-elemental ($\delta^{13}C$, $\delta^{81}Br$ and $\delta^{37}Cl$) isotope effects in biotic and abiotic degradation of 1-bromo-2-chloro-ethane. In Proceedings of the Isotopes 2019, The Cross-Disciplinary Conference on Stable Isotope Sciences, Raitenhaslach, Germany, 7–12 July 2019; pp. 114–115.

41. Adar, E.; Nativ, R. Isotopes as tracers in a contaminated fractured chalk aquitard. *J. Contam. Hydrol.* **2003**, *65*, 19–39. [CrossRef]

42. Nativ, R.; Adar, E.M.; Becker, A. Designing a monitoring network for contaminated ground water in fractured chalk. *Groundwater* **1999**, *37*, 38–47. [CrossRef]

43. Nativ, R.; Adar, E. *Evaluation of the Extent of Groundwater Contamination in the Ramat Hovav Industrial Park Area*; Concluding Report 1995–1997; Neot Hovav Industrial Council: Neot Hovav, Israel, 1997.

44. Brock, T.D.; Gustafson, J. Ferric Iron Reduction by Sulfur- and Iron-Oxidizing Bacteria. *Appl. Environ. Microbiol.* **1976**, *32*, 567–571.

45. Kodama, Y.; Watanabe, K. Isolation and characterization of a sulfur-oxidizing chemolithotroph growing on crude oil under anaerobic conditions isolation and characterization of a sulfur-oxidizing chemolithotroph growing on crude oil under anaerobic conditions. *Society* **2003**, *69*, 107–112.

46. Odum, E.P.; Finn, J.T.; Franz, E.H. Pertubation theory and the subsidy-stress gradient. *Bioscience* **1979**, *29*, 349–352. [CrossRef]

47. He, Z.; Zhang, P.; Wu, L.; Rocha, A.M.; Tu, Q.; Shi, Z.; Wu, B.; Qin, Y.; Wang, J.; Yan, Q.; et al. Microbial functional gene diversity predicts groundwater contamination and ecosystem functioning. *MBio* **2018**, *9*, e02435-17. [CrossRef]

48. Ning, Z.; Zhang, M.; He, Z.; Cai, P.; Guo, C.; Wang, P. Spatial pattern of bacterial community diversity formed in different groundwater field corresponding to electron donors and acceptors distributions at a petroleum-contaminated site. *Water* **2018**, *10*, 842. [CrossRef]

49. Atashgahi, S.; Lu, S.; Smidt, H. Overview of known organohalide respiring bacteria—Phylogenetic diversity and environmental distribution. In *Organohalide Respiring Bacteria*; Adrian, L., Loffler, F., Eds.; Springer: Berlin/Heidelberg, Germany, 2016.

50. Nijenhuis, I.; Nikolausz, M.; Köth, A.; Felföldi, T.; Weiss, H.; Drangmeister, J.; Grossmann, J.; Kästner, M.; Richnow, H.H. Assessment of the natural attenuation of chlorinated ethenes in an anaerobic contaminated aquifer in the Bitterfeld/Wolfen area using stable isotope techniques, microcosm studies and molecular biomarkers. *Chemosphere* **2007**, *67*, 300–311. [CrossRef]

51. Lovley, D.R.; Ueki, T.; Zhang, T.; Malvankar, N.S.; Shrestha, P.M.; Flanagan, K.A.; Aklujkar, M.; Butler, J.E.; Giloteaux, L.; Rotaru, A.-E.; et al. Geobacter: The Microbe electric's physiology, ecology, and practical applications. In *Advances in Microbial Physiology*; Academic Press: Cambridge, MA, USA, 2011; Volume 59, ISBN 9780123876614.

52. Iasur-Kruh, L.; Ronen, Z.; Arbeli, Z.; Nejidat, A. Characterization of an enrichment culture debrominating tetrabromobisphenol A and optimization of its activity under anaerobic conditions. *J. Appl. Microbiol.* **2010**, *109*, 707–715. [CrossRef] [PubMed]

53. Ben-Dov, E.; Brenner, A.; Kushmaro, A. Quantification of sulfate-reducing bacteria in industrial wastewater, by real-time polymerase chain reaction (PCR) using dsrA and apsA genes. *Microb. Ecol.* **2007**, *54*, 439–451. [CrossRef] [PubMed]

54. Yang, C.; Kublik, A.; Weidauer, C.; Seiwert, B.; Adrian, L. reductive dehalogenation of oligocyclic phenolic bromoaromatics by *Dehalococcoides mccartyi* Strain CBDB1. *Environ. Sci. Technol.* **2015**, *49*, 8497–8505. [CrossRef] [PubMed]

55. Futagami, T.; Morono, Y.; Terada, T.; Kaksonen, A.H.; Inagaki, F. Dehalogenation activities and distribution of reductive dehalogenase homologous genes in marine subsurface sediments. *Appl. Environ. Microbiol.* **2009**, *75*, 6905–6909. [CrossRef]

56. Kuppusamy, S.; Palanisami, T.; Megharaj, M.; Venkateswarlu, K.; Naisu, R. Ex-situ remediation technologies for environmental pollutants: A critical perspective. In *Reviews of Environmental Contamination and Toxicology*; Voogt, P., Ed.; Springer: Cham, Switzerland, 2015.

57. Rhee, S.K.; Fennell, D.E.; Häggblom, M.M.; Kerkhof, L.J. Detection by PCR of reductive dehalogenase motifs in a sulfidogenic 2-bromophenol-degrading consortium enriched from estuarine sediment. *FEMS Microbiol. Ecol.* **2003**, *43*, 317–324. [CrossRef]

58. Villemur, R.; Lanthier, M.; Beaudet, R.; Lépine, F. The *Desulfitobacterium* genus. *FEMS Microbiol. Rev.* **2006**, *30*, 706–733. [CrossRef]

59. Ronen, Z.; Abeliovich, A. Anaerobic-aerobic process for microbial degradation of tetrabromobisphenol A. *Appl. Environ. Microbiol.* **2000**, *66*, 2372–2377. [CrossRef]

60. Luo, J.; Kurt, Z.; Hou, D.; Spain, J.C. Modeling aerobic biodegradation in the capillary fringe. *Environ. Sci. Technol.* **2015**, *49*, 1501–1510. [CrossRef]

61. Maphosa, F.; Van Passel, M.W.J.; De Vos, W.M.; Smidt, H. Metagenome analysis reveals yet unexplored reductive dechlorinating potential of *Dehalobacter* sp. E1 growing in co-culture with *Sedimentibacter* sp. *Environ. Microbiol. Rep.* **2012**, *4*, 604–616.

62. Muyzer, G.; Stams, A.J.M. The ecology and biotechnology of sulphate-reducing bacteria. *Nat. Rev. Microbiol.* **2008**, *6*, 441–454. [CrossRef]

Article

Colloid Transport in a Single Fracture–Matrix System: Gravity Effects, Influence of Colloid Size and Density

Nikhil Bagalkot [1,*] and G. Suresh Kumar [2]

1   Department of Energy and Petroleum Engineering, University of Stavanger, 4036 Stavanger, Norway
2   Department of Ocean Engineering, Indian Institute of Technology Madras, Chennai 600036, India; gskumar@iitm.ac.in
*   Correspondence: nikhil.bagalkot@uis.no; Tel.: +47-4862-8290

Received: 5 October 2018; Accepted: 24 October 2018; Published: 27 October 2018

Abstract: A numerical model was developed to investigate the influence of gravitational force on the transport of colloids in a single horizontal fracture–matrix system. Along with major transport phenomena, prominence was given to study the mass flux at the fracture–matrix interface, and colloid penetration within the rock matrix. Results suggest that the gravitational force significantly alters and controls the velocity of colloids in the fracture. Further, it was shown that the colloid density and size play a vital part in determining the extent that gravity may influence the transport of colloids in both fracture and rock matrix. The mass flux transfer across the fracture–matrix interface is predominantly dependent on the colloidal size. As large as 80% reduction in penetration of colloids in the rock matrix was observed when the size of the colloid was increased from 50–600 nm. Similarly, the farther the density of colloid from that of the fluid in the fracture (water), then the higher the mitigation of colloids in the fracture and the rock matrix. Finally, a non-dimensional parameter "Rock Saturation Factor" has been presented in the present study, which can offer a straightforward approach for evaluating the extent of penetration of colloids within the rock matrix.

Keywords: fractured media; gravitational force; numerical method; rock penetration; colloid size; colloid transport

---

1. Introduction

Colloids are an integral part of fractured porous media. Studies [1,2] have established that colloids act as carriers, and enhance the migration of contaminants, which have leached into the subsurface water system, thereby, enhancing their sweep through the system. Significant advancements have been made in understanding the prediction of colloid behavior in fractured media. The knowledge imparted by these studies has been critical in understanding colloid advection, dispersion, adsorption, biodegradation and other transport phenomena [1–5]. However, apart from conventionally studied transport phenomena, an understanding of various forces acting on the colloids stability and migration is essential for predicting the near to actual spreading or migration of colloids. Forces such as the electric double layer, van der Waals, shear forces (lift and drag), and gravitational forces acting on colloids are the primary contributors to the stability of colloids [6–8]. In the present numerical experiment, forces such as lift, drag, electric double layer, and van der Waals forces acting on colloids are neglected, and the principal objective is to understand the influence of the gravitational force on the stability, and hence, the transport of the colloids in the fractured matrix-system. Equation (1) represents the force acting on the colloids due to gravitation (F_g) [6]. It can be noted from Equation (1) that the F_g is a function of the size and density of the colloids, and does not depend on the charge of the particle or its distance from another particle. Therefore, the influence of the double layer and van der Waals force arising due to these two factors on gravity can be eliminated. Hence, the assumption of selecting gravity as an isolated phenomenon influencing the stability of colloids would be valid.

The gravitational force is a potential driver of retention, thus has a strong influence on the velocity of the colloid through the fracture. The gravitational force acts as normal to the flow direction of the colloids, inducing a component of colloidal velocity normal to its flow direction [4]. Due to the normal component, the colloids drift towards the fracture wall, into the region of relatively low axial velocity.

Colloid density and size are the two parameters that contribute to the influence of the gravitational force on the transport of colloids in a fractured porous system. Stokes–Einstein's equation (Equation (4)) representing the diffusion of colloids show that diffusion is inversely a function of colloid size (diameter/radius). Further, the dispersion, which is a predominant phenomenon of transport in the fracture, is a function of velocity, which in turn is sensitive to the size of the colloid [9]. The axial velocity of colloid is sensitive to the size of colloid, under the gravitational force the reasonably large colloids tend to migrate towards the fracture wall due to reduced axial velocity [10]. Hence, it becomes critical to include the influence of colloidal size in the analysis of colloidal transport.

Few studies have been carried out that have examined the impact of the gravitational force on the transport of colloids in the fractured porous medium [11–13]. However, a majority of them consider gravity as an outcome of the change in potential energy (height) by carrying out either vertical or inclined column/fracture experiments. In such experiments, the gravity is a forced effect of the change in potential (height), which causes a change in the velocity of the colloid. The change in velocity of the colloid is not due to gravity alone, but may additionally be attributed to acceleration or deceleration of fluid (suspension) in which colloids are suspended, due to change in potential energy (height). In order to have a clear analysis of the impact of gravitational force on the colloids, the change in velocity of the fluid due to change in potential energy has to be isolated, and the experiment has to be carried out at zero potential energy change. Not many studies have considered the influence of gravity on the transport of colloids at zero potential energy change. The number of studies is even fewer when the system, in consideration for the analysis, is a multi-continuum fractured porous media. Some studies have come close enough, by analyzing the influence of external forces in a system that resembles that similar to fracture, sans the surrounding rock matrix [14]. Unni and Yang [15] and James and Chrysikopoulos [12] analyzed the influence of the external force on colloids flowing in a constant width channel, which was horizontal and inclined respectively. The channel represents the high permeability fracture zone, however, without the surrounding porous rock matrix (no mass flux transfer to the surrounding porous media). Jen and Li [16] in their study on performance assessment of high-level radioactive waste disposal in a fractured porous media proposed a theoretical model to include the external forces acting on the colloids in a fracture–matrix system. However, the primary focus of Jen and Li [16] was to understand the colloid assisted radionuclide transport and here a mathematical model for inclusion of gravity has been presented, and the detailed numerical analysis has not been carried out. Therefore, the current knowledge and the methods used in understanding the influence of the gravitational force on the stability are inadequate to accurately predict the colloid migration in fractured porous media under the influence of gravity. Additionally, the combined influence of gravitational force, size and density of colloids has received relatively minimal attention, especially in the multi continuum fracture–matrix system.

The focus of the current work is to investigate the influence of the gravitational force, along with different colloid size and density on the transport of colloids in a single horizontal multi-continuum fracture–matrix system. The influence of gravitational force on colloids is introduced through the velocity of the colloids into the transport equation. Apart from the analysis of major transport phenomena, prominence is given to analyzing the mass flux transfer at the fracture–matrix interface, and the diffusive transport in the porous rock matrix when gravitational force, different colloid size and density are taken into consideration. Finally, a novel non-dimensional parameter (Rock Saturation Factor) has been developed as a simple approach for evaluating the extent of penetration of colloids within the rock matrix.

2. Physical System

Figure 1 illustrates the conceptual model depicting the cross-section of a simplified single horizontal fracture–matrix system, which forms the fraction of the complete fracture network at large field scale. In Figure 1, b is the average half fracture aperture representing the fracture, B is the half fracture spacing, and L_f is the length of the fracture. Colloid enters into the fracture–matrix system from the left boundary (fracture inlet at left end, $X = 0$), then the injected colloids transport through the fracture, and some of them diffuse into the rock matrix. In Figure 1 the colloids, which are adsorbed onto the porous rock matrix or fracture, are termed as "immobile colloids" (filtered colloids onto the surface), and the colloids which are being transported are termed "mobile colloids". The fracture and matrix geometry repeat itself, and each fracture is symmetrical. Therefore, only one-half of a fracture and one-half of the rock matrix system is considered in the present analysis [17] (Figure 1).

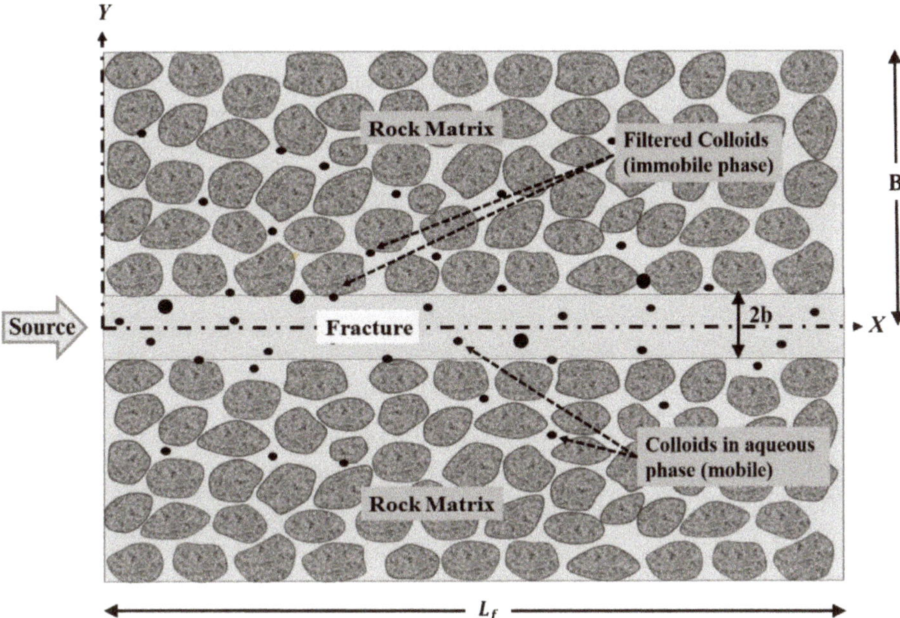

Figure 1. Conceptual model describing the cross-section of simplified single horizontal fracture surrounded by the porous rock matrix.

Assumptions made regarding the geometry and hydrodynamic properties of the system.

- Transport in the fracture is 1D with fluid flowing horizontally.
- Transport within the rock matrix is 1D with the transport of the fluid orthogonal to the direction of flow in the fracture.
- The fracture aperture is smaller than the length of the fracture ($2b << L_f$) [17].
- Transport along the fracture is faster compared to that within the porous rock matrix.
- Colloids can be transported in fracture and within rock matrix [4].
- Diffusion is the only transport mechanism within the rock matrix. Diffusion within the rock matrix is in the transverse direction (Y-axis) to that of the fracture axis (along with the X-axis) [4].
- Transverse diffusion and dispersion ensure complete mixing at all times across the fracture, and a concentration gradient does not occur across the width of the fracture [16].

In the modeling of the transport of colloids in the single horizontal fracture–matrix system, the following transport phenomena and forces were considered:

- Colloids in the aqueous phase were transported advectively and dispersed in the fracture.
- Colloids were filtered (adsorbed) onto the walls of the fracture and mobile colloids may attach on to the walls of the fracture [4].
- Irreversible deposition of colloids on rock surface may occur [3].
- Colloids diffuse into the porous rock matrix from the fracture [3].
- The clogging of colloids on the fracture–matrix interface is taken care of by the percentage of matrix diffusion of colloids (ε), which is constant throughout the analysis [4].
- Gravitational force act on colloids.

3. Mathematical Model

3.1. Model for the Inclusion of the Gravitational Force

The force due to gravity operates on the mass of the colloid, influencing the colloidal velocity along the fracture, and eventually, on the transport in the fracture, the gravitational acceleration acts transverse to the fracture axis. Equation (1) represents the force acting on the colloids due to gravitation (F_g) [6].

$$F_g = \frac{4}{3}\pi a_p{}^3(\rho_f - \rho_P)g, \tag{1}$$

where a_P the radius (size) of the colloid (m); g is the acceleration due to gravity (m/s^2); ρ_f and ρ_P are the density of the fluid (suspension) and the density of colloid in fluid respectively (kg/m^3). The effect of gravitational force on the transverse velocity of colloids is modeled based on the expression given by Jen and Li [16].

$$u_g = -2 \cdot a_P{}^2 \cdot (\rho_P - \rho_F)\frac{g}{9 \cdot \mu}, \tag{2}$$

where u_g (m/s) is the velocity of colloids in transverse direction due to gravitational force. It is clear from Equation (2) that the velocity due to the gravitational force is a function of the square of the size of the colloids (a_P).

The transverse velocity of colloids obtained from Equation (2) is substituted into Equation (3) [16], to obtain the velocity of the colloids in the longitudinal direction through the fracture. The first step in applying the model is to carry out the following integration:

$$E(y, a_P) = \frac{\int_0^b u_g(y, a_P)}{D_B}dy, \tag{3}$$

where E is the integral term representing the summation of transverse velocities. The limits (0 to b) represents the integration over the entire fracture aperture. D_B is the Brownian diffusion coefficient and is given by Stokes–Einstein relation:

$$D_B = \frac{K_B \cdot T}{6\pi \cdot a_P \cdot \mu}, \tag{4}$$

In the Equation (4), K_B (J K^{-1}) is Boltzmann's constant; T (K) is the temperature.

The expression that captures the effect of the transverse velocity on the longitudinal velocity of colloids through the fracture is represented by Equation (5).

$$V_c = \frac{\int_{-(b-a_P)}^{(b-a_P)} w(y)e^{E(y,a_P)}dy}{\int_{-(b-a_P)}^{(b-a_P)} e^{E(y,a_P)}dy}, \tag{5}$$

where b is the half fracture aperture, and $w(y)$ is the velocity distribution of the fluid in the fracture aperture (Equation (6)). Since a parallel plate model is assumed, the velocity distribution for a fully developed creeping flow will be parabolic and is incorporated into the model using Equation (6):

$$w(y) = \frac{3}{2}(V_f \cdot R)\left(1 - \frac{y^2}{b^2}\right), \tag{6}$$

where V_f (m/day) is the averaged fluid velocity through the fracture. Equation (5) computes the velocity of the colloids (V_c) in the fracture taking into account the influence of the gravitational force.

3.2. Governing Equations for the Transport in Fracture–Matrix System

A fractured porous media consists of a high permeability fracture surrounded by a region of relatively low permeability porous rock. The rock and fluid properties in both these regions of fluid flow are distinct. Therefore, mathematically each of these regions has to be considered as a separate continuum and represented by a separate mass transport equation. Then both of these equations are coupled to capture the mass flux transfer across the fracture–matrix interface. From the assumptions, and transport phenomena considered, the physics of the colloidal transport in a single fracture–matrix system is represented by two coupled partial differential equations derived from mass balance considerations. The partial differential equations governing the transport of colloids in fracture and the rock matrix have been adopted from Abdel-Salam and Chrysikopoulos [3] and Li et al. [4]. Equations (7) and (8) represent the transport of mobile and immobile colloids respectively in the fracture:

$$\frac{\partial}{\partial t}\left(C_f + \frac{\sigma_f}{b}\right) = -V_c \frac{\partial C_f}{\partial x} + D_f \frac{\partial^2 C_f}{\partial x^2} + \frac{\varepsilon \cdot \theta_m \cdot D_e}{b} \frac{\partial C_m}{\partial y}, \tag{7}$$

$$\frac{\partial \sigma_f}{\partial t} = \lambda_f \cdot V_c \cdot C_f \cdot b, \tag{8}$$

where x and y represent the longitudinal axis and transverse axis respectively, and t represents time (day). C_f (kg/m^3), σ_f (kg/m^2), and C_m (kg/m^3) are the concentration of mobile, immobile colloids, and concentration of colloids within the rock matrix respectively. D_f and D_e are the hydrodynamic dispersion coefficient (m^2/day) and diffusion coefficient (m^2/day) of rock matrix respectively. V_c is the velocity of colloids in fracture (m/day) which has been calculated using Equation (5). θ_m, ε, and λ_f are porosity of rock matrix, the percentage of matrix diffusion of colloids, and the filtration coefficient of colloids in fracture (m^{-1}) respectively. Experimental studies have shown that the release rate of the deposited particles is negligible for the case of the smooth parallel surface [18] and hence, colloid deposition can be assumed irreversible (Equation (8)).

Colloids migrate in the fracture, and some of them may be adsorbed on to the solid surface. The irreversible adsorption process considered in the present study can be represented by the Equation (8). In Equation (7), the first and second term on the RHS (Right Hand Side) represents the advection and dispersive transport phenomena of colloids in fracture respectively. The third term on RHS of Equation (7) represents the mass flux transfer of colloids from the fracture into the rock matrix. The rock matrix, which forms the wall of the fracture, is permeable, making it possible for the colloids to transport (perpendicular to fluid flow in fracture) into the rock matrix by process of diffusion. The loss of mass from a fracture is equal to the diffusive flux, which crosses the fracture–matrix interface, and represented by the coupling term of the Equation (7) (third term on RHS) (modeled on the Fick's first law) [17,19]. The pores of the rock matrix may block the colloids from entering the rock matrix from the fracture. The blockage may be small or total depending on the size of the colloids. Some of the filtered colloids on the fracture wall may further act as a hindrance for the mass flux transfer from fracture to the rock matrix. The parameter ε (percentage of diffusion flux of colloids into the rock matrix) in the final term of Equation (7) on RHS is a function of both sizes of colloids and filter colloids on the fracture wall that could hinder the mass flux transfer process. The value of ε varies from $\varepsilon = 0$ to 1, with $\varepsilon = 0$ being no flux transfer and $\varepsilon = 1$ being 100% flux transfer. In the present article, ε is constant (Table 1).

Equation (9) represents the diffusive transport of colloids within the porous rock matrix.

$$(1 + K_{dm})\frac{\partial C_m}{\partial t} = D_m \frac{\partial^2 C_m}{\partial y^2}, \tag{9}$$

where K_{dm} is the sorption partition coefficient for colloids within the rock matrix.

Equations (10)–(14) represent the set of initial and boundary conditions described for the colloidal transport within a single fracture–matrix system. Initial conditions are represented by Equation (10):

$$C_f(x, t = 0) = C_m(x, y \geq b, t = 0) = 0. \tag{10}$$

$$C_f(x = 0, t > 0) = C_o \tag{11}$$

Equation (7) is the boundary condition at fracture and matrix interface ($y = b$).

$$C_m(x, y = b, t > 0) = C_f(x, t > 0) \tag{12}$$

$$C_f(x = L_f, t > 0) = 0 \tag{13}$$

$$\frac{\partial C_m(x, y = B, t > 0)}{\partial y} = 0 \tag{14}$$

The data used to solve above numerical model is listed in Table 1.

Table 1. Transport parameters used for analysis of the transport of colloids in a single fracture–matrix system with variable fracture aperture.

Parameter	Value	Reference
Spatial grid size along the fracture, Δx (m)	0.006	Calculated
Time step size. Δt (days)	0.005	Calculated
Length of fracture, L (m)	20	Assumed
Thickness of rock matrix, B (m)	0.25	Assumed
Total time of simulation, T (days)	100	Assumed
Half fracture aperture (Parallel Plate Model), b (μm)	50	Assumed
Sorption partition coefficient for colloids in rock matrix, K_{dm}	0.2	[4]
Percentage of matrix diffusion flux of colloids, ε (0–1)	0.1	[4]
Porosity of rock matrix, θ	0.03	[4]
Hydrodynamic dispersion coefficient of colloids suspended in fracture, D_f(m^2/day)	0.013	[4]
Density of water, ρ_w(kg/m^3)	1000	
Average velocity of the colloids' particles, V_c (absence of gravity) (m/day)	1	Assumed
Fluid flow rate, Q (m^3/day)	50×10^{-6}	$Q = A \cdot V_c$
Filtration coefficient of the colloids in fracture, λ_f, (L/m)	0.3	[4]
Density of colloids, ρ_c (kg/m^3)	1500–5000	[4]
Radius of colloids, a_P(nm)	50–600	[4]
Boltzmann's constant, k_B(J/K)	1.38×10^{-23}	[4]

3.3. Rock Penetration Length and Rock Saturation Factor

The information of the extent of colloid penetration (diffusion) within the rock matrix from the fracture is vital to many areas of study. The extent of colloid penetration would help in accessing the level of contamination of colloid assisted radionuclides transport into the subsurface system. Nanoparticles which are one form of colloids are finding great applications in the field of petroleum (enhanced oil recovery), and the knowledge of the extent of penetration or spreading of nanoparticles within the rock matrix is vital. The higher the nanoparticles diffuse into the rock matrix, the better will be the oil recovery [20] from the fractured reservoir.

Effective and easy computation of the extent of penetration of colloids within the rock matrix is not an easy job. The transport of colloids (mass transfer) into the rock matrix from the fracture to the end of rock matrix is primarily attributed to two processes, the diffusion (Equation (4)), and the flux transfer at the fracture–matrix interface (coupling term, final term on the RHS of Equation (7)). It is evident from Equations (7)–(9) that a considerable number of parameters influence the diffusion or transport of particles from the fracture into the surrounding rock matrix (ε, D_m, D_f, V_c, b, λ_f). Therefore, it becomes tedious to accurately, and quickly evaluate the extent of penetration of colloids into the rock matrix. Hence, there is a need for a single measurable tool for determining the colloid or contaminant penetration within the rock matrix. The tool should be such that it considers all the parameters involved, and estimates the actual extent of diffusion of suspended particles within the rock matrix in a simple manner. One such measuring tool can be "Rock Saturation Factor" (RSF) developed in the present study. RSF effectively measures the extent of penetration/diffusion of suspended particles/colloids within the porous rock matrix (developed based on Bagalkot et al. [21]). RSF gives the information on the extent of the diffusion for various size and density of colloids or any suspended particle or solute within the rock matrix from the fracture. In Figure 2, B_o is the maximum rock penetration length (m), which is the maximum depth of penetration of colloids in rock matrix at a section of rock matrix, which is at 50% of L_{fo}. L_{fo} is the length measured along the fracture from the inlet to the point at which the concentration of colloid dies down to zero in the fracture, indicating the direct dependency of B_o on the value of L_{fo}. Equation (15) represents the Maximum rock penetration length.

$$B_o = f\left(L_{fo}\right). \tag{15}$$

Equation (16) shows the relationship of L_{fo} and other parameters considered in the analysis.

$$L_{fo} = f(\varepsilon, D_m, D_f, D_e, V_c, b, \lambda_f), \tag{16}$$

where $D = f(a_P)$ and $V_c = f(a_P, \rho_P, g)$. Substituting these two equations in Equation (16), we get.

$$L_{fo} = f(\varepsilon, a_p, \rho_P, g, D_f, D_m, b, \lambda_f). \tag{17}$$

Hence, L_{fo} not only dependent on the transport parameters, evidently, is a function of parameters (a_P, g, and ρ_P) that govern the gravitational force acting on colloids.

Using Equation (17) in Equation (15) gives a modified version of Equation (15) represented by Equation (18).

$$B_o = f(\varepsilon, a_p, \rho_P, g, D_f, D_m, b, \lambda_f). \tag{18}$$

From the relation presented by the Equation (18), it can be safely inferred that for the calculation of B_o all the parameters influencing colloid transport and the force of gravitation on the transport of colloids in a single horizontal fracture–matrix system has been considered. Hence, B_o is a reliable and suitable single measuring tool to quantify the extent of penetration of colloids in the porous rock matrix.

For a straightforward prediction of mitigation of contaminants, or the extent of penetration of suspended particles (nanoparticle, microbes, NAPLs or any suspended matter) a non-dimensional factor "Rock Saturation Factor" (RSF) is introduced in the present research. The work of Rock saturation factor gives the information about the extent of saturation of the rock at a section of rock matrix, which is at 50% of L_{fo} regarding colloids or any contaminant that has diffused into the rock matrix. The RSF is a non-dimensional form of maximum penetration length (B_o). Equation (19) represents the mathematical form of the RSF.

$$\text{RSF} = \frac{B_o}{B}, \tag{19}$$

where B_o (m) is the maximum depth of penetration of colloids within the rock matrix at a section of rock matrix which is at 50% of L_{fo} (Figure 2) and B (m) is the half fracture spacing. At $B_o = B$, the penetration (diffusion) of the colloids within the rock matrix is maximum. L_{fo} is the length measured along the fracture from the inlet to the point at which the concentration of colloid dies down to zero in the fracture (Figure 2). The schematics representing the penetration of colloids within the rock matrix, penetration length, and hence the RSF are shown in Figure 2. RSF = 1 indicates that the colloids/contaminants have penetrated the entire length of the rock matrix ($B = B_o$), signifying complete diffusion of colloids within the rock matrix. RSF = 0 symbolizes that the colloids were ineffective at penetrating the rock matrix, signifying no diffusion of colloids into the rock matrix. For the intermediate values of RSF (1 > RSF > 0) there is partial diffusion of colloids within the rock matrix (Figure 2).

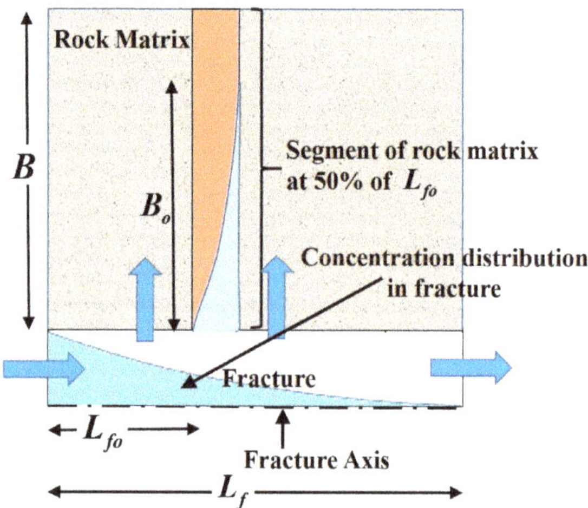

Figure 2. Schematic representation of parameters that are used to calculate "Rock Saturation Factor".

3.4. Numerical Model

The governing Equations (7)–(9) are solved for the given set of initial and boundary conditions (Equations (10)−(14)), to investigate the influence of gravity on the transport of colloids in a single coupled horizontal fracture–matrix system. A semi-implicit finite difference formulation is adopted to discretize the governing coupled partial differential equations. A first-order fully implicit backward (upwind) difference is employed to discretize the advection terms (first term on the RHS of Equation (7)). The Crank–Nicolson discretization scheme is used for the second order terms representing hydrodynamic dispersion (second term on the RHS (Right Hand Side) of Equation (7)) in fracture and diffusion in the rock matrix (first term on the RHS of Equation (10)). A two-point backwards differencing is used to discretize all the temporal terms (source terms). The iterations were carried out at each time step to satisfy the continuity at the fracture–matrix interface. A uniform grid spacing is adopted along the length of the fracture (X-axis), and a geometrically varying (increasing) grid space is applied to the rock matrix perpendicular to the flow in fracture (Y-axis). The grid size of the rock matrix increases geometrically, with the size of the first grid equal to half of the fracture aperture ($y_1 = b$). A smaller grid size is used in the first grid, which represents the interface of the fracture–matrix to capture the mass flux transfer occurring at the interface effectively.

Figure 3 provides the comparison of the present numerical model with that of analytical solution given by Abdel-Salam and Chrysikopoulos [3]. The present numerical model is in close agreement with the model presented by Abdel-Salam and Chrysikopoulos [3] establishing the credibility of the numerical model.

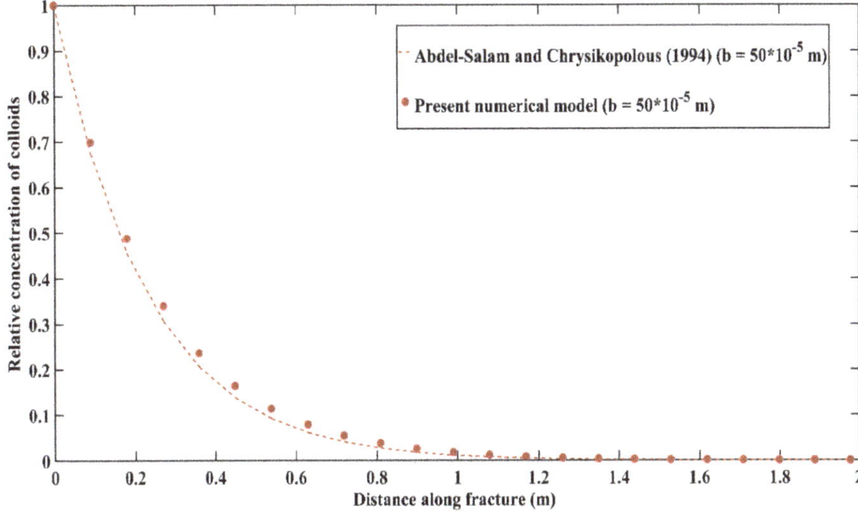

Figure 3. Verification of the present numerical model.

4. Results and Discussion

Figure 4 shows the dependency of the velocity of colloid in fracture (V_c) (Equation (5)) on the gravitational force, colloid density (1500, 3000, and 5000 kg/m^3) and colloid size (50–600 nm). Figure 4 also depicts the longitudinal velocity of colloid (V_c) if the gravitational force is neglected (hashed horizontal line). It may be observed in Figure 4 that when the gravitational force is applied the V_c is sensitive to change in both density and size of the colloid, indicating that the gravitational force is influential on the colloid velocity in the fracture. Further, for a fixed colloid density ($\rho_P > \rho_F$), the colloid velocity reduces exponentially, as the colloid size (radius) increases. At 1500 kg/m^3 the V_c diminishes by one order from 1m/day to around 0.1 m/day. The reduction in velocity due to gravity is amplified with the increase in the colloid density (1500 to 5000 kg/m^3). The observations in Figure 4 makes sense as previous studies have shown that the higher the density of the particles (colloids), the larger will be the effect of the gravity on the colloids [9,13,22]. When the colloid density (1500 kg/m^3) is near to that of fluid (suspension) (1000 kg/m^3), there exists a spectrum of a size of colloids (0–100 nm) over which there is a negligible influence of gravity and colloid velocity remains constant. Once the threshold is crossed the velocity of colloid drops exponentially with the increase in the colloid size. The size of the colloid does not exclusively influence the velocity of the colloid, but the density equally plays an important part. As the colloid density shifts away from the fluid density of 1000 kg/m^3 (3000 and 5000 kg/m^3), the range of spectrum becomes small (0–50 nm), indicating the increased influence of gravity. This behavior of colloid in the presence of gravity is consistent with the previously presented conceptual models [9,16], which predicted a more significant loss for the larger particles and negligible loss for smaller particle leading to faster transport.

Figure 5 depicts the spatial distribution of the relative concentration of colloids in the fracture for three different colloid densities (1500 kg/m^3, 3000 kg/m^3 and 5000 kg/m^3) and at a fixed colloid size of 300 nm (radius). There is a contrast in the relative concentration of the colloids in the fracture, among the considered density values, especially compared to the values that are near (1500 kg/m^3), and away (3000 and 5000 kg/m^3), from the density of the fluid (1000 kg/m^3). The distance traversed by colloid along fracture is more significant (>8m) for colloid density values (1500 kg/m^3) near to that of fluid density compared to the colloid densities (3000 kg/m^3 and 5000 kg/m^3) significantly higher than fluid density (<6 m). The influence of gravitational force on the transport of colloids in fracture becomes dominant as the density of colloid shifts away from that of the fluid density (1000 kg/m^3), and the gravity effects lead to more substantial retardation of colloids. As noted from Figure 4, for a colloid size of 300 nm, as the density is increased the velocity reduces considerably from 0.4 m/day for 1500 kg/m^3 to 0.08 m/day for 5000 kg/m^3 (80% reduction). Advection and dispersion phenomena in the fracture are closely related to the velocity of colloid, and an increase in colloid density will give rise to a reduction in velocity, leading to the weakening of advection and dispersion transport phenomena. Hence, a decrease in distance transported by colloid in fracture as density increases can be attributed to the decrease in the influence of advection and dispersion phenomena on the transport.

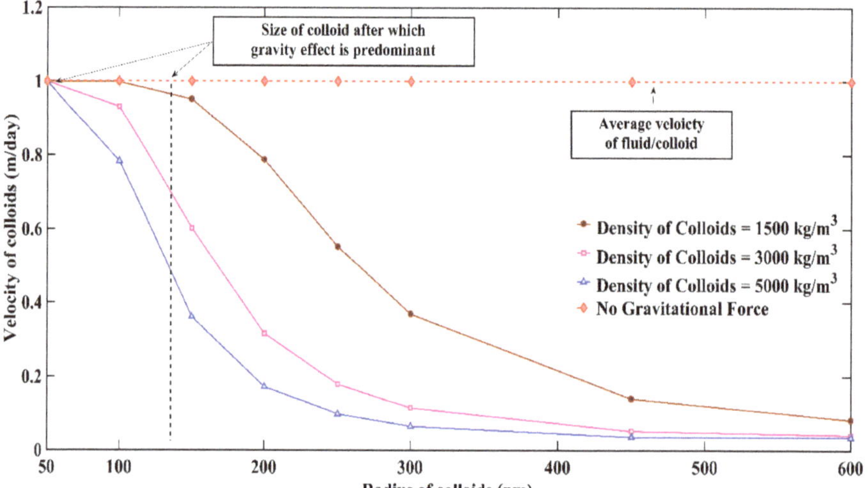

Figure 4. Influence of gravitational force on the colloidal velocity for different colloidal size (50–600 nm).

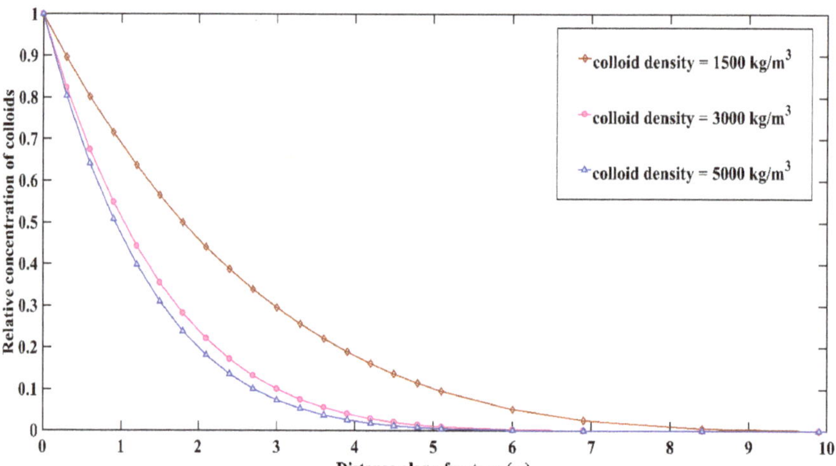

Figure 5. Spatial distribution of relative concentration of colloids along fracture at three different values of colloid densities (1500 kg/m^3, 3000 kg/m^3 and 5000 kg/m^3), at 300 nm colloid size.

Figure 6A–C show the contour representing the penetration of colloids within the rock matrix for 1500 kg/m^3, 3000 kg/m^3, and 5000 kg/m^3 respectively for 300 nm size colloid. Figure 6D shows the case when gravitational force is not considered at size of 300 nm. It may be observed from Figure 6C (colloid density of 5000 kg/m^3) that there is a diminished penetration (diffusion) of colloids (maximum

0.003 m along the rock matrix and maximum 2 m along the fracture). Slightly deeper penetration is observed when the density is brought down to 3000 kg/m^3 (maximum 0.0035 m along the rock matrix and around 3 m along the fracture) (Figure 6B). A significant difference in penetration is observed when the colloid density is reduced to 1500 kg/m^3 as observed in Figure 6A (above 0.0035 m along the rock matrix and around 9 m along fracture) when compared to other two colloidal densities (Figure 6B,C). The greater diffusion and hence penetration of colloids within the rock matrix for the case of 1500 kg/m^3 is due to the presence of a higher concentration of colloids in the fracture compared to higher density (3000 and 5000 kg/m^3). As seen in Figure 5 for the density of 1500 kg/m^3 greater length of the fracture–matrix interface is exposed to colloids (>8 m) as compared to colloids of densities of 3000 and 5000 kg/m^3 (<6 m) due to a more significant influence of the gravitational force. The longer length of colloids indicates the enhanced possibility of mass flux transfer of colloids into the rock matrix from fracture, which is precisely what is observed in Figure 6A. The high-density colloids (3000 and 5000 kg/m3) depreciate the diffusion within the rock matrix by approximately 80% when compared to the low-density colloids (1500 kg/m^3). Interestingly, when the gravitational effect is neglected (Figure 6D) the penetration of colloids is significantly higher than when the gravitational force is considered (Figure 6A–C). Due to the absence of gravity affect the velocity of the colloids would not be altered, as may be seen in Figure 4, this would affect the advection and mass flux at the interface, hence the observed difference. Further, comparing Figure 6D with Figure 6A–C, it may be concluded that for the present system, neglecting the gravitational force on colloids would lead over estimation in the assessment of penetration of colloids within the rock matrix.

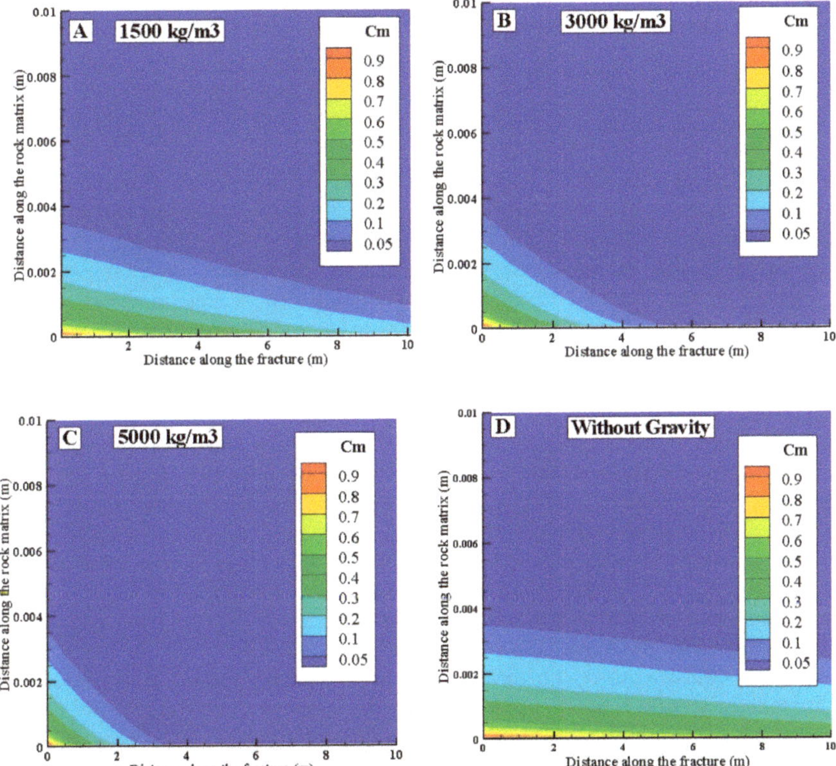

Figure 6. Contour plots depicting the diffusion of colloids within the rock matrix for three different values of colloid density 1500 (6A), 3000 (6B), and 5000 kg/m^3 (6C) and for no gravitational force (6D) at 300 nm colloid size.

From Equation (4) it may be observed that the diffusion coefficient (D_e) is independent of mass, but is inversely related to the size of the colloid. Consequently, a decrease in particle size will be accompanied by an increase in diffusion (when the rest of the parameters are fixed in Equation (4)). In the present article, the diffusion term (D_e) has been viewed from another perspective. Rather than examining the influence of the colloidal size on diffusion within the porous rock matrix, the influence of colloidal size on the mass flux transfer at the fracture–matrix interface will be investigated. The diffusion coefficient is one of the parameters present in the coefficient of the coupling term ($\varepsilon\theta_m D_e/b$), that controls the mass flux transfer at the fracture–matrix interface. The coupling term at-large controls the mass of colloids that can diffuse into the surrounding porous rock matrix from the high permeability fracture. The larger the value of the coefficient of coupling term, the higher the mass flux transfer at the interface, and subsequent enhanced diffusion (colloid penetration) within the rock matrix. Table 2 provides the values of the diffusion coefficient and the magnitude coupling term for different size of colloids (radius). In the analysis, the porosity of the rock matrix (θ_m), the percentage of matrix diffusion of colloids (ε), and the half fracture aperture (b) are fixed. It can be noted from Table 2 that the strength of the coupling term for 50 nm, 300 nm and 600 nm is 6.57×10^{-4} m/day, 1.31×10^{-4} m/day, and 0.58×10^{-4} m/day, respectively. Therefore, as the colloidal size increases the coupling term becomes weaker, indicating that there will be a diminished diffusion of colloids within the rock matrix for larger colloids in comparison with smaller.

Table 2. Measured coefficient of diffusion and coupling term (3rd term of RHS of Equation (7)) for various colloid size.

Size of Colloid (nm)	D_e (m²/day)	Coupling Coefficient ($\frac{\theta \cdot D_e}{b}$) (m/day)
50	3.29×10^{-7}	6.57×10^{-4}
300	0.65×10^{-7}	1.31×10^{-4}
600	0.27×10^{-7}	0.54×10^{-4}

Figure 7 shows the influence of the size of colloids on the extent penetration (diffusion) of colloids within the rock matrix for three sizes 50 nm (Figure 7A), 300 nm (Figure 7B) and 600 nm (Figure 7C), and at a fixed density of 3000 kg/m³. Further, Figure 7D shows the case when the gravitational force has been neglected for 600 nm colloid size. Comparing Figure 7A–C, it may be observed that there is a significant penetration of colloids for 50 nm (around 0.004 m along the rock matrix and around 6 m along the fracture) (Figure 7A) compared to 300 nm (Figure 7B), and 600 nm (Figure 7C). A significantly lower diffusion is witnessed for 300 nm (around 0.002 m along the rock matrix and just above 2 m along the fracture) and 600 nm (around 0.0015 m along the rock matrix and around 2 m along the fracture) (Figure 7B,C). The reason for the difference in diffusion of colloids for different size of colloid may be derived from data presented in Table 2. From Table 2, it may be confirmed that for smaller size colloids the coupling term is stronger and a weaker coupling term is observed for large sized colloids. Stronger coupling term implies a possibility of greater mass flux transport across the interface and subsequently, deeper penetration of colloids within rock matrix. The observed results of Figure 7 serve as further confirmation of this concept. The strength of the coupling term for 50 nm, 300 nm, and 600 nm are 6.57×10^{-4} m/day, 1.31×10^{-4} m/day, and 0.58×10^{-4} m/day, respectively. Hence, there is an approximately 90% reduction in the strength of the coupling term when size is increased from 50 nm (negligible gravity) to 600 nm (high gravity) and around 80% from 50 nm to 300 nm. Hence, a greater diffusion observed in the Figure 7A for 50 nm compared to 300 and 600 nm may be attributed to the presence of stronger coupling term for 50 nm colloid. Similar to Figure 6, it may be observed that even in Figure 7, there is a considerable difference in the cases where gravity effect is considered (Figure 7A–C) and when the gravity effect is neglected (Figure 7D). It may be observed that even at large colloid size of 600 nm neglecting gravity effect would show a deeper and more uniform penetration, which is not realistic. Therefore, it is critical to include the influence of gravitational force on colloid transport in fractured porous media.

Figure 8 depicts the spatial distribution of relative concentration of colloids along the fracture for three different values of colloid size (50 nm, 300 nm and 600 nm) at a fixed density of 3000 kg/m^3. At colloid density of 3000 kg /m^3, for a colloid size of 50 nm the colloid velocity is 1m/day (Figure 4), which is same as that of fluid and 0.1 m/day for 300 nm and 0.04 m/day for 600 nm. Hence, 50 nm due its smaller size is not influenced by gravity, while the same cannot be said for 300 nm and 600 nm size colloids. The impact of gravitational force is dominant as the size of colloids increases, causing a greater retardation for large sized colloids by increasing the degree of gravitational settling. It is also observed that the distance traversed by the colloids from fracture inlet increases with the decrease in colloidal size. Thus, the gravitational force significantly reduces the colloidal migration downstream for large sized colloids, which indirectly implies that the colloid-facilitated radionuclide may not get migrated to a large distance downstream as anticipated using the concept of colloids as a typical contaminant carrier.

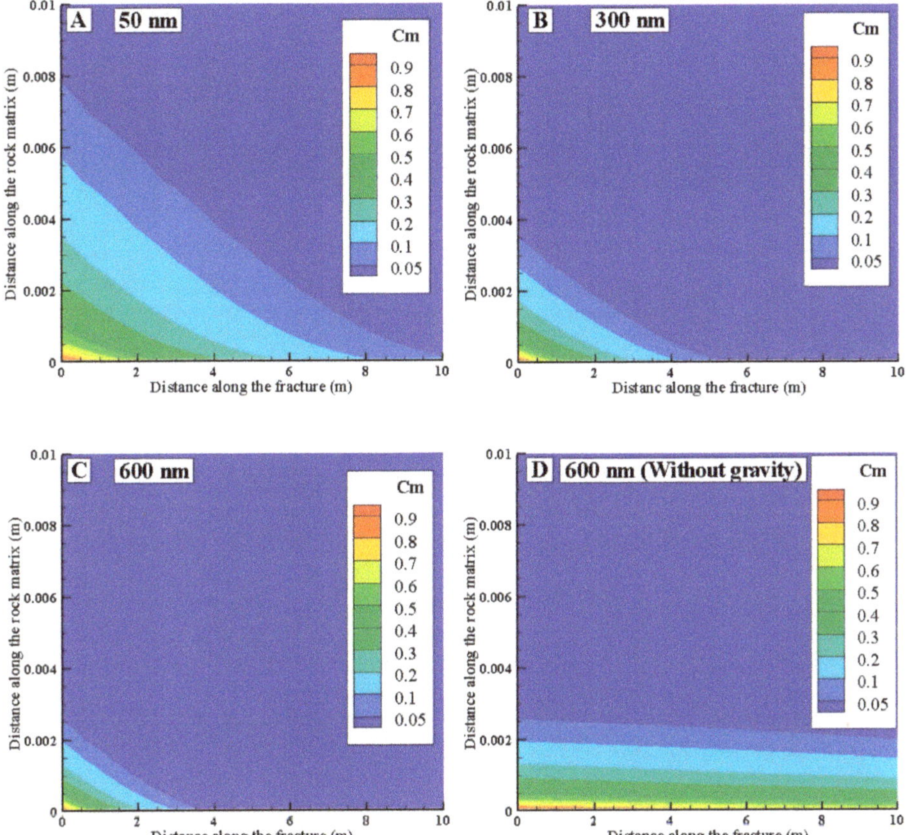

Figure 7. Contour plots depicting the diffusion of colloids within the rock matrix for three different values of colloid size 50 (7A), 300 (7B), and 600 nm (7C) at 3000 kg/m^3 colloid density and when there is no gravitational force (7D).

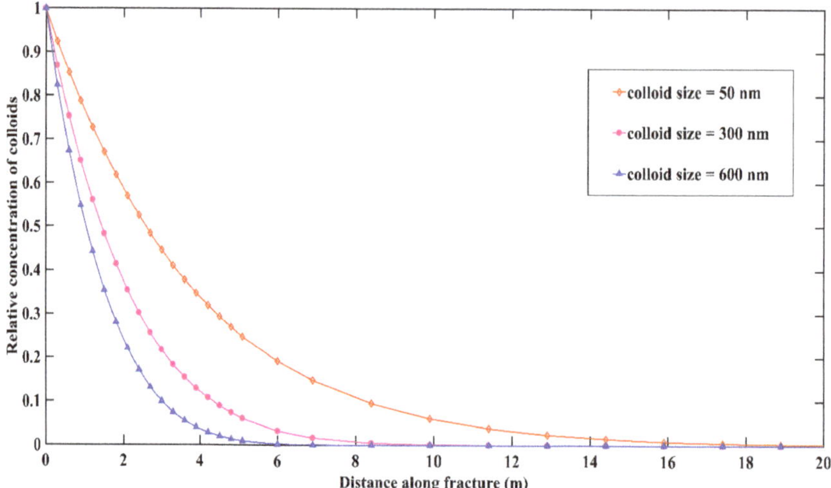

Figure 8. Spatial distribution of relative concentration of colloids along fracture at three different values of colloid size (50, 300, and 600 nm) and at 3000 kg/m^3 colloid density.

Table 3 shows the values of non-dimensional parameter rock saturation factor (RSF) developed in the present study to assess the extent of penetration of colloids within the rock matrix for three different colloid size and densities. The greater the value of RSF ($0 \leq$ RSF ≤ 1) the greater will the saturation of the rock matrix due to the diffusion of the colloids. For RSF = 1 implies 100% saturation of the rock matrix with the colloids or contaminants and for RSF = 0 implies no diffusion at all. The RSF data in Table 3 gives an insight into the influence of gravity, colloidal size and density put together with other influencing parameters (Equation (18)) on the extent of transport of colloids within the rock matrix. It can be noted from Table 3 that, as density increases from 1500 kg/m^3 to 5000 kg/m^3 except for 50 nm size for which there is negligible influence of gravity, there is a reduction RSF indicating lower saturation of the rock matrix by colloids for 300 and 600 nm size. Further, it can be noticed that there is hardly any penetration of colloids at 5000 kg/m^3 for 600 nm and 300 nm. At 5000 kg/m^3 a meager 7.5% (RSF = 0.075) of the rock matrix is saturated with 600 nm and 11.5% (RSF = 0.112) at 300 nm, whereas, for 50 nm around 21.5% (RSF = 0.214) of the rock matrix is saturated by colloids. Hence, at 5000 kg/m^3, there is approximately 65% for 600 nm and 45% for 300 nm lower diffusion is noted in comparison with 50 nm. A similar phenomenon is witnessed for 3000 kg/m^3 and 1500 kg/m^3 (Table 3) but at lesser impact. Few important observations can be made from Table 3. First, for 50 nm no change in the RSF value was noted at all densities, this could be attributed to the negligible influence of gravity for smaller size colloids. This shows that for 50 nm there is negligible influence of gravitational force as explained in the discussion of Figure 4. Second, for 300 nm or 600 nm as density increased from 1500 kg/m^3 to 5000 kg/m^3, the extent of saturation decreased significantly (approximately 28% for 300 nm (RSF = 0.156 to 0.112) and approximately 33% for 600 nm (RSF = 0.112 to 0.075)). This proves that the transport of larger size colloid is sensitive to the density and size of colloids as predicted by previous studies [9,16]. Therefore, there is an optimum range of colloid size present up to which the gravitational influence is negligible. The value of RSF will be helpful to gauge the extent of mitigation of hazardous chemical due to matrix diffusion, especially for the case of colloid assisted radionuclide transport (a possible practical application). RSF will further help in deciding the optimum size of the nanoparticle (petroleum EOR), for the nanoparticles to migrate extensively in rock matrix to create any impact. If the nanoparticles are too large and their density is greater than that of the fluid, then the nanoparticles will become unsuitable for Enhanced Oil Recovery applications.

Table 3. Measured Rock Saturation Factor values for different colloid density and size.

Colloid Size (nm)	Colloid Density (kg/m^3)		
	1500	3000	5000
	Rock Saturation Factor	**Rock Saturation Factor**	**Rock Saturation Factor**
50	0.214	0.214	0.214
300	0.156	0.122	0.112
600	0.112	0.100	0.075

5. Conclusions

A relatively new approach has been applied in the present study to understand the influence of gravitational force on the transport of colloids in a single fracture–matrix system using numerical experiments. The sensitivity of the transport phenomena on the size and density of colloids has been additionally included in the study. Based on the outcome of the developed numerical model the following conclusions can be drawn:

1. The gravitational force manifests a powerful control on the transport of colloidal transport by directly altering the velocity of the colloid through the fluid, and hence, altering advection and diffusion phenomena. The colloidal velocity reduces as the size of the colloid increases in the presence of the gravitational force. The reduction of colloidal velocity with the increase in the radius of colloid follows an exponential decreasing profile. This is further enhanced by an increase in colloidal density while a lower colloidal density will experience a minimum difference between the velocities of colloid and fluid. However, it was found that for a certain range of size (<50 nm) and for relatively low values of density (around 1500 kg/m^3) the velocity did not reduce indicating that influence of gravity would be negligible.

2. The density of colloid determines the magnitude of the impact of gravitational force on the transport of colloids. There is a limited colloid transport by the denser colloids along the fracture and within rock matrix. The influence of gravity becomes dominant as the density of colloid shifts away from that of the flowing fluid, and the gravitational influence a larger retardation of colloids for greater colloid density, enhancing the degree of gravitational settling. Roughly, 80% decrease in diffusion within the rock matrix and about 30% reduction in spatial spreading in fracture was witnessed as density increased from 1500 kg/m^3 to 5000 kg/m^3.

3. The mass flux transfer across the fracture–matrix interface is predominantly dependent on the size of the colloid. Large size colloids face a greater obstacle at the fracture–matrix interface compared to the small size. The mass flux transfer across the interface reduces by approximately 80% when colloid size increase from 50 nm to 300 nm, and around 90% when increased from 50 nm to 600 nm. Higher mass flux transfer for 50 nm further leads to enhanced diffusion within the rock matrix compared to a scant diffusion for 300 nm and 600 nm size colloids.

4. The developed non-dimensional parameter, Rock Saturation Factor (RSF) proved to be helpful in identifying the limits in terms of size and density of colloids that can penetrate deep into the rock matrix. RSF is an easy way to decide the optimum size that the colloids/suspended particle that can be transported within a fracture–matrix system to create a positive impact. The RSF will give an added advantage to the process of decision making with respect to size and density of particles before injecting in a fractured medium, especially in real-time applications such as petroleum and contamination remediation.

Author Contributions: N.B. and G.S.K. conceived and designed the numerical experiments. N.B. performed the numerical experiments and extracted and analyzed the data. N.B. wrote the paper.

Funding: This research received no external funding.

Acknowledgments: The authors acknowledge the Library University of Stavanger.

Conflicts of Interest: The authors declare no conflict of interest.

References

1. Champ, D.; Merritt, W.; Young, J. Potential for the rapid transport of plutonium in groundwater as demonstrated by core column studies. *MRS Online Proc. Lib. Arch.* **1981**, *11*. [CrossRef]
2. Baek, I.; Pitt, W. Colloid-facilitated radionuclide transport in fractured porous rock. *Waste Manag.* **1996**, *16*, 313–325. [CrossRef]
3. Abdel-Salam, A.; Chrysikopoulos, C.V. Analytical solutions for one-dimensional colloid transport in saturated fractures. *Adv. Water Resourc.* **1994**, *17*, 283–296. [CrossRef]
4. Li, S.-H.; Yang, H.-T.; Jen, C.-P. Modeling of colloid transport mechanisms facilitating migration of radionuclides in fractured media. *Nucl. Technol.* **2004**, *148*, 358–368. [CrossRef]
5. Silveira, C.S.; Alvim, A. *Numerical Modeling of Radionuclide Migration in Water-Saturated Planar Fracture: Study of Performance of Bentonite in the Far-Field Region*; International Atomic Energy Agency (IAEA): Vienna, Austria, 2011.
6. Elimelech, M. Effect of particle size on the kinetics of particle deposition under attractive double layer interactions. *J. Colloid Interface Sci.* **1994**, *164*, 190–199. [CrossRef]
7. Ryan, J.N.; Elimelech, M. Colloid mobilization and transport in groundwater. *Colloids Surf. A Physicochem. Eng. Asp.* **1996**, *107*, 1–56. [CrossRef]
8. Chang, Y.-I.; Ku, M.-H. Gravity-induced flocculation of non-brownian particles. *Colloids Surf. A Physicochem. Eng. Asp.* **2001**, *178*, 231–247. [CrossRef]
9. Cumbie, D.; McKay, L. Influence of diameter on particle transport in a fractured shale saprolite. *J. Contam. Hydrol.* **1999**, *37*, 139–157. [CrossRef]
10. James, S.C.; Chrysikopoulos, C.V. Monodisperse and polydisperse colloid transport in water-saturated fractures with various orientations: Gravity effects. *Adv. Water Resour.* **2011**, *34*, 1249–1255. [CrossRef]
11. Kretzschmar, R.; Barmettler, K.; Grolimund, D.; Yan, Y.-D.; Borkovec, M.; Sticher, H. Experimental determination of colloid deposition rates and collision efficiencies in natural porous media. *Water Resour. Res.* **1997**, *33*, 1129–1137. [CrossRef]
12. James, S.C.; Chrysikopoulos, C.V. Dense colloid transport in a bifurcating fracture. *J. Colloid Interface Sci.* **2004**, *270*, 250–254. [CrossRef] [PubMed]
13. Chrysikopoulos, C.V.; Syngouna, V.I. Effect of gravity on colloid transport through water-saturated columns packed with glass beads: Modeling and experiments. *Environ. Sci. Technol.* **2014**, *48*, 6805–6813. [CrossRef] [PubMed]
14. Abhishek, R.; Bagalkot, N.; Kumar, G.S. Effect of transverse forces on velocity of nanoparticles through a single fracture in a fractured petroleum reservoir. *Int. J. Oil Gas Coal Technol.* **2015**, *12*, 379–395. [CrossRef]
15. Unni, H.; Yang, C. Brownian dynamics simulation and experimental study of colloidal particle deposition in a microchannel flow. *J. Colloid Interface Sci.* **2005**, *291*, 28–36. [CrossRef] [PubMed]
16. Jen, C.-P.; Li, S.-H. Effects of hydrodynamic chromatography on colloid-facilitated migration of radionuclides in the fractured rock. *Waste Manag.* **2001**, *21*, 499–509. [CrossRef]
17. Sudicky, E.; Frind, E. Contaminant transport in fractured porous media: Analytical solutions for a system of parallel fractures. *Water Resour. Res.* **1982**, *18*, 1634–1642. [CrossRef]
18. Bowen, B.D.; Epstein, N. Fine particle deposition in smooth parallel-plate channels. *J. Colloid Interface Sci.* **1979**, *72*, 81–97. [CrossRef]
19. Grisak, G.; Pickens, J.-F. Solute transport through fractured media: 1. The effect of matrix diffusion. *Water Resour. Res.* **1980**, *16*, 719–730. [CrossRef]
20. Sivasankar, P.; Kumar, G.S. Numerical modelling of enhanced oil recovery by microbial flooding under non-isothermal conditions. *J. Pet. Sci. Eng.* **2014**, *124*, 161–172. [CrossRef]

21. Bagalkot, N.; Zare, A.; Kumar, G.S.J.E. Influence of fracture heterogeneity using linear congruential generator (lcg) on the thermal front propagation in a single geothermal fracture-rock matrix system. *Energies* **2018**, *11*, 916. [CrossRef]

22. James, S.C.; Chrysikopoulos, C.V. Analytical solutions for monodisperse and polydisperse colloid transport in uniform fractures. *Colloids Surf. A Physicochem. Eng. Asp.* **2003**, *226*, 101–118. [CrossRef]

Article

Interplays between State and Flux Hydrological Variables across Vadose Zones: A Numerical Investigation

Zhaoxin Wang [1], Tiejun Wang [1,2,*] and Yonggen Zhang [1,2]

[1]   Institute of Surface-Earth System Science, Tianjin University, Weijin Road 92, Tianjin 300072, China; wangzhaoxin@tju.edu.cn (Z.W.); ygzhang@tju.edu.cn (Y.Z.)
[2]   Tianjin Key Laboratory of Earth Critical Zone Science and Sustainable Development in Bohai Rim, Tianjin University, Weijin Road 92, Tianjin 300072, China
*   Correspondence: tiejun.wang@tju.edu.cn

Received: 17 May 2019; Accepted: 16 June 2019; Published: 20 June 2019

Abstract: Knowledge of both state (e.g., soil moisture) and flux (e.g., actual evapotranspiration (ET_a) and groundwater recharge (GR)) hydrological variables across vadose zones is critical for understanding ecohydrological and land-surface processes. In this study, a one-dimensional process-based vadose zone model with generated soil hydraulic parameters was utilized to simulate soil moisture, ET_a, and GR. Daily hydrometeorological data were obtained from different climate zones to drive the vadose zone model. On the basis of the field phenomenon of soil moisture temporal stability, reasonable soil moisture spatiotemporal structures were reproduced from the model. The modeling results further showed that the dependence of ET_a and GR on soil hydraulic properties varied considerably with climatic conditions. In particular, the controls of soil hydraulic properties on ET_a and GR greatly weakened at the site with an arid climate. In contrast, the distribution of mean relative difference (MRD) of soil moisture was still significantly correlated with soil hydraulic properties (most notably residual soil moisture content) under arid climatic conditions. As such, the correlations of MRD with ET_a and GR differed across different climate regimes. In addition, the simulation results revealed that samples with average moisture conditions did not necessarily produce average values of ET_a and GR (and vice versa), especially under wet climatic conditions. The loose connection between average state and flux hydrological variables across vadose zones is partly because of the high non-linearity of subsurface processes, which leads to the complex interactions of soil moisture, ET_a, and GR with soil hydraulic properties. This study underscores the importance of using soil moisture information from multiple sites for inferring areal average values of ET_a and GR, even with the knowledge of representative sites that can be used to monitor areal average moisture conditions.

Keywords: soil moisture; groundwater recharge; evapotranspiration; vadose zone; soil hydraulic property; climate

---

1. Introduction

Estimation of state (e.g., soil moisture) and flux (e.g., actual evapotranspiration (ET_a) and groundwater recharge (GR)) hydrological variables across vadose zones are of great importance for many research and application purposes, and especially for understanding ecohydrological and land-surface processes in Earth's critical zones [1–3]. Due to the high non-linearity and extreme complexity of hydrological systems, those state and flux variables tend to display significant spatiotemporal variability. For example, a wealth of field and numerical studies has shown that there was considerable variability in soil moisture at different spatiotemporal scales, as a result of its complex interactions with surrounding environments [4,5]. Likewise, flux hydrological variables, such as ET_a and GR, may also vary substantially across landscapes, largely because of the marked spatial variations in soil texture, vegetation, and climatic conditions [1,6–9]. As such, accurate estimation of those state and flux variables presents grand challenges to scientific communities [10–13]. Given the tight connections between state and flux variables [12,14,15], it is thus imperative to elucidate the interplays among those variables to improve their estimation across vadose zones.

To enhance the representativeness of estimation of state and flux hydrological variables across vadose zones, various techniques have been developed [6,16,17]. Notably, in the field of soil hydrology, the method of temporal stability analysis (*TSA*) initially proposed by Vachaud et al. [18] has been widely adopted to identify representative locations for monitoring soil moisture conditions [19]. Based on field observations, Vachaud et al. [18] found that there was a temporal persistence in the spatial structure of soil moisture. Specifically, Vachaud et al. [18] showed that soil moisture levels at certain locations in a field were closer to the areal average moisture conditions, whereas other locations consistently displayed either drier or wetter conditions than the areal average moisture conditions. As such, Vachaud et al. [18] suggested that it was feasible to apply the *TSA* method, which is based on the metric of mean relative difference (*MRD*) of soil moisture (an indicator of the relative wetness at a location as compared to the areal average wetness), to seek representative sites for monitoring areal average moisture conditions (e.g., sites with *MRD* values closest to zero). Since the seminal work of Vachaud et al. [18], the use of the *TSA* method has also been extended to optimize soil moisture monitoring networks, diagnose soil moisture spatiotemporal patterns, fill missing soil moisture data, and improve the practice of water resources management [20–24].

Despite its wide applications, the implication of the *TSA* method for understanding flux hydrological variables across vadose zones has not been adequately addressed. In general, the availability of field measurements of ET_a and GR is extremely limited due to the high operational and maintenance costs of measurement equipment [3,25]. As an effective alternative to resolve this issue, soil moisture data have been used to infer ET_a and GR, such as through inverse modeling [15,26,27]. Given the significant spatial variability in ET_a and GR, it naturally leads to the question as to whether soil moisture data from representative locations identified by the *TSA* method can be used to infer areal average values of ET_a and GR. To answer this question, it requires the knowledge of the relationships of soil moisture with ET_a and GR. Although previous modeling studies have revealed that soil moisture, ET_a, and GR were affected by similar factors (e.g., soil, vegetation, and climate), the dependence of those variables on environmental factors might vary considerably [9,28,29]. Therefore, further investigation into the interplays among soil moisture, ET_a, and GR is still warranted.

The primary goals of this study were two-fold: (1) to investigate the relationships of soil moisture with ET_a and GR under different climatic conditions and (2) to examine whether soil samples with average moisture conditions can also produce average hydrological fluxes, such as ET_a and GR, across vadose zones. To this end, a commonly used process-based vadose zone model with generated soil hydraulic parameters was utilized to simulate soil moisture, ET_a, and GR, as this type of modeling approach has been widely adopted for understanding hydrological processes across vadose zones (e.g., Vereecken et al. [30]; Wang et al. [31]; Martinez et al. [28]; also see the review by Vereecken et al. [5]). Based on the simulation results, the distribution patterns of *MRD*, ET_a, and GR were first discussed, and then the correlations of *MRD* with ET_a and GR were examined to understand the interdependence of those variables on climatic conditions and soil hydraulic properties. Finally, with the aid of the *TSA* method, the feasibility of using representative locations of soil moisture for estimating areal average values of ET_a and GR was evaluated.

2. Methods and Materials

2.1. Model Setup and Data Inputs

The widely used Hydrus-1D model [32] was employed in this study to simulate soil moisture, ET_a, and GR, due to the accuracy of its numerical algorithm [33]. The Hydrus-1D model is a one-dimensional process-based vadose zone model, which utilizes the Richards equation to simulate soil water movement in vadose zones:

$$\frac{\partial \theta}{\partial t} = \frac{\partial}{\partial x}\left[K(h)\left(\frac{\partial h}{\partial x}\right) - K(h)\right] - S(h) \tag{1}$$

where θ (L^3/L^3) is volumetric soil moisture content, t (T) is time, x (L) is spatial coordinate, h (L) is pressure head, K (L/T) is hydraulic conductivity, and S (1/T) is a sink/source term (e.g., root water uptake). To be concise, only brief information on the model setup is given here, and detailed mathematical descriptions of the simulated processes can be found in Simunek et al. [32]. Specifically, the length of the simulated soil columns was set to be 3 m with a total of 301 spatial nodes evenly distributed at 1 cm intervals. Additional numerical experiments showed that further increases in the number of the spatial nodes did not improve the simulation results. At the surface, an atmospheric boundary condition was chosen for calculating actual soil evaporation (E_a), which primarily depended on potential soil evaporation (E_p) and a predefined pressure head at the surface [34]. Surface runoff was allowed to occur when precipitation (P) intensity was greater than the soil infiltration capacity or the surface soil layer became saturated. A unit hydraulic gradient condition was imposed at the lower boundary of the simulated soil columns. As such, GR is defined here as the amount of water that passes downward across the lower boundary [6,31].

The method of Feddes et al. [35] was used to compute root water uptake, which depended on potential transpiration (T_p), soil matric potential, and root density distribution. Note that root water uptake was assumed to be equal to actual transpiration (T_a) in this study. Daily potential transpiration T_p was obtained by partitioning daily potential evapotranspiration (ET_p) into T_p and E_p according to Beer's law [36]:

$$E_p = ET_p \times e^{-k \times LAI} \tag{2}$$

$$T_p = ET_p - E_p \tag{3}$$

where k is an extinction coefficient (0.5 in this study) and LAI is leaf area index. Following Wang and Zlotnik [37], LAI data were retrieved from the MODIS_MOD15 dataset [38] based on the geographical coordinates of the study sites. The exponential model of Jackson et al. [39] was adopted to describe the root density distribution according to the vegetation types at the study sites. Finally, the sum of E_a and T_a was set to be equal to ET_a.

Instead of generating synthetic data to drive the Hydrus-1D model [28,40], five years of daily hydrometeorological data from four study sites with contrasting hydroclimatic conditions across China were obtained from the National Centers for Environmental Information [41], including Hangzhou, Guiyang, Tianjin, and Yinchuan. Specifically, the climates at Hangzhou and Guiyang are humid and sub-humid, respectively. By comparison, Tianjin is located in a semi-arid climate zone, while Yinchuan is located in an arid climate zone. First, daily P and maximum and minimum air temperatures were retrieved for each site. Given the data availability, the Hargreaves equation was used to calculate daily ET_p based on maximum and minimum air temperatures [42]. The summary of mean annual P ($\overline{P}$) and mean annual ET_p ($\overline{ET_p}$) during the study period is provided in Table 1 for the four sites. Based on the land cover of the study sites, forest was assumed at Hangzhou and Guiyang, while grass and shrub were assumed at Tianjin and Yinchuan, respectively. Physiological parameter values for different vegetation types, which were used for computing root water uptake, were obtained from literature [43–45]. Finally, to minimize the impact of initial conditions on the simulation results, all simulations were repeated for four times (i.e., a total of 20 years) until the simulated soil moisture profiles were in equilibrium with atmospheric forcings, and the simulation results from the last repetition were used in the analysis.

Table 1. Mean annual precipitation ($\overline{P}$), potential evapotranspiration ($\overline{ET_p}$), and aridity index ($\overline{ET_p/P}$) at the study sites.

Location	Latitude	Longitude	$\overline{P}$ (cm/year)	$\overline{ET_p}$ (cm/year)	$\overline{ET_p/P}$(-)
Hangzhou	30.23°N	120.17°E	150.9 (127.4–172.9) *	111.8 (108.0–115.1)	0.74 (0.62–0.89)
Guiyang	26.58°N	106.73°E	103.8 (73.5–137.1)	101.7 (95.9–104.6)	0.98 (0.74–1.40)
Tianjin	39.10°N	117.17°E	54.3 (35.6–75.7)	104.1 (99.9–107.2)	1.92 (1.37–2.81)
Yinchuan	38.47°N	106.20°E	20.8 (16.6–29.2)	108.6 (106.8–110.5)	5.22 (3.74–6.43)

* The range of annual values is given in the parentheses.

2.2. Generation of Soil Hydraulic Parameters

To solve Equation (1), the constitutive relations among θ, h, and K are required. In this study, the van Genuchten–Mualem (VGM) model [46,47] was utilized:

$$\theta(h) = \begin{cases} \theta_r + \dfrac{\theta_s - \theta_r}{(1+|\alpha h|^n)^m}, & h < 0 \\ \theta_s, & h \geq 0 \end{cases} \tag{4}$$

$$K(h) = K_S \times S_e^{\,l} \times (1 - (1 - S_e^{1/m})^m)^2 \tag{5}$$

where θ_r (L^3/L^3) and θ_s (L^3/L^3) are residual and saturated soil moisture content, respectively, K_S (L/T) is saturated hydraulic conductivity, and S_e (-) is effective saturation degree that is defined as $(\theta - \theta_r)/(\theta_s - \theta_r)$. The parameters α, n, and l are shape factors: α(1/L) is related to the inverse of air entry pressure, n (-) is a measure of pore size distributions with $m = 1 - 1/n$, and l (-) is a measure of pore tortuosity and connectivity.

In previous modeling studies, because of the lack of field data, synthetic soil hydraulic parameter datasets have been routinely used to characterize the spatial variability in soil hydraulic properties for simulating soil moisture, ET_a, and GR [9,28,48,49]. Here, the method of Zhang and Schaap [50] was utilized to generate the VGM parameters for loamy sand, sandy loam, and silt loam. This method is built on the assumption of either normal (θ_r and θ_s) or lognormal (α, n, and K_s) distributions of the VGM parameters. The validity of this assumption has been verified by Zhang and Schaap [50]. Based on a Monte Carlo procedure, new samples were drawn from the parameter (normal or lognormal) distributions with means and standard deviations given by Zhang and Schaap [50] for each soil texture class. In addition, following Wang et al. [31] and Zhang et al. [51], the parameter l was fixed to be 0.5 in all simulations. In this study, a total of 200 samples were generated for each soil texture class and the statistical summary of the generated VGM parameters is provided in Table 2. Note that soil profiles with uniform soil textures were used in this study. The main purpose of using uniform soil profiles is to promote general understanding, and thus reach general conclusions regarding the impact of soil hydraulic properties on subsurface hydrological processes. By comparison, the use of non-uniform soil profiles might lead to site-specific conclusions (as done in the majority of field-scale soil moisture studies), which, however, deserves future investigation based on numerical approaches.

Table 2. Average values of soil hydraulic parameters for the van Genuchten–Mualem model generated for different soil textural classes.

Soil Texture	θr (cm^3 cm^{-3})	θs (cm^3 cm^{-3})	α (1/cm)	n (-)	K_s (cm/day)
Loamy sand	0.064 (0.035) *	0.386 (0.063)	0.036 (0.039)	1.908 (0.580)	342.5 (864.1)
Sandy loam	0.076 (0.047)	0.381 (0.068)	0.040 (0.075)	1.673 (0.359)	125.4 (334.1)
Silt loam	0.098 (0.057)	0.432 (0.077)	0.006 (0.007)	1.790 (0.475)	62.8 (180.9)

* Standard deviation values are given in parenthesis.

2.3. Temporal Stability Analysis of Soil Moisture

Given its wide applications for analyzing soil moisture spatiotemporal patterns [19,52,53], the *TSA* method was adopted in this study to examine the relationships of soil moisture with ET_a and *GR*. The *TSA* method is based on the metric of relative difference (*RD*) of soil moisture, which is defined as:

$$RD_{ij} = \frac{\theta_{ij} - \overline{\theta}_j}{\overline{\theta}_j} \tag{6}$$

where θ_{ij} is soil moisture content at location i and time j, and $\overline{\theta}_j$ is the spatial average of soil moisture content at time j. With a series of *RD* over time, *MRD* at location i is given as:

$$MRD_i = \frac{1}{m} \sum_{j=1}^{m} RD_{ij} \tag{7}$$

where m is the number of observations over the study period. By definition, the term *MRD* characterizes the wetness condition at a location relative to the areal average wetness condition of the study field over the study period. Therefore, *MRD* can be used to characterize the spatial distribution of soil moisture (average state) and the sites with *MRD* values closest to zero can be selected as representative locations for monitoring field average moisture conditions [18,19].

The standard deviation of *RD* (*SDRD*) is another useful metric for quantifying soil moisture temporal dynamics and defined as:

$$SDRD_i = \left[\frac{1}{m-1} \sum_{j=1}^{m} \left(RD_{ij} - MRD_i\right)^2 \right]^{0.5} \tag{8}$$

According to its definition, sites with lower *SDRD* values indicate higher temporal stability of soil moisture than those with higher *SDRD* values. Note that instead of using soil moisture storage in this study, daily volumetric soil moisture content at 25 cm extracted from the last repetition of the simulations was analyzed. Since soil moisture storage is equivalent to depth-weighted soil moisture content and similar simulation results were obtained from other depths (e.g., 10, 50, and 100 cm), the conclusions made in this study essentially remain the same based on depth-specific simulation data. In addition, the use of depth-specific data is to be consistent with the framework set for soil moisture monitoring networks (e.g., Nebraska Mesonet, Oklahoma Mesonet, and Soil Climate Analysis Network), which report soil moisture data at different soil depths.

3. Results and Discussion

3.1. Comparison of Hydroclimatic Conditions across the Study Sites

The statistical summary of $\overline{P}$ and $\overline{ET_p}$ during the study period is reported in Table 1 for the study sites. Among the four sites, there was a significant gradient in $\overline{P}$ increasing from 20.8 cm/year at

Yinchuan to 150.9 cm/year at Hangzhou; whereas, the spatial variations in $\overline{ET_p}$ were much smaller with a range from 101.7 cm/year at Guiyang to 111.8 cm/year at Hangzhou. As a result, a wide range of mean annual aridity index (i.e., $\overline{ET_p}/\overline{P}$) existed across the sites, varying from 0.74 at Hangzhou to 5.22 at Yinchuan. Given the diversity of climate regimes among the selected sites (e.g., from arid to humid climates), mean annual ratios of ET_a (defined as $\overline{ET_a}/\overline{P}$) and GR ($\overline{GR}/\overline{P}$) were used in the following analysis to ensure the comparability of the simulated patterns of ET_a and GR at different sites. With regard to the interannual variability, there were also pronounced variations in annual P at all sites; however, the interannual variability in ET_p was less significant (Table 1). It should be noted that climate seasonality is another important climatic factor that affects ET_a and GR [40,54]. Therefore, to examine the climate seasonality at the study sites, mean monthly P and ET_p during the five-year study period are plotted in Figure 1. It can be seen from Figure 1 that the climates at the sites exhibited similar seasonal patterns with highest P and ET_p occurring during summer months. As such, it is reasonable to argue that climate seasonality did not exert a significant impact on the comparison of the distribution patterns of the simulated $\overline{ET_a}/\overline{P}$ and $\overline{GR}/\overline{P}$ across the sites in this study.

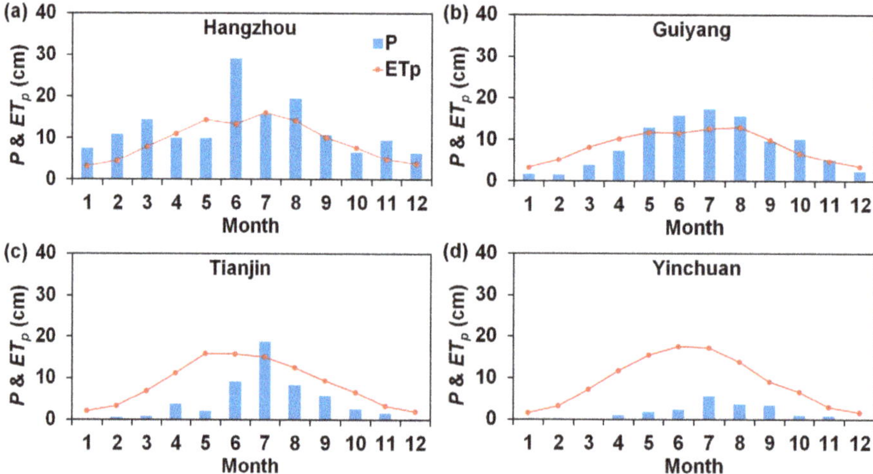

Figure 1. Mean monthly precipitation (P) and potential evapotranspiration (ET_p) during the study period at (**a**) Hangzhou, (**b**) Guiyang, (**c**) Tianjin, and (**d**) Yinchuan.

3.2. Simulated Patterns of MRD

To investigate the relationships of soil moisture with ET_a and GR on the basis of the *TSA* method, it is necessary to first examine whether simulated soil moisture data could mimic the field phenomenon of soil moisture temporal stability as shown in previous field studies. To this end, Figure 2 shows the ranked *MRD* with associated *SDRD* obtained for different soil textures at the study sites. As previously explained, only soil moisture data simulated at 25 cm were used to compute *MRD* and *SDRD* in Figure 2. In general, the simulated *MRD* patterns were similar to those derived from field observations at various spatiotemporal scales [20–24]. In particular, Figure 2 shows that certain samples displayed consistently wetter soil moisture conditions (i.e., with larger *MRD* values) than others. More importantly, the ranges of both *MRD* and *SDRD* were in line with previously reported values obtained at watershed and regional scales [24,53,55]. For example, the range of *MRD* for loamy sand was from −0.765 to 1.287 at Hangzhou, from −0.797 to 1.324 at Guiyang, from −0.819 to 1.237 at Tianjin, and from −0.912 to 1.166 at Yinchuan (Table 3). It should be emphasized that there were variations in the statistical characteristics of *MRD* and *SDRD* produced by different modeling studies [28,29], which could be largely attributed to the differences in the synthetic soil hydraulic parameter datasets and hydrometeorological conditions used for simulations.

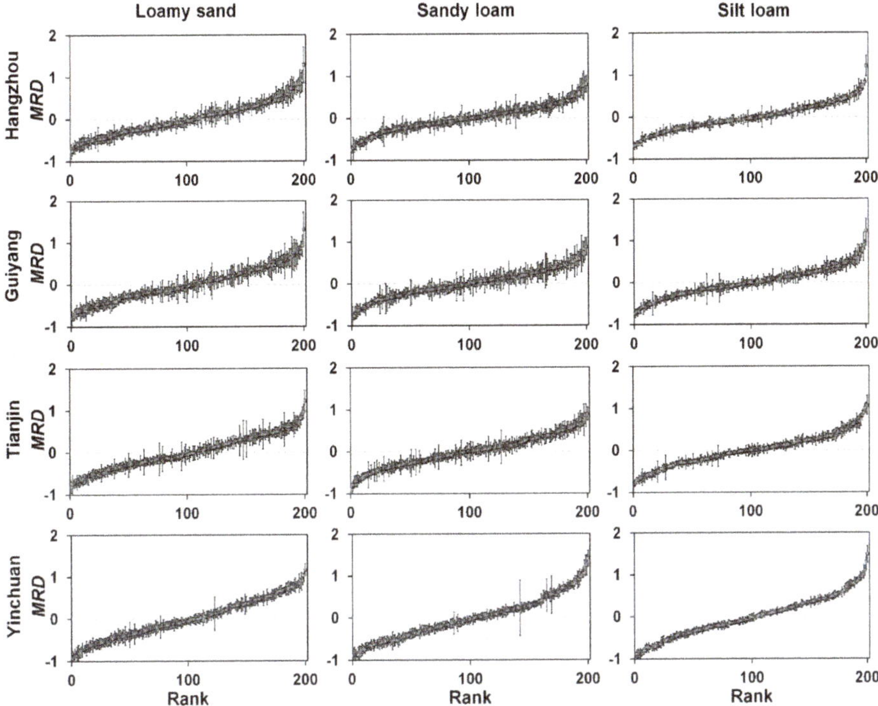

Figure 2. Ranked mean relative difference (*MRD*) of soil moisture with one standard deviation (vertical bars) for different soil textures at the study sites. Simulated soil moisture data at 25 cm were used to compute *MRD*.

Table 3. Statistical summary of mean relative difference (*MRD*) and standard deviation of relative difference (*SDRD*) of soil moisture. Simulated soil moisture data at 25 cm were used to compute *MRD* and *SDRD*.

Location	Loamy Sand			Sandy Loam			Silt Loam		
	Range of *MRD*	Range of *SDRD*	Mean *SDRD*	Range of *MRD*	Range of *SDRD*	Mean *SDRD*	Range of *MRD*	Range of *SDRD*	Mean *SDRD*
Hangzhou	−0.765–1.287	0.046–0.428	0.139	−0.741–0.799	0.041–0.402	0.131	−0.690–1.210	0.022–0.264	0.085
Guiyang	−0.797–1.324	0.037–0.486	0.158	−0.789–0.873	0.049–0.430	0.159	−0.732–1.230	0.027–0.370	0.112
Tianjin	−0.819–1.237	0.043–0.462	0.133	−0.977–0.891	0.053–0.397	0.134	−0.793–1.064	0.032–0.301	0.091
Yinchuan	−0.912–1.166	0.035–0.397	0.114	−0.889–1.410	0.043–0.664	0.111	−0.890–1.485	0.024–0.199	0.073

It is worth noting that inconsistent findings have been reported from previous studies regarding the environmental conditions and factors that mainly control soil moisture temporal stability. For example, there still exists a debate on whether soil moisture is temporally more stable under wetter or drier conditions [29,52,56]. In this study, the average soil moisture temporal stability as indicated by mean *SDRD* values varied noticeably among the sites (Table 3). It appears that soil moisture tended to be temporally more stable under both very wet (e.g., Hangzhou) and dry (e.g., Yinchuan) climatic conditions, while soil moisture temporal stability was lowest under intermediate climatic conditions (e.g., Guiyang). This result is in general agreement with the conclusion made by Lawrence and

Hornberger [48], who showed that the observed soil moisture variability peaked in temperature regions with intermediate moisture conditions. Soil moisture levels remain high in humid regions due to the constant P inputs, which results in less temporal dynamics of soil moisture. In contrast, because of infrequent rainfall events in arid regions, soil moisture is also temporally less dynamic. Thus, it leads to the lowest soil moisture temporal stability at the sites with intermediate climatic conditions (e.g., in sub-humid regions).

Table 3 also reveals that mean $SDRD$ values differed for different soil textures under the same climatic condition. For example, mean $SDRD$ values were comparably smaller for silt loam than those for loamy sand and sandy loam at all sites. Based on a theoretical framework, Vereecken et al. [30] showed that the variability in soil hydraulic parameters could significantly affect soil moisture variability. From Table 2, it can be seen that the lower mean $SDRD$ values for silt loam were mostly due to the smaller variability in the parameters α and K_S. It, thus, corroborates our previous conjecture that the statistical characteristics of MRD and $SDRD$ produced by synthetic modeling approaches are partly dependent on the variability in soil hydraulic parameters generated for simulations. In summary, the results from the TSA demonstrate the viability of using synthetic modeling approaches for reproducing reasonable spatiotemporal structures of soil moisture as observed in the fields.

3.3. Distribution Patterns of $\overline{ET_a}/\overline{P}$ and $\overline{GR}/\overline{P}$

Following Wang et al. [31], histograms of simulated $\overline{ET_a}/\overline{P}$ and $\overline{GR}/\overline{P}$, as shown in Figure 3, are used to discern the differences in the distribution patterns of $\overline{ET_a}/\overline{P}$ and $\overline{GR}/\overline{P}$ under different climatic conditions and for different soil textures. For the same soil texture, Figure 3 reveals that the ranges of $\overline{ET_a}/\overline{P}$ and $\overline{GR}/\overline{P}$ varied noticeably among the study sites. In particular, the distribution of $\overline{ET_a}/\overline{P}$ increased towards the higher end of the range between 0 and 100% with increasing climate dryness, and the distribution of $\overline{GR}/\overline{P}$ exhibited an opposite trend. This modeling result suggests that $\overline{ET_a}/\overline{P}$ becomes progressively larger while $\overline{GR}/\overline{P}$ grows gradually less important from humid to arid sites, which is consistent with the general understanding of the climate effect on ET_a and GR [57]. Rainfall water infiltrated into soils needs to first satisfy the atmospheric demand for ET_a before becoming GR. Thus, the opportunity for the infiltrated water to be evapotranspired is much higher under drier climatic conditions due to the higher atmospheric demands for evapotranspiration, resulting in the increasing trend in $\overline{ET_a}/\overline{P}$ with increasing climate dryness.

Although the respective ranges were similar for $\overline{ET_a}/\overline{P}$ and $\overline{GR}/\overline{P}$ under the same climatic conditions, there were noticeable variations in the distribution patterns of $\overline{ET_a}/\overline{P}$ and $\overline{GR}/\overline{P}$ for different soil textures (Figure 3). In the order of loamy sand, sandy loam, and silt loam, there were proportionally more samples with higher $\overline{ET_a}/\overline{P}$ (or lower $\overline{GR}/\overline{P}$) values towards the higher (or lower) end of the distribution ranges, most notably at the sites of Hangzhou, Guiyang, and Tianjin. It reflects the controls of soil hydraulic properties on ET_a and GR processes. Due to the higher infiltration capacities (e.g., Table 2 shows that K_s becomes gradually smaller from loamy sand to sandy loam and silt loam), coarser soils tend to produce lower $\overline{ET_a}/\overline{P}$ and higher $\overline{GR}/\overline{P}$ ratios [40,58]. Interestingly, at the site of Yinchuan with very dry climatic conditions, the contrast in the distribution patterns of $\overline{ET_a}/\overline{P}$ and $\overline{GR}/\overline{P}$ was less obvious among different soil textures. Regardless of soil textural differences, most of P becomes ET_a in arid regions due to the limited water supplies, which greatly weakens the dependence of ET_a and GR on soil hydraulic properties. It, thus, suggests that the controls of soil hydraulic properties on ET_a and GR vary with climatic conditions.

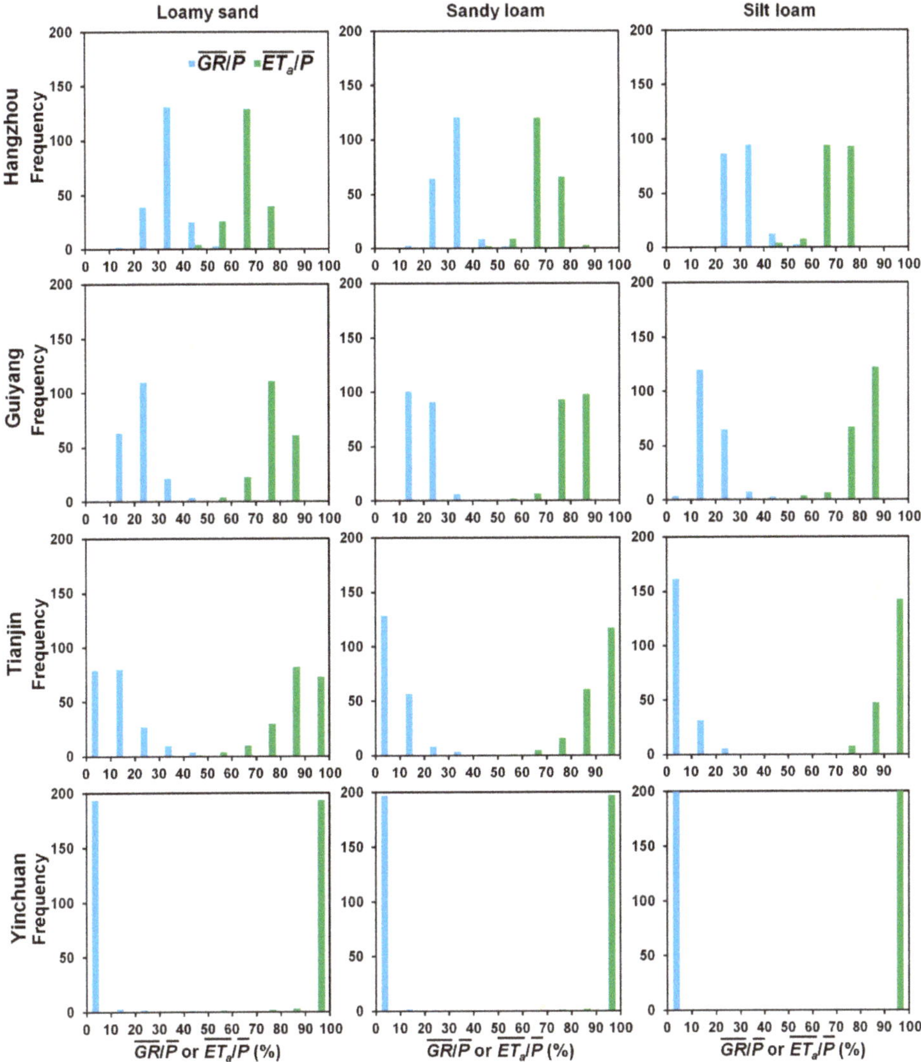

Figure 3. Distributions of simulated mean annual ratios of actual evapotranspiration ($\overline{ET_a}$) and groundwater recharge ($\overline{GR}$) over precipitation ($\overline{P}$).

3.4. Relationships of MRD with $\overline{ET_a}/\overline{P}$ and $\overline{GR}/\overline{P}$

To investigate the interplays between soil moisture and flux hydrological variables across vadose zones on the basis of the *TSA* method, the relationships between *MRD* and $\overline{ET_a}/\overline{P}$ are plotted in Figure 4 for the study sites. Note that due to the symmetry of the distributions of $\overline{ET_a}/\overline{P}$ and $\overline{GR}/\overline{P}$ (Figure 3), the relationships between *MRD* and $\overline{GR}/\overline{P}$ are not shown here for brevity. In addition, the non-parametric Spearman's rank correlation test was also carried out, and the correlation coefficients (r_p) of *MRD* with $\overline{ET_a}/\overline{P}$ and $\overline{GR}/\overline{P}$ are reported in Table 4. Overall, there existed positive correlations between *MRD* and $\overline{ET_a}/\overline{P}$, while negative ones between *MRD* and $\overline{GR}/\overline{P}$ (Table 4). The positive dependence of ET_a on soil moisture levels has been widely observed in the field [12,59]. Meanwhile, finer textured soils tend to have larger water holding capacities and lower *GR* [31], which leads to higher soil moisture levels and *MRD* values for finer soils, and thus the negative correlation between *MRD* and *GR*.

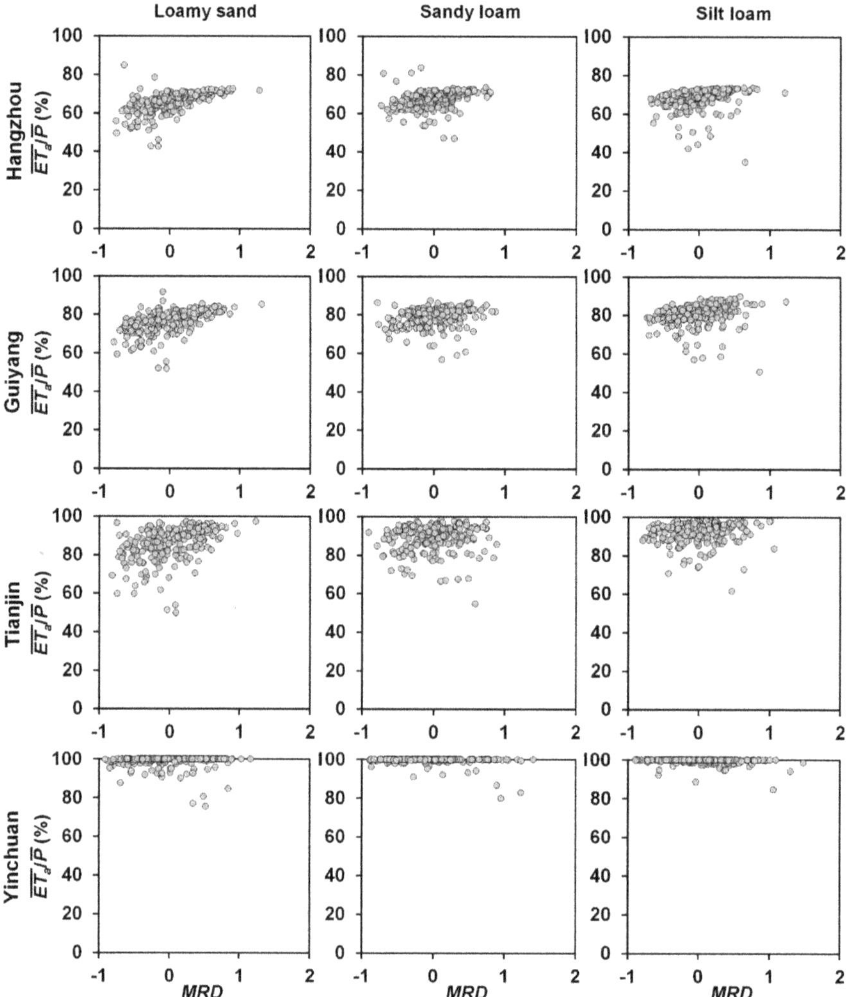

Figure 4. Relationships of mean relative difference (*MRD*) of soil moisture with mean annual ratio of actual evapotranspiration ($\overline{ET_a}$) over precipitation ($\overline{P}$).

More importantly, Figure 4 reveals that the ranges of *MRD* were similar at all sites for different soil textures; whereas, the ranges of $\overline{ET_a}/\overline{P}$ varied substantially across the sites. Specifically, the positive correlation between *MRD* and $\overline{ET_a}/\overline{P}$ was much stronger under wetter climatic conditions (e.g., Hangzhou and Guiyang) than under drier climatic conditions (e.g., Tianjin and Yinchuan; Table 4). In addition to climatic controls, soil hydraulic properties also play significant roles in affecting soil moisture levels, and thus *MRD*. As observed by Wang and Franz [9], the spatial distributions of *MRD* in Utah and the southeast United States were largely determined by soil hydraulic properties (most notably θ_r). Physically speaking, soil moisture levels cannot reach below θ_r under natural conditions. As such, regardless of climatic conditions, soil moisture levels might still vary considerably due to the variability in soil hydraulic properties (Figure 4). In contrast, as previously illustrated, the dependence of ET_a on soil hydraulic properties greatly weakened under drier climatic conditions, leading to the smaller variability in $\overline{ET_a}/\overline{P}$ for different samples (e.g., at Yinchuan). To further demonstrate the varying dependence of MRD and $\overline{ET_a}/\overline{P}$ on soil hydraulic properties with climatic conditions, r_p of MRD and $\overline{ET_a}/\overline{P}$ with different soil hydraulic parameters was computed for sandy loam, and the results are reported in Table 5. It is clear from Table 5 that the control of θ_r on MRD greatly strengthened with increasing climate dryness, while the impacts of soil hydraulic properties on $\overline{ET_a}/\overline{P}$ along the gradient of increasing climate dryness was less obvious. Therefore, owing to the differences in the dependence of MRD and $\overline{ET_a}/\overline{P}$ on soil hydraulic properties, the correlation between MRD and $\overline{ET_a}/\overline{P}$ varied across different climate regimes. Moreover, Table 4 shows that the correlation between MRD and $\overline{ET_a}/\overline{P}$ was also dependent on soil texture. It appears that correlations between MRD and $\overline{ET_a}/\overline{P}$ were stronger for coarser soils (e.g., loamy sand) under the same climatic conditions, probably due to the tighter hydrological connection between vadose zone and land surface processes.

Table 4. Spearman's rank correlation coefficients of mean relative difference (MRD) of soil moisture with mean annual ratios of actual evapotranspiration ($\overline{ET_a}$) and groundwater recharge ($\overline{GR}$) over precipitation ($\overline{P}$).

Location	Loamy Sand		Sandy Loam		Silt Loam	
	$\overline{ET_a}/\overline{P}$	$\overline{GR}/\overline{P}$	$\overline{ET_a}/\overline{P}$	$\overline{GR}/\overline{P}$	$\overline{ET_a}/\overline{P}$	$\overline{GR}/\overline{P}$
Hangzhou	**0.679**	**−0.668**	**0.431**	**−0.429**	**0.461**	**−0.462**
Guiyang	**0.630**	**−0.623**	**0.288**	**−0.279**	**0.378**	**−0.393**
Tianjin	**0.482**	**−0.478**	0.123	−0.145 *	0.188 **	**−0.258**
Yinchuan	**0.255**	**−0.297**	0.227 **	**−0.241**	0.089	−0.158 *

* $p < 0.05$; ** $p < 0.01$; values in bold with $p < 0.001$.

Table 5. Spearman's rank correlation coefficients of soil hydraulic parameters for sandy loam with mean relative difference (MRD) of soil moisture and mean annual ratio of actual evapotranspiration ($\overline{ET_a}$) over precipitation ($\overline{P}$).

Soil Hydraulic Parameters	Hangzhou		Guiyang		Tianjin		Yinchuan	
	MRD	$\overline{ET_a}/\overline{P}$	MRD	$\overline{ET_a}/\overline{P}$	MRD	$\overline{ET_a}/\overline{P}$	MRD	$\overline{ET_a}/\overline{P}$
θ_r	**0.411**	**−0.372**	**0.573**	**−0.369**	**0.734**	**−0.356**	**0.920**	−0.018
θ_s	**0.497**	**0.481**	**0.398**	**0.422**	0.219 **	**0.351**	0.026	−0.095
α	−0.012	−0.176 *	0.026	**−0.307**	0.038	**−0.443**	0.061	**−0.493**
n	**−0.568**	**−0.270**	**−0.578**	**−0.351**	**−0.548**	**−0.363**	**−0.294**	−0.156 *
K_s	**−0.357**	**−0.406**	**−0.263**	**−0.301**	−0.109	−0.172 *	0.008	**0.286**

* $p < 0.05$; ** $p < 0.01$; values in bold with $p < 0.001$.

One of the main aims of the *TSA* method is to identify representative locations for measuring areal average moisture conditions [18]. Here, the criterion of $MRD < \pm 0.05$ was adopted to select representative samples with moisture conditions close to the average [60]. To examine whether the selected representative samples could also produce $\overline{ET_a}/\overline{P}$ values close to the average $\overline{ET_a}/\overline{P}$ of all samples (denoted as $\overline{ET_a}/\overline{P}^*$ hereafter), histograms of $\overline{ET_a}/\overline{P}$ associated with $MRD < \pm 0.05$ are shown in Figure 5 for different soil textures at the study sites. It can be seen from Figure 5 that the resulting number of representative samples with $MRD < \pm 0.05$ were different across the sites and for different soil textures. More importantly, although MRD only changed from −0.05 to 0.05 for the selected representative samples, associated $\overline{ET_a}/\overline{P}$ generally exhibited much wider ranges when compared to $\overline{ET_a}/\overline{P}^*$, except for at the site of Yinchuan. Due to the arid climate at Yinchuan, the spread of associated $\overline{ET_a}/\overline{P}$ was significantly reduced; thus, $\overline{ET_a}/\overline{P}^*$ corresponded well with the majority of associated

$\overline{ET_a}/\overline{P}$. Nevertheless, Figure 5 suggests that representative soil samples with moisture conditions close to the average as identified by the *TSA* method did not necessarily lead to representative soil samples for inferring $\overline{ET_a}/\overline{P}^*$, especially under wet climatic conditions.

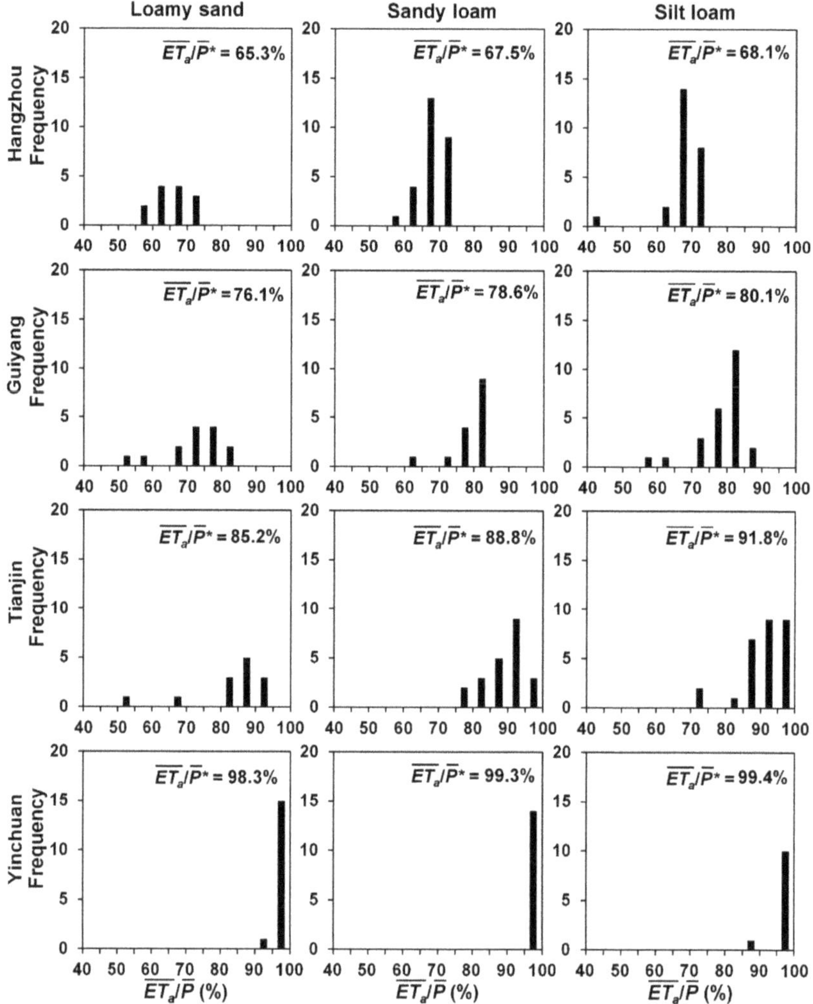

Figure 5. Distributions of mean annual ratio of actual evapotranspiration ($\overline{ET_a}$) over precipitation ($\overline{P}$). Samples with *MRD* < ±0.05 are used here to plot the distributions of $\overline{ET_a}/\overline{P}$. $\overline{ET_a}/\overline{P}^*$ is the average value of $\overline{ET_a}/\overline{P}$ from all samples.

With the aid of the *TSA* concept, the criterion of $(\overline{ET_a}/\overline{P} - \overline{ET_a}/\overline{P}^*) < \pm 0.01$ was used here for choosing representative samples with $\overline{ET_a}/\overline{P}$ close to $\overline{ET_a}/\overline{P}^*$. Using the same method as above, a similar result can be obtained, which shows that in spite of the use of the narrow range of $\overline{ET_a}/\overline{P}$ for selecting representative samples, the distributions of associated *MRD* with $(\overline{ET_a}/\overline{P} - \overline{ET_a}/\overline{P}^*) < \pm 0.01$ were much more scattered. It suggests that samples with average flux conditions might exhibit significant variability in soil moisture conditions.

To further reveal the correlations between average state and flux hydrological variables, an example is given in Figure 6, which shows the relationships of MRD and $\overline{ET_a}/\overline{P}$ with different soil hydraulic parameters for sandy loam at Guiyang. In the figure, samples with $MRD < \pm 0.05$ are highlighted. Part of the reason for the loose connection between average state and flux variables can be attributed to the complex interactions of MRD and $\overline{ET_a}/\overline{P}$ with soil hydraulic parameters, due to the high non-linearity of subsurface processes. Figure 6 shows that MRD from the representative samples was almost irrespective of θ_r, but associated $\overline{ET_a}/\overline{P}$ exhibited a clear negative relationship with θ_r. It is clear that samples with average moisture conditions generally do not correspond with samples with average $\overline{ET_a}/\overline{P}$ values, especially for the study sites with wet climatic conditions.

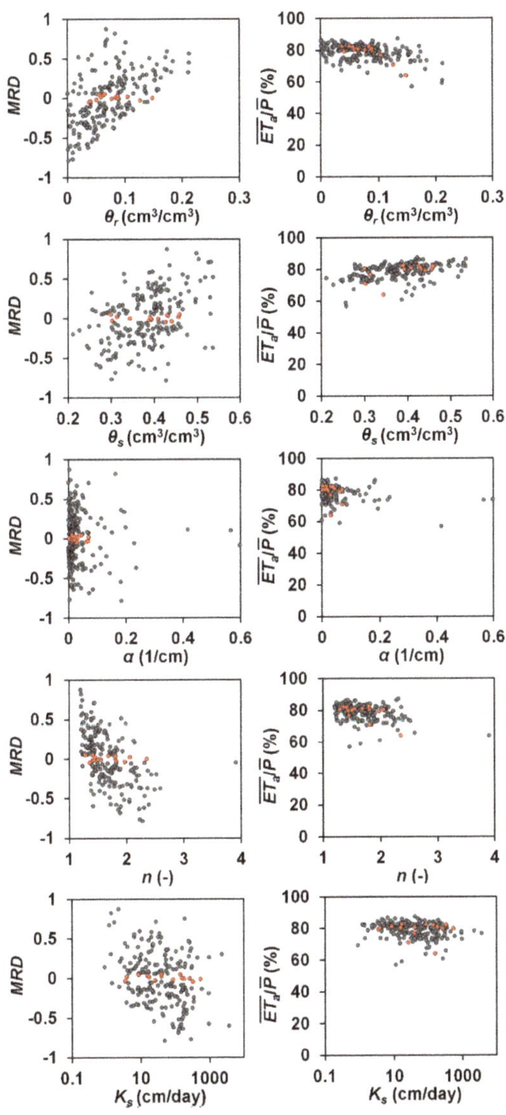

Figure 6. Relationships of MRD and $\overline{ET_a}/\overline{P}$ with different soil hydraulic parameters for sandy loam at Guiyang. Samples with $MRD < \pm 0.05$ are highlighted in red.

4. Conclusions

A synthetic modeling approach with generated soil hydraulic parameters was utilized in this study to investigate the interplays between state and flux hydrological variables across vadose zones in different climate zones. On the basis of the *TSA* method, reasonable soil moisture spatiotemporal structures were reproduced from the model. The simulation results revealed that although *MRD*, ET_a, and *GR* were affected by both climate and soil, their dependence on soil hydraulic properties varied considerably with climatic conditions. In particular, the controls of soil hydraulic properties on ET_a and *GR* greatly weakened under arid climatic conditions due to the limited water supplies, whereas the distribution of *MRD* still largely depended on soil hydraulic properties under the same climatic conditions. As such, the correlations of *MRD* with ET_a and *GR* varied with climatic conditions. Moreover, the modeling results showed that samples with average moisture conditions did not necessarily correspond with those with average flux conditions, which resulted from the complex interactions of soil moisture, ET_a, and *GR* with soil hydraulic properties. Although the *TSA* method has been widely used for selecting representative locations for monitoring areal average moisture conditions, the results of this study underscore the importance of using soil moisture information from multiple sites to infer areal average values of ET_a and *GR*, even with the knowledge of representative soil moisture monitoring sites.

Author Contributions: Conceptualization, T.W.; formal analysis, Z.W., T.W. and Y.Z.; data curation, Z.W. and Y.Z.; writing—original draft preparation, T.W.; writing—review and editing, Z.W.

Acknowledgments: The authors would like to thank the National Centers for Environmental Information for providing the hydrometeorological data used in this study. This study was supported by the National Natural Scientific Foundation of China (U1612441) and the National Key R&D Program of China (2016 YFA0601002). The authors would like to acknowledge the financial support from the Tianjin University, and T.W. also acknowledges the financial support from the Thousand Talent Program for Young Outstanding Scientists for this study.

Conflicts of Interest: The authors declare no conflict of interest.

References

1. Scanlon, B.R.; Keese, K.E.; Flint, A.L.; Flint, L.E.; Gaye, C.B.; Edmunds, W.M.; Simmers, I. Global synthesis of groundwater recharge in semiarid and arid regions. *Hydrol. Process.* **2006**, *20*, 3335–3370. [CrossRef]
2. Robinson, D.A.; Campbell, C.S.; Hopmans, J.W.; Hornbuckle, B.K.; Jones, S.B.; Knight, R.; Ogden, F.; Selker, J.; Wendroth, O. Soil moisture measurement for ecological and hydrological watershed-scale observatories: A review. *Vadose Zone J.* **2008**, *7*, 358–389. [CrossRef]
3. Wang, K.; Dickinson, R.E. A review of global terrestrial evapotranspiration: Observation, modeling, climatology, and climatic variability. *Rev. Geophys.* **2012**, *50*. [CrossRef]
4. Seneviratne, S.I.; Corti, T.; Davin, E.L.; Hirschi, M.; Jaeger, E.B.; Lehner, I.; Orlowsky, B.; Teuling, A.J. Investigating soil moisture–climate interactions in a changing climate: A review. *Earth-Sci. Rev.* **2010**, *99*, 125–161. [CrossRef]
5. Vereecken, H.; Huisman, J.A.; Pachepsky, Y.; Montzka, C.; van der Kruk, J.; Bogena, H.; Weihermüller, L.; Herbst, M.; Martinez, G.; Vanderborght, J. On the spatio-temporal dynamics of soil moisture at the field scale. *J. Hydrol.* **2014**, *516*, 76–96. [CrossRef]
6. Keese, K.E.; Scanlon, B.R.; Reedy, R.C. Assessing controls on diffuse groundwater recharge using unsaturated flow modeling. *Water Resour. Res.* **2005**, *41*. [CrossRef]
7. Szilagyi, J.; Zlotnik, V.A.; Gates, J.B.; Jozsa, J. Mapping mean annual groundwater recharge in the Nebraska Sand Hills, USA. *Hydrogeol. J.* **2011**, *19*, 1503–1513. [CrossRef]
8. Trambauer, P.; Dutra, E.; Maskey, S.; Werner, M.; Pappenberger, F.; van Beek, L.P.H.; Uhlenbrook, S. Comparison of different evaporation estimates over the African continent. *Hydrol. Earth Syst. Sci.* **2014**, *18*, 193–212. [CrossRef]
9. Wang, T.; Franz, T.E.; Zlotnik, V.A. Controls of soil hydraulic characteristics on modeling groundwater recharge under different climatic conditions. *J. Hydrol.* **2015**, *521*, 470–481. [CrossRef]
10. Daly, E.; Porporato, A. A review of soil moisture dynamics: From rainfall infiltration to ecosystem response. *Environ. Eng. Sci.* **2005**, *22*, 9–24. [CrossRef]

11. Kirchner, J.W. Getting the right answers for the right reasons: Linking measurements, analyses, and models to advance the science of hydrology. *Water Resour. Res.* **2006**, *42*. [CrossRef]

12. Vivoni, E.R.; Moreno, H.A.; Mascaro, G.; Rodriguez, J.C.; Watts, C.J.; Garatuza-Payan, J.; Scott, R.L. Observed relation between evapotranspiration and soil moisture in the North American monsoon region. *Geophys. Res. Lett.* **2008**, *35*. [CrossRef]

13. Peters-Lidard, C.D.; Clark, M.; Samaniego, L.; Verhoest, N.E.C.; van Emmerik, T.; Uijlenhoet, R.; Achieng, K.; Franz, T.E.; Woods, R. Scaling, similarity, and the fourth paradigm for hydrology. *Hydrol. Earth Syst. Sci.* **2017**, *21*, 3701–3713. [CrossRef] [PubMed]

14. Jung, M.; Reichstein, M.; Ciais, P.; Seneviratne, S.I.; Sheffield, J.; Goulden, M.L.; Bonan, G.; Cescatti, A.; Chen, J.; de Jeu, R.; et al. Recent decline in the global land evapotranspiration trend due to limited moisture supply. *Nature* **2010**, *467*, 951. [CrossRef]

15. Wang, T.; Franz, T.E.; Yue, W.; Szilagyi, J.; Zlotnik, V.A.; You, J.; Chen, X.; Shulski, M.D.; Young, A. Feasibility analysis of using inverse modeling for estimating natural groundwater recharge from a large-scale soil moisture monitoring network. *J. Hydrol.* **2016**, *533*, 250–265. [CrossRef]

16. Crow, W.T.; Berg, A.A.; Cosh, M.H.; Loew, A.; Mohanty, B.P.; Panciera, R.; de Rosnay, P.; Ryu, D.; Walker, J.P. Upscaling sparse ground-based soil moisture observations for the validation of coarse-resolution satellite soil moisture products. *Rev. Geophys.* **2012**, *50*. [CrossRef]

17. Liu, S.; Xu, Z.; Song, L.; Zhao, Q.; Ge, Y.; Xu, T.; Ma, Y.; Zhu, Z.; Jia, Z.; Zhang, F. Upscaling evapotranspiration measurements from multi-site to the satellite pixel scale over heterogeneous land surfaces. *Agric. For. Meteorol.* **2016**, *230–231*, 97–113. [CrossRef]

18. Vachaud, G.; Passerat De Silans, A.; Balabanis, P.; Vauclin, M. Temporal stability of spatially measured soil water probability density function1. *Soil Sci. Soc. Am. J.* **1985**, *49*, 822–828. [CrossRef]

19. Vanderlinden, K.; Vereecken, H.; Hardelauf, H.; Herbst, M.; Martínez, G.; Cosh, M.H.; Pachepsky, Y.A. Temporal stability of soil water contents: A review of data and analyses. *Vadose Zone J.* **2012**, *11*. [CrossRef]

20. Pachepsky, Y.A.; Guber, A.K.; Jacques, D. Temporal persistence in vertical distributions of soil moisture contents. *Soil Sci. Soc. Am. J.* **2005**, *69*, 347–352. [CrossRef]

21. Starr, G.C. Assessing temporal stability and spatial variability of soil water patterns with implications for precision water management. *Agric. Water Manag.* **2005**, *72*, 223–243. [CrossRef]

22. Guber, A.K.; Gish, T.J.; Pachepsky, Y.A.; van Genuchten, M.T.; Daughtry, C.S.T.; Nicholson, T.J.; Cady, R.E. Temporal stability in soil water content patterns across agricultural fields. *Catena* **2008**, *73*, 125–133. [CrossRef]

23. Brocca, L.; Melone, F.; Moramarco, T.; Morbidelli, R. Spatial-temporal variability of soil moisture and its estimation across scales. *Water Resour. Res.* **2010**, *46*. [CrossRef]

24. Wang, T.; Liu, Q.; Franz, T.E.; Li, R.; Lang, Y.; Fiebrich, C.A. Spatial patterns of soil moisture from two regional monitoring networks in the United States. *J. Hydrol.* **2017**, *552*, 578–585. [CrossRef]

25. Scanlon, B.R.; Healy, R.W.; Cook, P.G. Choosing appropriate techniques for quantifying groundwater recharge. *Hydrogeol. J.* **2002**, *10*, 18–39. [CrossRef]

26. Turkeltaub, T.; Kurtzman, D.; Bel, G.; Dahan, O. Examination of groundwater recharge with a calibrated/validated flow model of the deep vadose zone. *J. Hydrol.* **2015**, *522*, 618–627. [CrossRef]

27. Foolad, F.; Franz, T.E.; Wang, T.; Gibson, J.; Kilic, A.; Allen, R.G.; Suyker, A. Feasibility analysis of using inverse modeling for estimating field-scale evapotranspiration in maize and soybean fields from soil water content monitoring networks. *Hydrol. Earth Syst. Sci.* **2017**, *21*, 1263–1277. [CrossRef]

28. Martínez, G.; Pachepsky, Y.A.; Vereecken, H. Temporal stability of soil water content as affected by climate and soil hydraulic properties: A simulation study. *Hydrol. Process.* **2014**, *28*, 1899–1915. [CrossRef]

29. Wang, T. Modeling the impacts of soil hydraulic properties on temporal stability of soil moisture under a semi-arid climate. *J. Hydrol.* **2014**, *519*, 1214–1224. [CrossRef]

30. Vereecken, H.; Kamai, T.; Harter, T.; Kasteel, R.; Hopmans, J.; Vanderborght, J. Explaining soil moisture variability as a function of mean soil moisture: A stochastic unsaturated flow perspective. *Geophys. Res. Lett.* **2007**, *34*. [CrossRef]

31. Wang, T.; Zlotnik, V.A.; Šimunek, J.; Schaap, M.G. Using pedotransfer functions in vadose zone models for estimating groundwater recharge in semiarid regions. *Water Resour. Res.* **2009**, *45*. [CrossRef]

32. Simunek, J.; Sejna, M.; Saito, H.; Sakai, M.; van Genuchten, M.T. *The HYDRUS-1 D Software Package for Simulating the One-Dimensional Movement of Water, Heat, and Multiple Solutes in Variably-Saturated Media*; Version 4.17; Depart. of Environ. Sci., University of California Riverside: Riverside, CA, USA, 2013; p. 308.

33. Zlotnik, V.A.; Wang, T.; Nieber, J.L.; Šimunek, J. Verification of numerical solutions of the Richards equation using a traveling wave solution. *Adv. Water Resour.* **2007**, *30*, 1973–1980. [CrossRef]

34. Neuman, S.P.; Feddes, R.A.; Bresler, E. *Finite Element Simulation of Flow in Saturated-Unsaturated Soils Considering Water Uptake by Plants*; Third Annual Report; Proj. A10-SWC-77; Hydraul. Eng. Lab., Technion: Haifa, Israel, 1974.

35. Feddes, R.A.; Kowalik, P.J.; Neuman, S.P. *Simulation of Field Water Use and Crop Yield*; John Wiley and Sons: New York, NY, USA, 1978.

36. Ritchie, J.T. Model for predicting evaporation from a row crop with incomplete cover. *Water Resour. Res.* **1972**, *8*, 1204–1213. [CrossRef]

37. Wang, T.; Zlotnik, V.A. A complementary relationship between actual and potential evapotranspiration and soil effects. *J. Hydrol.* **2012**, *456–457*, 146–150. [CrossRef]

38. Myneni, R.B.; Hoffman, S.; Knyazikhin, Y.; Privette, J.L.; Glassy, J.; Tian, Y.; Wang, Y.; Song, X.; Zhang, Y.; Smith, G.R.; et al. Global products of vegetation leaf area and fraction absorbed PAR from year one of MODIS data. *Remote Sens. Environ.* **2002**, *83*, 214–231. [CrossRef]

39. Jackson, R.B.; Canadell, J.; Ehleringer, J.R.; Mooney, H.A.; Sala, O.E.; Schulze, E.D. A global analysis of root distributions for terrestrial biomes. *Oecologia* **1996**, *108*, 389–411. [CrossRef]

40. Small, E.E. Climatic controls on diffuse groundwater recharge in semiarid environments of the southwestern United States. *Water Resour. Res.* **2005**, *41*. [CrossRef]

41. Available online: http://www.ncdc.noaa.gov/cdo-web/ (accessed on 19 June 2019).

42. Hargreaves, G.H.; Samani, Z.A. Estimation potential evapotranspiration. *J. Irrig. Drain. Eng. ASCE* **1982**, *108*, 223–230.

43. Gutiérrez-Jurado, H.A.; Vivoni, E.R.; Harrison, J.B.J.; Guan, H. Ecohydrology of root zone water fluxes and soil development in complex semiarid rangelands. *Hydrol. Process.* **2006**, *20*, 3289–3316. [CrossRef]

44. Vogel, T.; Dohnal, M.; Dusek, J.; Votrubova, J.; Tesar, M. Macroscopic modeling of plant water uptake in a forest stand involving root-mediated soil water redistribution. *Vadose Zone J.* **2013**, *12*. [CrossRef]

45. Wesseling, J.G. *Meerjarige Simulatie van Grondwaterstroming voor Verschillende Bodemprofielen, Grondwatertrappen en Gewassen Met Het Model SWATRE*. Rapport 152; Wageningen, DLO-Staring Centrum. 1991, p. 63. Available online: https://library.wur.nl/WebQuery/wurpubs/fulltext/304226 (accessed on 19 June 2019).

46. Mualem, Y. A new model for predicting the hydraulic conductivity of unsaturated porous media. *Water Resour. Res.* **1976**, *12*, 513–522. [CrossRef]

47. Van Genuchten, M.T. A closed-form equation for predicting the hydraulic conductivity of unsaturated soils. *Soil Sci. Soc. Am. J.* **1980**, *44*, 892–898. [CrossRef]

48. Lawrence, J.E.; Hornberger, G.M. Soil moisture variability across climate zones. *Geophys. Res. Lett.* **2007**, *34*. [CrossRef]

49. Teuling, A.J.; Hupet, F.; Uijlenhoet, R.; Troch, P.A. Climate variability effects on spatial soil moisture dynamics. *Geophys. Res. Lett.* **2007**, *34*. [CrossRef]

50. Zhang, Y.; Schaap, M.G. Weighted recalibration of the Rosetta pedotransfer model with improved estimates of hydraulic parameter distributions and summary statistics (Rosetta3). *J. Hydrol.* **2017**, *547*, 39–53. [CrossRef]

51. Zhang, Y.; Schaap, M.G.; Guadagnini, A.; Neuman, S.P. Inverse modeling of unsaturated flow using clusters of soil texture and pedotransfer functions. *Water Resour. Res.* **2016**, *52*, 7631–7644. [CrossRef]

52. Martínez-Fernández, J.; Ceballos, A. Temporal stability of soil moisture in a large-field experiment in Spain. *Soil Sci. Soc. Am. J.* **2003**, *67*, 1647–1656. [CrossRef]

53. Wang, T.; Franz, T.E. Field observations of regional controls of soil hydraulic properties on soil moisture spatial variability in different climate zones. *Vadose Zone J.* **2015**, *14*. [CrossRef]

54. Yokoo, Y.; Sivapalan, M.; Oki, T. Investigating the roles of climate seasonality and landscape characteristics on mean annual and monthly water balances. *J. Hydrol.* **2008**, *357*, 255–269. [CrossRef]

55. Cosh, M.H.; Jackson, T.J.; Moran, S.; Bindlish, R. Temporal persistence and stability of surface soil moisture in a semi-arid watershed. *Remote Sens. Environ.* **2008**, *112*, 304–313. [CrossRef]

56. Zhao, Y.; Peth, S.; Wang, X.Y.; Lin, H.; Horn, R. Controls of surface soil moisture spatial patterns and their temporal stability in a semi-arid steppe. *Hydrol. Process.* **2010**, *24*, 2507–2519. [CrossRef]

57. Zhang, L.; Dawes, W.R.; Walker, G.R. Response of mean annual evapotranspiration to vegetation changes at catchment scale. *Water Resour. Res.* **2001**, *37*, 701–708. [CrossRef]

58. Wang, T.; Istanbulluoglu, E.; Lenters, J.; Scott, D. On the role of groundwater and soil texture in the regional water balance: An investigation of the Nebraska Sand Hills, USA. *Water Resour. Res.* **2009**, *45*. [CrossRef]

59. Chen, X.; Rubin, Y.; Ma, S.; Baldocchi, D. Observations and stochastic modeling of soil moisture control on evapotranspiration in a Californian oak savanna. *Water Resour. Res.* **2008**, *44*. [CrossRef]

60. Hu, W.; Shao, M.; Han, F.; Reichardt, K.; Tan, J. Watershed scale temporal stability of soil water content. *Geoderma* **2010**, *158*, 181–198. [CrossRef]

Article

Sensitivity of Potential Groundwater Recharge to Projected Climate Change Scenarios: A Site-Specific Study in the Nebraska Sand Hills, USA

Zablon Adane [1,2], **Vitaly A. Zlotnik** [1], **Nathan R. Rossman** [1,3], **Tiejun Wang** [4] **and Paolo Nasta** [1,5,*]

[1] Department of Earth and Atmospheric Sciences, University of Nebraska-Lincoln, Lincoln NE 68588, USA; zablon@huskers.unl.edu (Z.A.); vzlotnik1@unl.edu (V.A.Z.); Nathan.Rossman@hdrinc.com (N.R.R.)

[2] The World Resources Institute, Washington DC 20002, USA

[3] HDR, Omaha, NE 68106, USA

[4] Institute of Surface-Earth System Science, Tianjin University, Tianjin 300072, China; tiejun.wang@tju.edu.cn

[5] Department of Agriculture, Division of Agricultural, Forest and Biosystems Engineering, University of Napoli Federico II, Via Università n. 100, 80055 Portici (Napoli), Italy

[*] Correspondence: paolo.nasta@unina.it

Received: 5 April 2019; Accepted: 30 April 2019; Published: 6 May 2019

Abstract: Assessing the relationship between climate forcings and groundwater recharge (GR) rates in semi-arid regions is critical for water resources management. This study presents the impact of climate forecasts on GR within a probabilistic framework in a site-specific study in the Nebraska Sand Hills (NSH), the largest stabilized sand dune region in the USA containing the greatest recharge rates within the High Plains Aquifer. A total of 19 downscaled climate projections were used to evaluate the impact of precipitation and reference evapotranspiration on GR rates simulated by using HYDRUS 1-D. The analysis of the decadal aridity index (AI) indicates that climate class will likely remain similar to the historic average in the RCP2.6, 4.5, and 6.0 emission scenarios but AI will likely decrease significantly under the worst-case emission scenario (RCP8.5). However, GR rates will likely decrease in all of the four emission scenarios. The results show that GR generally decreases by ~25% under the business-as-usual scenario and by nearly 50% in the worst-case scenario. Moreover, the most likely GR values are presented with respect to probabilities in AI and the relationship between annual-average precipitation and GR rate were developed in both historic and projected scenarios. Finally, to present results at sub-annual time resolution, three representative climate projections (dry, mean and wet scenarios) were selected from the statistical distribution of cumulative GR. In the dry scenario, the excessive evapotranspiration demand in the spring and precipitation deficit in the summer can cause the occurrence of wilting points and plant withering due to excessive root-water-stress. This may pose significant threats to the survival of the native grassland ecology in the NSH and potentially lead to desertification processes if climate change is not properly addressed.

Keywords: scenario-based projections; HYDRUS 1-D; aridity index; water balance; grassland; root water stress

---

Highlights

(1)   Multi-decadal averages of groundwater recharge rates will likely decrease in all emission scenarios despite the region maintaining its historic climate classification as semi-arid.

(2)   In the dry climate scenario analysis, summer is the most sensitive season with a projected reduction in precipitation by 25% and in groundwater recharge rates by 60%.

(3)   The dry climate scenario indicates a notable increase in root water stress and a greater likelihood of desiccation of grass and desertification.

1. Introduction

Socioeconomic drivers have historically determined and will continue to dictate the ever-increasing rate of groundwater depletion around the world. Water consumption in most countries has considerably increased over recent decades due to population and economic growth, with irrigation water use being the leading consumer. In addition to overexploitation, groundwater recharge (GR) regimes are affected by climate change and variability, which threatens environmental sustainability and the livelihoods of billions around the world. The dire water scarcity issues make it imperative to assess the consequences of climate change on water resources at global, continental, and local scales with particular emphasis on GR rates that feed shallow aquifers [1–4]. Groundwater levels of many aquifers have been showing a decreasing trend over the last few decades due to excessive groundwater extraction for irrigation that surpasses GR and replenishing rates [5,6]. The vulnerability of groundwater resources emphasizes the need for stakeholders and decision-makers to have reliable information regarding the relationship between climate and GR rates, particularly in semi-arid regions where water scarcity issues are well documented [6].

The assessment of climate change can vary in the type of impact, scale, and severity. The negative consequences of climate change in some settings can prove to be pronounced and include severe droughts [7], sea-level rise [8], wildfires [9], extreme heat waves [10], floods and inundation [11], reduction of biodiversity [12], and many others.

The impact of climate change on the water budget in semi-arid regions primarily occurs via changes to potential evapotranspiration (ET) rates and precipitation (P) regime [13]. The rise in temperatures can cause increases in potential ET [14]. Similarly, climate change has been reported to impact the amount and seasonality of P regime [15] which can alter surface runoff and streamflow [16–19], soil moisture [20], and GR rates [21–23].

Climate change can impact GR rates by altering the availability of water through changes in P and temperature, as well as in the composition of vegetation [24]. While on the one hand some studies have found increases in GR rates in areas experiencing permafrost thaws due to temperature rise and increase in P [25,26], on the other hand, several studies have estimated decreased recharge as a result of reduced P or greater temperatures [23,27,28]. Further, Bouraoui et al. (1999) [29] reported considerable reductions in GR near Grenoble, France, predominantly due to increases in evaporation during the recharge season. The extent to which GR is affected by climate warming will also depend on the local soils and vegetation in addition to climate [30].

The objective of this paper is to examine the relationship between historic and projected future climate forcings (in the form of precipitation, reference evapotranspiration, and aridity index) and the GR rates in a site-specific study in the Nebraska Sand Hills (NSH). This groundwater-dependent grassland ecosystem represents the largest stabilized sand dune formation in the western hemisphere. One of the largest grass-stabilized dune regions in the world, the NSH provides the greatest recharge rates within the High Plains Aquifer under semi-arid climate conditions. This paper is a follow-up of four studies in the NSH region [31–34]. The remainder of this paper is organized as follows: Section 2 presents a brief description of the study area, Nebraska National Forest (NNF), located in the NSH and modeling methodology based on HYDRUS-1D. Section 3 discusses the analysis of climate change impacts on GR rates by comparing the historic and projected water balance at a decadal, annual and sub-annual resolution within a probabilistic framework. Section 4 provides a concluding summary of the analysis presented herein.

2. Study Area and Modeling Approach

The study site is located in a century-old experimental forest (Nebraska National Forest, near the town of Halsey, Nebraska, USA) surrounded by mixed-grass prairie in the south-central part of the NSH (Figure 1). The NSH landscape is comprised mainly of eolian sand dunes that were deposited as recently as a few thousand years ago [35]. The soil is approximately 92%–97% sand [36], which contributes to the greatest recharge rates in the High Plains Aquifer [37]. The native vegetation of the NSH region consists of mixed-prairie grassland including little bluestem (*Schizachyrium scoparium*), switchgrass (*Panicum virgatum*), sand dropseed (*Sporobolus cryptandrus*), and Kentucky bluegrass (*Poa pratensis*), and is suitable to the historic land uses of ranching and cattle grazing. A weather station (Longitude −100°15′00″, Latitude 41°9′00″, elevation 824 m) recorded daily values of precipitation, wind speed, air temperature, relative humidity and solar radiation from 1990 to 2012. The climate is semi-arid continental with annual precipitation (P) ranging from 36.0 to 82.0 cm yr^{-1}, and with reference crop evapotranspiration (ET$_0$, referred to a well-watered hypothetical reference alfalfa-crop) ranging from 128.5 to 166.7 cm yr^{-1}. We remind that with the availability of all meteorological data, we used the well-known standard Penmann-Monteith equation for computing ET$_0$ as recommended by the Food and Agriculture Organization (FAO), in Paper no. 56 [38]. Overland flow is negligible due to the high infiltration capacity of the sandy soils. Other details are described by Adane and Gates (2015) [39]. The native grassland plot (Longitude −100°21′16″, Latitude 41°50′41″, elevation 860 m), denoted with the acronym G, represents the natural grassland ecosystem in the NSH [32,33].

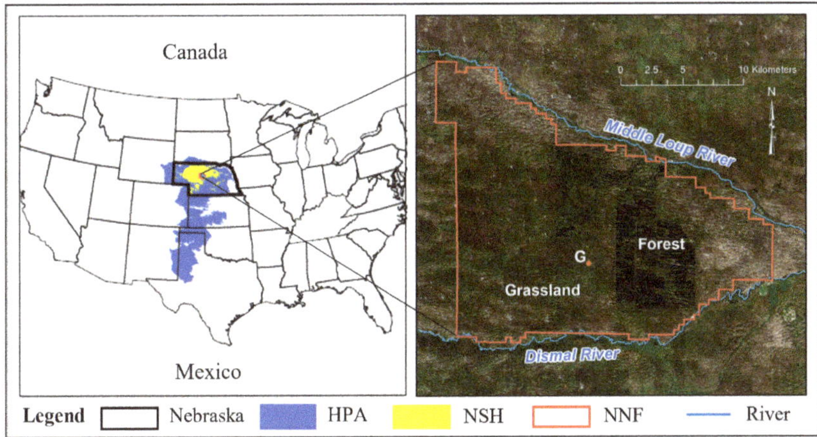

Figure 1. Location of the study area—Nebraska National Forest, (NNF), Nebraska Sand Hills (NSH), High Plains Aquifer (HPA), experimental grassland plot site (G), and Middle Loup and Dismal rivers.

The effect of climate change on potential GR rates was assessed using a scenario-based approach by considering historic (1950–2000) and plausible climate projections (2000–2100). Both historic and projected climate data contained daily values of P, and minimum and maximum temperatures obtained from the High Plains Regional Climate Center (HPRCC). In this case, la ack of data for wind speed, relative humidity, and solar radiation made it a necessity to use the temperature-based Hargreaves equation [40] to estimate reference crop evapotranspiration (ET$_0$). The formula only requires minimum and maximum temperature data while the extraterrestrial radiation term is estimated by using study site latitude and day of the year. Nonetheless, the performance of the Hargreaves equation to estimate ET$_0$ has considerable limitations given the limited input dataset [41]. For this reason, the performance of Hargreaves formula was tested against the Penman-Monteith equation, by using the meteorological data at Halsey weather station. In doing so, ET$_0$-values based on solely temperature data were compared with those based on a complete set of meteorological data from 1990 till 2012. The results indicate that the Hargreaves equation underestimates annual average ET$_0$ (119.2 cm yr^{-1}) compared to ET$_0$ values estimated by Penman-Monteith equation (145.5 cm yr^{-1}) [42]. The root mean square error (RMSE) and R^2-value calculated on the daily ET$_0$-values are 0.11 cm d^{-1} and 0.88, respectively. Archibald and Walters (2014) [43] found similar and acceptable performance at Mead station in eastern Nebraska (Latitude 41°17′00″). A total of 19 climate models were used in this study (Table 1).

Table 1. Selected 19 global climate models used in this study and corresponding institutions.

	Acronym	Institution
1	BCC_CSM	Beijing Climate Center (China)
2	CANESM	Canadian Centre for Climate Modelling and Analysis (Canada)
3	CCSM	University Corporation for Atmospheric Research (UCAR) (USA)
4	CESM_BGC	NCAR Earth System Laboratory (USA)
5	CNRM_CM	National Centre for Meteorological Research
6	CSIRO	Commonwealth Scientific and Industrial Research Organization (Australia)
7	GFDL_CM	Geophysical Fluid Dynamics Laboratory (USA)
8	GFDL_G1	Geophysical Fluid Dynamics Laboratory (USA)
9	GFDL_M1	Geophysical Fluid Dynamics Laboratory (USA)
10	INMCM	Institute of Numerical Mathematics (Russia)
11	IPSL_CM	Institute Pierre-Simon Laplace (France)
12	IPSL_MR	Institute Pierre-Simon Laplace (France)
13	MIROC_ESM	Japan Agency for Marine-Earth Science and Technology (Japan)
14	MIRCO_ESM	Japan Agency for Marine-Earth Science and Technology (Japan)
15	MICROC	Japan Agency for Marine-Earth Science and Technology (Japan)
16	MPI_LR	Max Planck Institute (Germany)
17	MPI_MR	Max Planck Institute (Germany)
18	MRI_GCM	Meteorological Research Institute (Japan)
19	NORESM	Norwegian Climate Center (Norway)

Each climate model considered four plausible scenarios based on future greenhouse gas concentration trajectories (Representative Concentration Pathways, RCPs = 2.6, 4.5, 6.0, and 8.5). The four RCPs represent possible ranges of radiative forcing values in the year 2100 relative to pre-industrial values (+2.6, +4.5, +6.0, and +8.5 W/m^2, respectively). RCP 2.6 assumes that global annual greenhouse gas emissions peak between 2010 and 2020 and decline substantially thereafter (optimistic scenario). Emissions in RCP 4.5 and RCP 6.0 peak around 2040 and 2080 then decline, respectively while the RCP 8.5 scenarios assume emissions continue to rise throughout the 21st Century (worst-case scenario). A total of 76 climate projections (four emission scenarios over 19 climate models) were expected, but only 65 were available because some climate models did not provide time series of daily P, or minimum and maximum temperatures under RCPs 4.5 and 6.0 emission scenarios. The aridity index, AI is described as the annual P over the annual ET_0 ($AI = P/ET_0$) and is commonly used for climate classification [44–47]. This climate indicator was widely used for evaluating the effects of climate change on overall water balance [48–50].

The water balance in the soil-plant-atmosphere system was numerically evaluated using HYDRUS 1-D [51]. Numerical simulations at daily temporal resolution utilized available P data and the calculated ET_0 from the Hargreaves equation.

The model setup in HYDRUS 1-D requires vegetation parameters, which were determined by direct measurements or were obtained from tabulated values in the scientific literature [33]. It is worth noting that considering net P (P_{net}) instead of total precipitation accounts for precipitation that was intercepted by foliage cover and does not reach the soil surface. The leaf area index (LAI) determines the amount of precipitation interception that is subtracted from P in order to obtain P_{net} falling on the soil surface [52].

The crop coefficient (K_c) of grass converts ET_0 into crop potential evapotranspiration (ET_p). The LAI is used for partitioning ET_p into potential evaporation (E_p) and potential transpiration (T_p) [33]. Therefore, P_{net} and E_p represent the system-dependent time-variable water fluxes of the upper boundary condition whilst T_p determines the potential root water uptake and depends on maximum root depth (z_r) and root distribution (assumed to be linear along the soil profile by varying from maximum at the soil surface to minimum at z_r). Both E_p and T_p are reduced by water stress conditions into actual evaporation (E_a) and actual transpiration (T_a) depending on soil moisture storage status. A set of five unknown soil hydraulic parameters (SHPs) belonging to the water retention curve and the hydraulic conductivity curve [53], denoting residual soil water content, saturated soil water content, two water retention shape parameters, and saturated hydraulic conductivity, were calibrated and validated in the study presented by Adane et al. (2018) [33]. The depth of the soil profile was assumed to be 220 cm following Adane et al. (2018) [33], and the lower boundary condition is set to free drainage representing potential GR, as the depth to the water table is at least 4 m in the NNF [31]. The assessment of the lag time will provide an estimate of the actual GR to the water table, rather than just draining through the root zone [31]. Initial conditions were set in terms of pressure head values along the soil profile to impose hydraulic equilibrium.

The total number of numerical simulations in HYDRUS-1D included:

(a)   one numerical simulation under historic climate conditions (1950–2000);

(b)   sixty-five numerical simulations under future climate projections (2000–2100).

Actual transpiration (T_a), actual evaporation (E_a), and GR rates represent the model outputs that were stored at daily time resolution and subsequently aggregated to monthly and annual sums.

In this study, the Student's t-distribution is used as an alternative to the normal distribution when sample sizes are small to estimate the expected value of P, ET_0, ET_p, AI, E_a, T_a and GR. The Student's t-distribution is used to construct the 95% confidence interval ($\alpha = 0.05$) around the sample mean $\bar{x}$ in each emission scenario to infer the expected value (μ) of a continuously distributed population:

$$\mu = \bar{x} \pm t_{\alpha,n-1} \frac{s}{\sqrt{n}} \tag{1}$$

where n is the number of observations (degrees of freedom are therefore defined as $n - 1$), and s is the sample standard deviation, and $t_{\alpha,n-1}$ is student's t-value at 95% confidence interval and $n - 1$ degrees of freedom.

3. Results and Discussion

This Section is divided into three sub-sections. The first sub-section presents the comparison between historic and future decadal-average GR rates in response to varying climate forcings. The second part presents the annual average GR rates. Finally, in the third sub-section, three representative scenarios were selected out of 65 available climate projections to compare results at sub-annual time resolution.

3.1. Multi-Decadal-Average Results of Historical and Projected ET_a and GR

Multi-decadal-average values (referring to the average of annual values of 2000–2100) of climate forcings (P, ET_0, ET_p), aridity index (AI) and model output (ET_a and GR) for historic and climate projections (16 projections in RCP2.6, 18 projections in RCP4.5, 12 projections in RCP6.0, 19 projections in RCP8.5) are listed in Table A1 (in the Appendix A). Given that the climate models were relatively few, the observations were assumed to be distributed normally (P, ET_0, AI, ET_p, ET_a) or lognormally (GR).

The historical average values of all 50 years (1950–2000) are presented in the first row of Table A1. The historical aridity index value was AI = 0.49 corresponding to the semi-arid class ($0.2 < \text{AI} \le 0.5$) based on aridity index classification reported in Spinoni et al. (2015) [54]. The simulated historical data indicate that ET_p and ET_a are 69% and 40% of the ET_0, respectively. Considering negligible losses to precipitation interception and overland flow, baseline GR was approximately 16% of the historic P. This result agrees with previous GR estimates in the Nebraska Sand Hills [33,39,55,56] that on average estimated approximately 10 cm yr^{-1} under native grassland conditions. However, spatial differences may be important across the Sand Hills as estimated net GR rates based on remote-sensing and other ancillary climate data. For instance, data presented by Szilagyi et al. (2011) [57] for the period 2000–2009 showed that GR rates varied spatially from approximately −0.3 to 12.5 cm yr^{-1} where the negative values indicate that in some parts of the NNF annual ET was greater than P. Detailed comparisons of historical GR rates in the Sand Hills in general and NNF, in particular, are available in Adane and Gates (2015) [39].

In all emission scenarios evaluated, the projected rise in temperature generated a corresponding rise in ET_0, resulting in projected multi-decadal-averages of 123.8 ± 1.8 cm yr^{-1}, 126.5 ± 1.9 cm yr^{-1}, 127.3 ± 2.3 cm yr^{-1}, and 149.8 ± 8.7 cm yr^{-1}, for RCP2.6, RCP4.5, RCP6.0 and RCP8.5, respectively (see Equation (1)). The mean values correspond to an 8.5%, 10.9%, 11.6%, and 31.4% increase, respectively, when compared to the historic value (114.1 cm yr^{-1}). It is also worth noting that the uncertainty (given by the confidence interval around the mean quantified by Equation (1)) in ET_0 rise also increased with increasing CO_2 concentration projected, with the greatest uncertainty observed under the worst-case climate scenario (RCP8.5). Further, P is characterized by a slight increase (about 1% for all four emission scenarios) when compared to the decadal-average historic value (55.7 cm yr^{-1}). This result agrees with the conclusion by Lauffenburger et al. (2018) [58] that the projected change in P between 1990 and 2050 in western Nebraska (Northern High Plains) is a statistically non-significant increase.

Due to the substantial differences in the rates of change between ET_0 and P, the use of a synthetic climate measure such as the aridity index (AI) is effective to evaluate recharge rates between the historic and climate change projections. Figure 2a shows the box plots of the projected multi-decadal-average AI-values obtained from the 19 climate models for each emission scenario. The AI values resulting from the analysis were 0.47 ± 0.014, 0.45 ± 0.015, 0.46 ± 0.021, and 0.40 ± 0.029, for RCP2.6, RCP4.5, RCP6.0 and RCP8.5, respectively (see Equation (1)). These results suggest that the region is projected to become more arid (with decreasing AI-values) compared to the historical baseline (AI = 0.49) but still remaining in the same semi-arid class [59]. The greatest change in aridity is likely to occur only under the RCP8.5 climate projection scenario.

The ANOVA test at 5% significance level rejected the hypothesis that the mean AI values of the four groups were equal to each other given that F-value = 12.75 was larger than the critical F-value. However, if the AI-values of the RCP8.5 were removed, the ANOVA test would accept the hypothesis that the mean values of the RCP2.6, RCP4.5 and RCP6.0 group were significantly equal. This statistical analysis suggests that the change in aridity index under the climate change scenarios was only significant under the worst-case scenario.

The analysis also indicates that all climate scenarios caused a decline in GR rates compared to the historical baseline (9.02 cm yr^{-1}). Projections of climate change induced a general decrease in the simulated potential GR, resulting in future period multi-decadal-averages of 7.15 ± 0.60 cm yr^{-1}, 6.87 ± 0.71 cm yr^{-1}, 6.95 ± 0.74 cm yr^{-1}, and 5.51 ± 0.78 cm yr^{-1}, for RCP2.6, RCP4.5, RCP6.0 and RCP8.5, respectively (see Equation (1)). The mean values represent decreases ranging from 21% in RCP2.6 to 39% in RCP8.5 in terms of percentage values (see box plots in Figure 2b). Nevertheless, box plots denote high uncertainty around the median value, with a coefficient of variation spanning from 18% to 30% for RCP2.6 and RCP8.5, respectively.

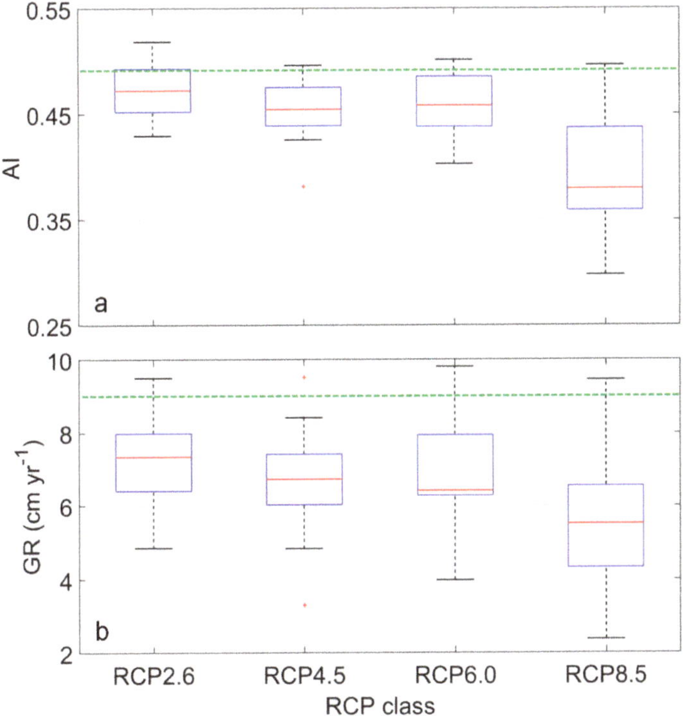

Figure 2. Box plots of annual-averages (**a**) aridity index (AI) values and (**b**) groundwater recharge (GR) values classified according to the four RCPs. The horizontal dashed green line depicts the historic value.

Lauffenburger et al. (2018) [58] reported that median annual recharge for the low (+1.0 °C, 478 ppm CO_2) and high (+2.4 °C, 567 ppm CO_2) global warming scenarios in western Nebraska rangeland are projected to decrease by about 9% and 94% relative to historical values, respectively.

The decrease in potential GR rates is due to a substantial increase in ET losses. This climatic setting also poses questions on probabilities related to plant survival, covered in Section 3.3.

3.2. Annual-Average Results of Historical and Projected AI, ET_a, and GR

The model simulations comprise 50 years of historical and 100 years of 65 climate projections of climate forcings (P, ET_0, ET_p), aridity index (AI), and model output ET_a and GR. The resulting annual-averages covering the projected 100-year simulation period are presented within a probabilistic framework shown in empirical histograms of AI values and corresponding GR (Figure 3). The normal and lognormal distributions (red lines) were fit on annual AI values and GR rates (blue histograms).

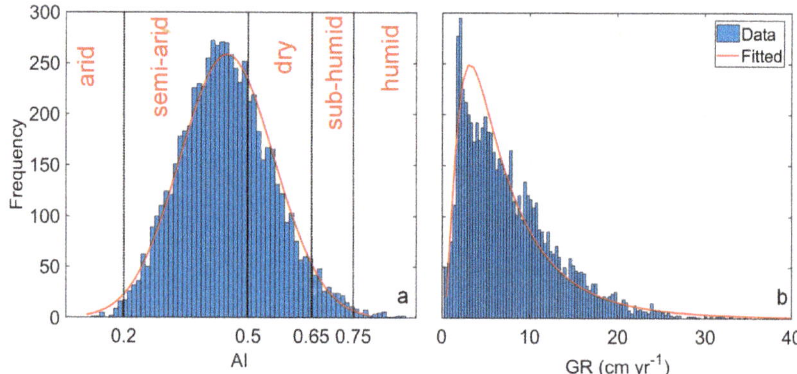

Figure 3. Frequency distribution (blue histograms) and fitted distribution (red lines) for (**a**) normally distributed aridity index, classified into arid, semi-arid, dry, sub-humid, and humid classes; and (**b**) lognormally distributed potential groundwater recharge rates.

The probabilities for each climate class and corresponding median potential GR rates are reported in Table 2. Approximately 69% of the total aridity index values fall in the semi-arid climate class and are associated with the median potential GR rate (GR = 5.19 cm) that is significantly lower than the historic GR (9.02 cm yr^{-1}, see Table A1). The probabilities of the climate shifting from semi-arid to dry (0.5 < AI ≤ 0.65) are approximately 26% and would produce a median potential GR of 9.64 cm yr^{-1}, which is slightly greater (4.6%) than the historic value (9.02 cm yr^{-1}). The combined probabilities attributed to extreme conditions such as arid, sub-humid, and humid climates were very low (~5%). The median-annual GR rates for the extreme climate conditions were far from the baseline rates.

Table 2. Probabilities of annual AI-values and associated median annual potential groundwater recharge (GR) rates.

Value	Class	Probability	Median GR
		%	cm yr^{-1}
AI ≤ 0.2	Arid	0.56	1.87
0.2 < AI ≤ 0.5	Semi-arid	68.58	5.19
0.5 < AI ≤ 0.65	Dry	26.24	9.64
0.65 < AI ≤ 0.75	Sub-humid	3.76	14.19
A > 0.75	Humid	0.86	21.93

The coarse sandy soils in the Nebraska Sand Hills are not conducive to overland flow, thus the water balance of the soil profile is likely controlled by vertical hydrological processes, namely infiltration and ET. Moreover, precipitation is one of the main drivers of GR under semi-arid climate conditions [60]. The relationship between annual P and GR was quantified through regression lines and the corresponding coefficient of determination, namely the R-squared (R^2), for evaluating its goodness of fit for both the historic and projected data group. The data are presented in Table 3 and visualized through a scatter plot in Figure 4, where the circles are colored according to the aridity index values.

Table 3. Regression lines and corresponding goodness of fit (R^2) for historical data and the four climate projections (RCP2.6, RCP4.5, RCP6.0, and RCP8.5).

Scenario	Regression Line	R^2
Historic	GR = 0.335P − 9.06	0.407
RCP2.6	GR = 0.2767P − 7.87	0.351
RCP4.5	GR = 0.2829P − 8.37	0.371
RCP6.0	GR = 0.2794P − 8.33	0.351
RCP8.5	GR = 0.2419P − 7.64	0.349

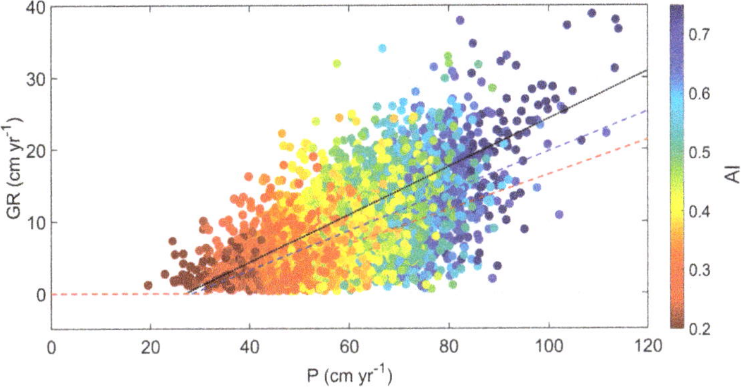

Figure 4. Scatter plot representing the relationship between annual-average P and GR with circle colors determined by the aridity index. The regression lines are showed for historic data (dotted black line), RCP2.6 (dashed blue line) and RCP8.5 (dashed red line).

Even though the regressions are clearly affected by heteroscedasticity (with higher spread around the regression line for higher AI-values), the results presented in Figure 4 and Table 3 represent a good opportunity to roughly estimate potential GR rates from projected P with annual temporal resolution. Given the R^2-values listed in Table 3, P is able to explain about 30%–40% of GR temporal variability. It is apparent that all projected regression lines lie below the historic line (dotted black line), which means that GR rates will likely decrease, especially under the worst-case scenario (RCP8.5, dashed red line), as can be observed from the slopes of the regression lines reported in Table 3.

3.3. Selection of Three Representative Climate Projections for Assessing the Relation between Precipitation and Groundwater Recharge Rates at Sub-Annual Time Scale

Detailed results can be explored at sub-annual resolution. Nonetheless, it was not practical to analyze simulations pertaining to all climate projections. Therefore, for the purpose of encompassing the full spectrum of most probable future climate projections, three representative climate scenarios were selected following the procedure recommended by Rossman et al. (2018) [34]. The temporal dynamics of all cumulative GR rates, colored according to RCP, are shown in Figure 5a. The corresponding frequency distribution of final cumulative GR rates recorded for the year 2100 with sample mean, $\bar{x}$, and standard deviation, s, is displayed in Figure 5b. The dry, mean, and wet GR projections selected based on this analysis were CSIRO (RCP8.5), GFDL_M1 (RCP2.6), and MPI_MR (RCP4.5), respectively (Table A1), based on the closest GR rates identified using the $\bar{x} - s$, $\bar{x}$ and $\overline{x + s}$ distribution criteria, respectively (Figure 5b).

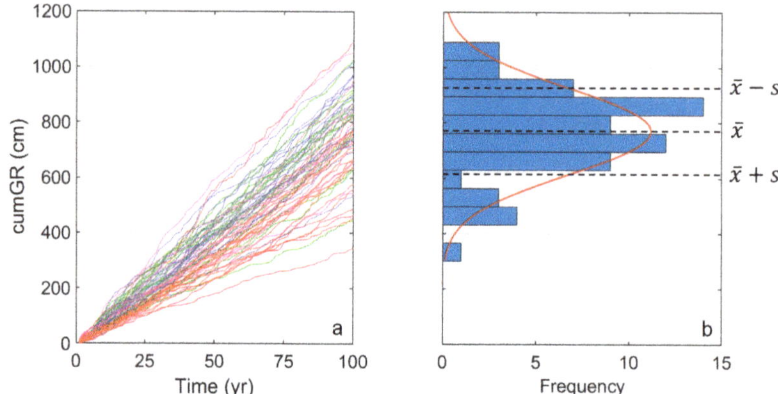

Figure 5. (**a**) Dynamics of cumulative GR rate projections from 2000 to 2100 under the four emission scenarios (blue, green, magenta and red lines represent RCP2.6, RCP4.5, RCP6.0 and RCP8.5, respectively); (**b**) frequency distribution (blue histograms) of the final cumulative GR-values with corresponding fitted normal distribution (red line) with mean ($\bar{x}$) and standard deviation (s) that are used to select three representative climate projections (dry, mean and wet given by $\bar{x} - s$, $\bar{x}$, and $\bar{x} + s$, respectively)

The annual time series of ET_a, P, ET_0, and potential GR rates of the three representative scenarios are shown in Figure 6. While the projected multi-decadal-averages of P are 54.0 cm yr^{-1}, 59.4 cm yr^{-1}, 58.9 cm yr^{-1}, in dry, mean, wet scenarios, respectively, the three ET_0 time series are characterized by time-variant increasing rates as predicted in other semi-arid environments [61]. Thus, the corresponding AI time series (not shown) suggests that the climate might suffer from an increasing aridity even though the progressive shift does not change the projected climate class from the historical designation (semi-arid). The results of the analyses are consistent with Oglesby et al. (2015) [62] that suggested that while the form and seasonality of P will change in Nebraska, the annual amount of P may not vary substantially. Similarly, the report forecasts that Nebraska will likely experience temperature increases ranging from 4 to 5 °F under low emission scenario and 8 to 9 °F in the high emission scenario. Both emission scenarios would result in greater aridity index given that P would remain relatively stable.

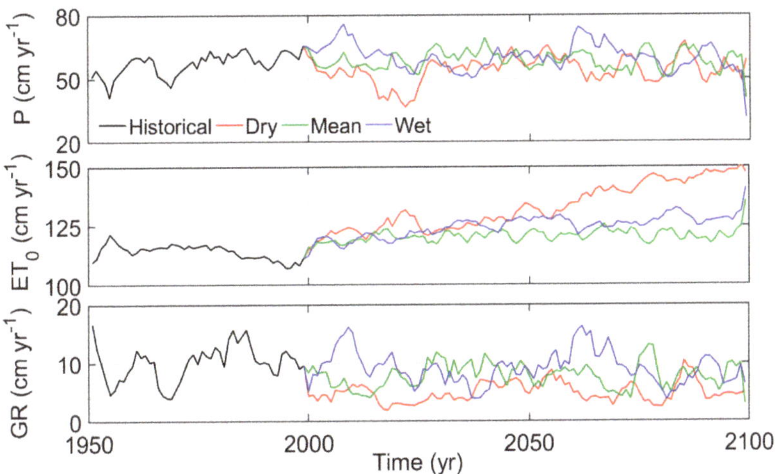

Figure 6. Historic and projected trends (dry, mean and wet projections) in annual averages of precipitation (P), reference evapotranspiration (ET$_0$), and potential groundwater recharge (GR) rates.

Annual-average groundwater recharge in the wet, mean, and dry projections are 8.40 cm yr^{-1}, 6.74 cm yr^{-1}, and 3.94 cm yr^{-1}, representing −7%, −25% and −56% changes relative to the historical baseline (9.02 cm yr^{-1}). Crosbie et al. (2013) [56] estimated a baseline historical recharge of 7.8 cm yr^{-1} for the Northern High Plains. The selected three future climate scenarios, namely dry, median, and wet correspond to 10%, 50%, and 90% exceedances for low (+1.0 °C, 478 ppm CO_2), medium (+1.7 °C, 523 ppm CO_2), and high (+2.4 °C, 567 ppm CO_2) global warming scenarios. The study reported that GR was projected to be 5.93 cm yr^{-1} (−24%), 8.42 cm yr^{-1} (+8%), and 10.30 cm yr^{-1} (+32%) under the dry, median, wet scenarios, respectively. Rossman et al. (2018) [34] also found an increase in projected GR under both the median and wet projections at a rate of 0.27 cm yr^{-1} (+5%) and 2.21 cm yr^{-1} (+30%), and a decrease under the dry projection by 1.50 cm yr^{-1} (-40%) averaged over 2010–2099 relative to the baseline averaged over 2000–2009. However, it is worth noting the differences in the baseline historical recharge rates used between this analysis (9.02 cm yr^{-1}) for NNF and Rossman et al. 2018 [34] (7.0 cm yr^{-1}) for NSH. The difference in projected recharge rates may also have occurred due to simulation scales (spatial and temporal) and differences in the modeling approach. Spatial resolution differences can create discrepancies in the historical averages of climate forcings and the resulting parameters such as ET$_a$ and GR. While the focus of this study is on the NNF site, Crosbie et al. (2013) [56] and Rossman et al. (2018) [34] provide averaged values over Northern High Plains and the NSH, and make use of WAVES, a one-dimensional point-scale model that simulates the fluxes of mass and energy between the atmosphere, vegetation, and soil systems [63] and the variable infiltration capacity (VIC) macroscale hydrology model [64], respectively. Further, differences can also arise from the increased number of climate models used in this analysis, as opposed to the number of climate models used in Crosbie et al. (2013) [56] and Rossman et al. (2018) [34]. Despite these differences, GR rates are nearly equivalent to those reported in Crosbie et al. (2013) [56] when only looking at results derived from the same climate models. Discrepancies will continue to exist when comparing GR rates derived from averaging all climate projections over the entire NHP and NSH spatial extents. To describe and test the entirety of reasons for differences among previous studies was not an objective of this investigation, but it could be a focus for future research. In semi-arid regions, the impact of precipitation on GR is more important when analyzing the seasonal distribution rather than the annual sums [65]. The distribution of monthly sums of precipitation, actual ET, and potential GR highlights interesting aspects. Historic precipitation, actual ET and potential GR follow similar temporal trends with maxima in May (P = 9.07 cm), June (ET$_a$ = 9.82 cm), and June (GR = 1.30 cm). Minima occur in winter months likely related to negligible precipitation, low temperatures, and freezing conditions. The projected monthly distributions under the mean and wet scenarios roughly reproduce similar trends for P, while the dry scenario generates precipitation lower than the historic baseline during the summer seasons and higher than the historic baseline during the winter seasons. The dry climate projection indicates that summer months produce 17.4 cm yr^{-1} of P, which is a 25% reduction in summer month P compared to the historic value (23.1 cm yr^{-1}). This significant water deficit is reflected in a 60% loss in GR rates in the summer, which is the major season for groundwater replenishment in this region. Spring is the next most important season for GR. However, GR was also substantially reduced in April and May months due to the projected excessive actual ET demand (+25%) in the dry scenario. Outside of the summer and spring months of the year, the values for the water balance components observed under the historic time period and the dry scenario projection were similar.

The water deficit detected by using monthly sums (Figure 7) poses the risk of threatening plant survival in this site. In this study, we applied time-invariant parameters for root water uptake as we assumed no change in grassland cover in HYDRUS-1D [17]. Prolonged root water stress is known to cause plant wilting and subsequent withering. Increasing chances of root water stress induced by a change in climate will have consequences to the native grassland ecology of the Nebraska Sand Hills. In this study, daily values of root water stress were explored in the historical data and the mean, wet, and dry scenario projections. Frequency distribution of decimal logarithmic values (18,262 daily values) of pressure head at root depth was used to explore the impact of climate change on grassland root water stress. The results were fitted by using a nonparametric kernel probability smoothing distribution.

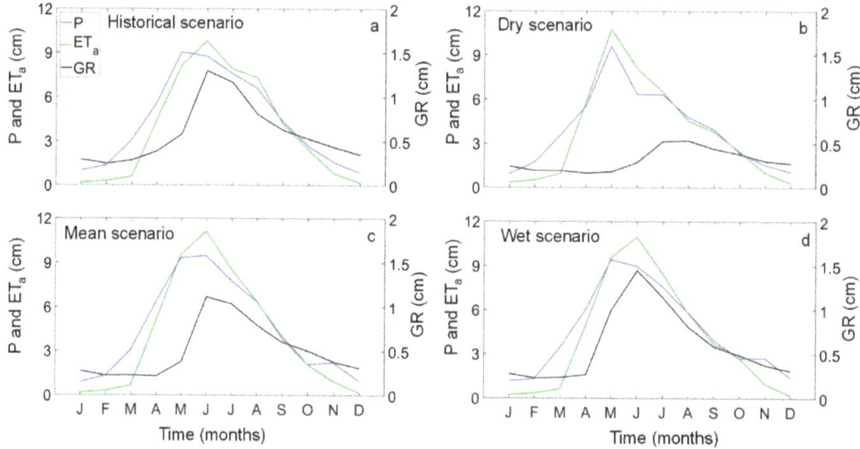

Figure 7. Monthly averages of precipitation (blue line), actual evapotranspiration (green), and potential groundwater recharge (GR; black line) under (**a**) historical records, (**b**) dry, (**c**) mean and (**d**) wet climate projections. P and ET$_a$ are referred to in the primary y-axis while GR is referred to in the secondary y-axis.

The root water stress analysis was performed within a probabilistic framework, where the probabilities associated with no-stress, stress, and wilting point were considered (Table 4). The analysis indicates that the historic conditions resulted in equivalent percentages of root water stress and no stress conditions (about 49% and 47%, respectively) as well as low probabilities (4.3%) to reach wilting points. While the root water stress probabilities under the mean and wet climate projections were equivalent to the historic condition, the wilting point probabilities under the projections nearly double from 4.3% to 8.5% (Table 4). In contrast, the dry scenario analysis indicates that while the stress probability remains equivalent, the no stress probability is considerably reduced from 48.8% to 35.1%. Of particular interest is also a significant increase in the probability of wilting point (Figure 8).

The wilting point probability under the dry climate scenario increased about four times compared to the historic condition. This analysis indicates that the conditions that would be induced by the dry scenario will increase the risk for grass to die and increase the chance of desertification of this region substantially.

Table 4. Statistical probabilities (%) of no root water stress, root water stress, and wilting point conditions under historic records and dry, mean, wet climate projections.

Scenario	No Stress	Stress	Wilting Point
	%	%	%
Historic	48.8	46.9	4.3
Dry	35.1	46.4	18.5
Mean	46.1	45.4	8.5
Wet	46.9	44.8	8.3

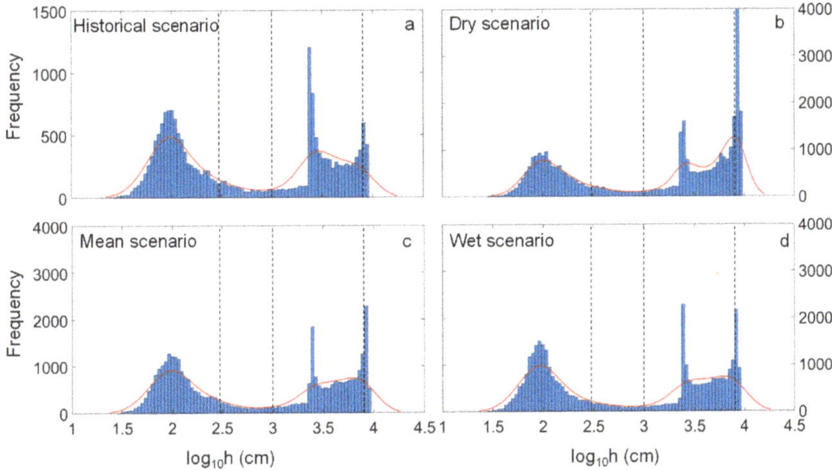

Figure 8. Frequency distribution (blue histograms) and fitted nonparametric kernel probabilistic distribution (red line) of matric pressure head (in $\log_{10}$) under (**a**) historical records, (**b**) dry, (**c**) mean and (**d**) wet climate projections. Vertical dashed lines indicate the Feddes' parameters controlling root water uptake.

4. Conclusions

The study examined the relationship between historic (1950–2000) and projected future (2000–2100) climate forcings and the GR rates in a site-specific study in the Nebraska National Forest of the Nebraska Sand Hills region. Toward this objective, we considered 65 projections of GR that included four greenhouse gas emissions scenarios. These climate inputs were used in the local-scale numerical model of soil-atmosphere-vegetation dynamics based on HYDRUS-1D. The results have been averaged over decadal, annual, and sub-annual time scales within a probabilistic framework accounting for uncertainty in climate projections. Dry, mean, and wet GR recharge scenarios were defined from the statistical parameters of cumulative GR from daily model runs. The historic climate in the NSH is classified as semi-arid with annual-average potential GR rate of 9.02 cm yr^{-1}. Projections of decadal precipitation appear to remain quasi-steady with negligible (1%) increase while increasing forecasts of ET$_0$ range from 8.6% to 31.4%. These main factors are shown to cause a decrease in decadal GR rates ranging from 21% to 39% in extreme cases. It is also worth noting that the Hargreaves approach to estimate ET$_0$ can inject uncertainty in recharge estimation. In this analysis, the results indicate that GR will likely be overestimated if ET_0 is estimated using the Hargreaves equation. In contrast, GR is likely to be even lower if ET_0 is estimated by using the Penman-Monteith equation.

The results highlight the need to consider adaptation and mitigation measures to respond to projected decreases in GR, which can exacerbate groundwater depletion in the High Plains Aquifer. The outcomes also underscore the need for exploring measures to combat the projected potential increases in wilting probability (~8.3% to 18.5% compared to 4.3% historical baseline), to protect the native grassland ecosystem, and to prevent desertification processes.

Author Contributions: For research articles with several authors, a short paragraph specifying their individual contributions must be provided. The following statements should be used "conceptualization, Z.A., P.N., N.R., T.W. and V.Z.; methodology, Z.A. and P.N.; software, P.N. and Z.A.; validation, P.N., Z.A., N.R. and T.W.; formal analysis, Z.A., P.N., N.R., T.W. and V.Z.; investigation, Z.A. and P.N.; resources, V.Z.; data curation, P.N.; writing—original draft preparation, P.N. and Z.A.; writing—review and editing, Z.A., P.N., N.R., T.W. and V.Z.; visualization, X.X.; supervision, V.Z. and P.N.; project administration, V.Z.; funding acquisition, V.Z.

Funding: The authors acknowledge the financial support provided by the NSF IGERT program (DGE-0903469), the Nebraska Geological Society's Yatkola-Edwards scholarship and the American Association of Petroleum Geologists.

Conflicts of Interest: The authors declare no conflict of interest.

Appendix A

Table A1. Multi-decadal-average values of P, ET_0, AI, ET_p, ET_a, and GR for historic data and different RCP classes.

Data Sets	Model	P	ET_0	AI	ET_p	ET_a	GR
		cm yr^{-1}	cm yr^{-1}	-	cm yr^{-1}	cm yr^{-1}	cm yr^{-1}
Historic data 1950–2000	NA	55.7	114.1	0.49	78.5	45.8	9.02
	BCC_CSM	56.9	124.0	0.46	85.3	49.2	8.0
	CANESM	58.0	127.9	0.45	87.6	48.5	6.9
	CCSM	59.8	123.1	0.49	84.5	49.2	8.2
	CESM_BGC	n.a.	n.a.	n.a.	n.a.	n.a.	n.a.
	CNRM_CM	n.a.	n.a.	n.a.	n.a.	n.a.	n.a.
	CSIRO	56.6	125.3	0.45	86.6	46.0	7.2
	GFDL_CM	66.3	127.4	0.52	87.8	55.4	7.5
	GFDL_G1	55.7	122.2	0.46	84.3	47.8	5.8
RCP 2.6	GFDL_M1	59.4	119.6	0.50	82.0	48.0	6.7
2000–2100	INMCM	n.a.	n.a.	n.a.	n.a.	n.a.	n.a.
	IPSL_CM	57.9	123.3	0.47	86.4	49.1	5.9
	IPSL_MR	56.2	124.6	0.45	85.3	46.9	7.0
	MIROC_ESM	59.3	129.0	0.46	87.3	46.8	9.5
	MIRCO_ESM	60.6	128.4	0.47	88.9	49.5	7.4
	MICROC	53.9	125.6	0.43	85.5	47.3	4.9
	MPI_LR	56.4	120.7	0.47	83.0	48.6	6.3
	MPI_MR	61.9	120.4	0.51	82.8	50.6	9.2
	MRI_GCM	58.6	117.1	0.50	80.6	50.2	5.9
	NORESM	62.7	122.2	0.51	84.2	51.6	8.0
	BCC_CSM	56.2	126.6	0.44	87.7	47.2	7.4
	CANESM	58.3	129.8	0.45	89.3	48.4	7.2
	CCSM	57.5	126.8	0.45	88.2	46.9	6.2
	CESM_BGC	59.6	125.1	0.48	85.9	48.1	7.7
	CNRM_CM	58.2	124.3	0.47	85.5	47.4	8.5
	CSIRO	55.7	128.6	0.43	88.0	46.5	6.3
	GFDL_CM	n.a.	n.a.	n.a.	n.a.	n.a.	n.a.
	GFDL_G1	59.5	123.7	0.48	85.3	49.8	6.0
RCP 4.5	GFDL_M1	60.7	120.9	0.50	82.7	50.7	6.0
2000–2100	INMCM	54.5	121.7	0.45	83.8	44.8	7.5
	IPSL_CM	53.7	127.6	0.42	89.4	47.3	4.8
	IPSL_MR	54.1	128.3	0.42	87.7	45.7	6.8
	MIROC_ESM	60.3	131.6	0.46	89.7	49.0	9.5
	MIRCO_ESM	60.8	133.0	0.46	90.2	49.5	7.4
	MICROC	49.4	132.8	0.37	89.7	44.1	3.3
	MPI_LR	56.5	124.2	0.45	85.6	47.5	6.7
	MPI_MR	58.9	124.8	0.47	85.8	48.9	8.4
	MRI_GCM	58.7	119.8	0.49	82.4	49.8	5.6
	NORESM	59.1	127.1	0.46	87.5	50.8	8.4

Table A1. *Cont.*

Data Sets	Model	P	ET$_0$	AI	ET$_p$	ET$_a$	GR
RCP 6.0 2000–2100	BCC_CSM	61.7	126.1	0.49	87.0	49.1	9.8
	CANESM	n.a.	n.a.	n.a.	n.a.	n.a.	n.a.
	CCSM	56.2	126.6	0.44	87.2	47.5	6.3
	CESM_BGC	n.a.	n.a.	n.a.	n.a.	n.a.	n.a.
	CNRM_CM	n.a.	n.a.	n.a.	n.a.	n.a.	n.a.
	CSIRO	n.a.	n.a.	n.a.	n.a.	n.a.	n.a.
	GFDL_CM	65.3	128.5	0.51	88.6	55.2	8.2
	GFDL_G1	59.0	124.2	0.47	85.2	50.9	6.4
	GFDL_M1	59.3	121.9	0.49	83.7	49.3	7.2
	INMCM	n.a.	n.a.	n.a.	n.a.	n.a.	n.a.
	IPSL_CM	55.7	126.5	0.44	88.2	46.6	6.3
	IPSL_MR	52.5	129.1	0.41	88.2	43.6	6.4
	MIROC_ESM	60.1	130.8	0.46	88.8	48.6	7.7
	MIRCO_ESM	59.6	131.6	0.45	89.3	48.5	8.9
	MICROC	52.4	128.4	0.41	88.1	45.9	4.0
	MPI_LR	n.a.	n.a.	n.a.	n.a.	n.a.	n.a.
	MPI_MR	n.a.	n.a.	n.a.	n.a.	n.a.	n.a.
	MRI_GCM	58.2	118.5	0.49	81.8	50.6	5.8
	NORESM	57.4	126.1	0.46	86.5	48.4	6.4
		cm yr^{-1}	cm yr^{-1}	-	cm yr^{-1}	cm yr^{-1}	cm yr^{-1}
RCP 8.5 2000–2100	BCC_CSM	60.8	163.2	0.37	89.7	48.5	9.5
	CANESM	62.4	133.5	0.47	112.6	52.6	6.8
	CCSM	61.0	165.1	0.37	110.1	51.5	5.5
	CESM_BGC	60.7	128.0	0.47	107.6	52.1	5.6
	CNRM_CM	56.7	128.7	0.44	108.4	49.2	5.2
	CSIRO	54.0	131.8	0.41	111.7	48.0	3.9
	GFDL_CM	67.8	131.4	0.52	110.8	58.4	6.6
	GFDL_G1	59.0	161.8	0.36	109.2	52.2	4.8
	GFDL_M1	60.2	158.1	0.38	107.1	51.8	6.4
	INMCM	53.2	125.6	0.42	108.3	46.5	5.1
	IPSL_CM	54.7	164.9	0.33	113.3	47.8	4.0
	IPSL_MR	49.8	166.9	0.30	112.7	44.3	3.4
	MIROC_ESM	60.3	170.5	0.35	112.8	50.6	7.8
	MIRCO_ESM	61.5	170.4	0.36	113.5	52.4	6.1
	MICROC	51.8	167.8	0.31	111.7	47.3	2.4
	MPI_LR	54.8	130.9	0.42	110.4	48.6	4.2
	MPI_MR	61.2	127.7	0.48	108.1	50.8	6.6
	MRI_GCM	61.0	153.7	0.40	104.0	52.9	5.6
	NORESM	57.7	165.9	0.35	111.5	50.2	5.3

References

1. Eckhardt, K.; Ulbrich, U. Potential impacts of climate change on groundwater recharge and streamflow in a central European low mountain range. *J. Hydrol.* **2003**, *284*, 244–252. [CrossRef]
2. Ng, G.-H.C.; McLaughlin, D.; Entekhabi, D.; Scanlon, B.R. Probabilistic analysis of the effects of climate change on groundwater recharge. *Water Resour. Res.* **2010**, *46*, 1–18. [CrossRef]
3. Kim, J.H.; Jackson, R.B. A Global Analysis of Groundwater Recharge for Vegetation, Climate, and Soils. *Vadose Zone J.* **2012**, *11*. [CrossRef]
4. Nasta, P.; Gates, J.B.; Wada, Y. Impact of climate indicators on continental-scale potential groundwater recharge in Africa. *Hydrol. Process.* **2016**, *30*, 3420–3433. [CrossRef] [PubMed]
5. Terrell, B.L.; Johnson, P.N.; Segarra, E. Ogallala aquifer depletion: Economic impact on the Texas high plains. *Water Policy* **2002**, *4*, 33–46. [CrossRef]
6. Scanlon, B.R.; Keese, K.E.; Flint, A.L.; Flint, L.E.; Gaye, C.B.; Edmunds, W.M.; Simmers, I. Global synthesis of groundwater recharge in semiarid and arid regions. *Hydrol. Process.* **2006**, *20*, 3335–3370. [CrossRef]
7. Cheng, L.; Hoerling, M.; Aghakouchak, A.; Livneh, B.; Quan, X.W.; Eischeid, J. How has human-induced climate change affected California drought risk? *J. Clim.* **2016**, *29*, 111–120. [CrossRef]
8. Carson, M.; Köhl, A.; Stammer, D.; Slangen, A.A.B.; Katsman, C.A.; van de Wal, R.S.W.; Church, J.; White, N. Coastal sea level changes, observed and projected during the 20th and 21st century. *Clim. Chang.* **2016**, *134*, 269–281. [CrossRef]
9. Turco, M.; Llasat, M.C.; von Hardenberg, J.; Provenzale, A. Climate change impacts on wildfires in a Mediterranean environment. *Clim. Chang.* **2014**, *125*, 369–380. [CrossRef]

10. Mori, N.; Yasuda, T.; Mase, H.; Tom, T.; Oku, Y. Projection of Extreme Wave Climate Change under Global Warming. *Hydrol. Res. Lett.* **2010**, *4*, 15–19. [CrossRef]

11. Arnell, N.W.; Gosling, S.N. The impacts of climate change on river flood risk at the global scale. *Clim. Chang.* **2016**, *134*, 387–401. [CrossRef]

12. Bellard, C.; Bertelsmeier, C.; Leadley, P.; Thuiller, W.; Courchamp, F. Impacts of climate change on the future of biodiversity. *Ecol. Lett.* **2012**, *15*, 365–377. [CrossRef] [PubMed]

13. Gunkel, A.; Shadeed, S.; Hartmann, A.; Wagener, T.; Lange, J. Model signatures and aridity indices enhance the accuracy of water balance estimations in a data-scarce Eastern Mediterranean catchment. *J. Hydrol. Reg. Stud.* **2015**, *4*, 487–501. [CrossRef]

14. Milly, P.C.D.; Dunne, K.A. Potential evapotranspiration and continental drying. *Nat. Clim. Chang.* **2016**, *6*, 946. [CrossRef]

15. Trenberth, K.E. Changes in precipitation with climate change. *Clim. Res.* **2011**, *47*, 123–138. [CrossRef]

16. Graham, L.P.; Andreáasson, J.; Carlsson, B. Assessing climate change impacts on hydrology from an ensemble of regional climate models, model scales and linking methods—A case study on the Lule River basin. *Clim. Chang.* **2007**, *81*, 293–307. [CrossRef]

17. Kling, H.; Fuchs, M.; Paulin, M. Runoff conditions in the upper Danube basin under an ensemble of climate change scenarios. *J. Hydrol.* **2012**, *424–425*, 264–277. [CrossRef]

18. Huang, J.; Zhang, Z.; Feng, Y.; Hong, H. Hydrologic response to climate change and human activities in a subtropical coastal watershed of southeast China. *Reg. Environ. Chang.* **2013**, *13*, 1195–1210. [CrossRef]

19. López-Moreno, J.I.; Zabalza, J.; Vicente-Serrano, S.M.; Revuelto, J.; Gilaberte, M.; Azorin-Molina, C.; Morán-Tejeda, E.; García-Ruiz, J.M.; Tague, C. Impact of climate and land use change on water availability and reservoir management: Scenarios in the Upper Arago'n River, Spanish Pyrenees. *Sci. Total Environ.* **2014**, *493*, 1222–1231. [CrossRef]

20. Le Roux, P.C.; Aalto, J.; Luoto, M. Soil moisture's underestimated role in climate change impact modelling in low-energy systems. *Glob. Chang. Biol.* **2013**, *19*, 2965–2975. [CrossRef]

21. Goderniaux, P.; Brouyère, S.; Fowler, H.J.; Blenkinsop, S.; Therrien, R.; Orban, P.; Dassargues, A. Large scale surface-subsurface hydrological model to assess climate change impacts on groundwater reserves. *J. Hydrol.* **2009**, *373*, 122–138. [CrossRef]

22. Herrmann, F.; Baghdadi, N.; Blaschek, M.; Deidda, R.; Duttmann, R.; La Jeunesse, I.; Sellami, H.; Vereecken, H.; Wendland, F. Simulation of future groundwater recharge using a climate model ensemble and SAR-image based soil parameter distributions—A case study in an intensively-used Mediterranean catchment. *Sci. Total Environ.* **2016**, *543*, 889–905. [CrossRef]

23. Meixner, T.; Manning, A.H.; Stonestrom, D.A.; Allen, D.M.; Ajami, H.; Blasch, K.W.; Brookfield, A.E.; Castro, C.L.; Clark, J.F.; Gochis, D.J.; et al. Implications of projected climate change for groundwater recharge in the western United States. *J. Hydrol.* **2016**, *534*, 124–138. [CrossRef]

24. Taylor, R.G.; Scanlon, B.; Döll, P.; Rodell, M.; Van Beek, R.; Wada, Y.; Longuevergne, L.; Leblanc, M.; Famiglietti, J.S.; Edmunds, M.; et al. Ground water and climate change. *Nat. Clim. Chang.* **2013**, *3*, 322. [CrossRef]

25. Kitabata, H.; Nishizawa, K.; Yoshida, Y.; Maruyama, K. Permafrost Thawing in Circum-Arctic and Highlands under Climatic Change Scenario Projected by Community Climate System Model (CCSM3). *SOLA* **2006**, *2*, 53–56. [CrossRef]

26. Nyenje, P.M.; Batelaan, O. Estimating the effects of climate change on groundwater recharge and baseflow in the upper Ssezibwa catchment, Uganda. *Hydrol. Sci. J.* **2009**, *54*, 713–726. [CrossRef]

27. Crosbie, R.S.; McCallum, J.L.; Walker, G.R.; Chiew, F.H.S. Modelling climate-change impacts on groundwater recharge in the Murray-Darling Basin, Australia. *Hydrogeol. J.* **2010**, *18*, 1639–1656. [CrossRef]

28. Goodarzi, M.; Abedi-Koupai, J.; Heidarpour, M.; Safavi, H.R. Evaluation of the Effects of Climate Change on Groundwater Recharge Using a Hybrid Method. *Water Resour. Manag.* **2016**, *30*, 133–148. [CrossRef]

29. Bouraoui, F.; Vachaud, G.; Li, L.Z.X.; Le Treut, H.; Chen, T. Evaluation of the impact of climate changes on water storage and groundwater recharge at the watershed scale. *Clim. Dyn.* **1999**, *15*, 153–161. [CrossRef]

30. Green, T.R.; Taniguchi, M.; Kooi, H.; Gurdak, J.J.; Allen, D.M.; Hiscock, K.M.; Treidel, H.; Aureli, A. Beneath the surface of global change: Impacts of climate change on groundwater. *J. Hydrol.* **2011**, *405*, 532–560. [CrossRef]

31. Rossman, N.R.; Zlotnik, V.A.; Rowe, C.M.; Szilagyi, J. Vadose zone lag time and potential 21st century climate change effects on spatially distributed groundwater recharge in the semi-arid Nebraska Sand Hills. *J. Hydrol.* **2014**, *519*, 656–669. [CrossRef]
32. Adane, Z.; Nasta, P.; Gates, J.B. Links between soil hydrophobicity and groundwater recharge under plantations in a sandy grassland setting, Nebraska Sand Hills, USA. *For. Sci.* **2017**, *63*, 388–401. [CrossRef]
33. Adane, Z.A.; Nasta, P.; Zlotnik, V.; Wedin, D. Impact of grassland conversion to forest on groundwater recharge in the Nebraska Sand Hills. *J. Hydrol. Reg. Stud.* **2018**, *15*, 171–183. [CrossRef]
34. Rossman, N.R.; Zlotnik, V.A.; Rowe, C.M. Using cumulative potential recharge for selection of GCM projections to force regional groundwater models: A Nebraska Sand Hills example. *J. Hydrol.* **2018**, *561*, 1105–1114. [CrossRef]
35. Miao, X.; Mason, J.A.; Johnson, W.C.; Wang, H. High-resolution proxy record of Holocene climate from a loess section in Southwestern Nebraska, USA. *Palaeogeogr. Palaeoclimatol. Palaeoecol.* **2007**, *245*, 368–381. [CrossRef]
36. Wang, T.; Wedin, D.; Zlotnik, V.A. Field evidence of a negative correlation between saturated hydraulic conductivity and soil carbon in a sandy soil. *Water Resour. Res.* **2009**. [CrossRef]
37. Scanlon, B.R.; Faunt, C.C.; Longuevergne, L.; Reedy, R.C.; Alley, W.M.; McGuire, V.L.; McMahon, P.B. Groundwater depletion and sustainability of irrigation in the US High Plains and Central Valley. *Proc. Natl. Acad. Sci. USA* **2012**, *109*, 9320–9325. [CrossRef]
38. Allen, R.G. FAO Irrigation and Drainage Paper Crop by. *Irrig. Drain.* **1998**, *300*, 300.
39. Adane, Z.A.; Gates, J.B. Determining the impacts of experimental forest plantation on groundwater recharge in the Nebraska Sand Hills (USA) using chloride and sulfate. *Hydrogeol. J.* **2015**, *23*, 81–94. [CrossRef]
40. Hargreaves, G.H.; Allen, R.G. History and Evaluation of Hargreaves Evapotranspiration Equation. *J. Irrig. Drain. Eng.* **2003**, *129*, 53–63. [CrossRef]
41. Pelosi, A.; Medina, H.; Villani, P.; D'Urso, G.; Chirico, G.B. Probabilistic forecasting of reference evapotranspiration with a limited area ensemble prediction system. *Agric. Water Manag.* **2016**, *178*, 106–118. [CrossRef]
42. Droogers, P.; Allen, R.G. Estimating reference evapotranspiration under inaccurate data conditions. *Irrig. Drain. Syst.* **2002**, *16*, 33–45. [CrossRef]
43. Archibald, J.A.; Walter, M.T. Do energy-based PET models require more input data than temperature-based models?—An evaluation at four humid fluxnet sites. *J. Am. Water Resour. Assoc.* **2014**, *50*, 497–508. [CrossRef]
44. Huo, Z.; Dai, X.; Feng, S.; Kang, S.; Huang, G. Effect of climate change on reference evapotranspiration and aridity index in arid region of China. *J. Hydrol.* **2013**, *492*, 24–34. [CrossRef]
45. Liu, X.; Zhang, D.; Luo, Y.; Liu, C. Spatial and temporal changes in aridity index in northwest China: 1960 to 2010. *Theor. Appl. Climatol.* **2013**, *112*, 307–316. [CrossRef]
46. Nastos, P.T.; Politi, N.; Kapsomenakis, J. Spatial and temporal variability of the Aridity Index in Greece. *Atmos. Res.* **2013**, *119*, 140–152. [CrossRef]
47. Zhang, K.X.; Pan, S.M.; Zhang, W.; Xu, Y.H.; Cao, L.G.; Hao, Y.P.; Wang, Y. Influence of climate change on reference evapotranspiration and aridity index and their temporal-spatial variations in the Yellow River Basin, China, from 1961 to 2012. *Quat. Int.* **2015**, *380*, 75–82. [CrossRef]
48. Arora, V.K. The use of the aridity index to assess climate change effect on annual runoff. *J. Hydrol.* **2002**, *265*, 164–177. [CrossRef]
49. Nkomozepi, T.; Chung, S.O. The effects of climate change on the water resources of the Geumho River Basin, Republic of Korea. *J. Hydro-Environ. Res.* **2014**, *8*, 358–366. [CrossRef]
50. Gudmundsson, L.; Greve, P.; Seneviratne, S.I. The sensitivity of water availability to changes in the aridity index and other factors—A probabilistic analysis in the Budyko space. *Geophys. Res. Lett.* **2016**, *43*, 6985–6994. [CrossRef]
51. Šimůnek, J.; van Genuchten, M.T.; Šejna, M. Recent developments and applications of the HYDRUS computer software packages. *Vadose Zone J.* **2016**, *15*, 25. [CrossRef]
52. Nasta, P.; Gates, J.B. Plot-scale modeling of soil water dynamics and impacts of drought conditions beneath rainfed maize in Eastern Nebraska. *Agric. Water Manag.* **2013**, *128*, 120–130. [CrossRef]
53. Van Genuchten, M.T. A Closed-form Equation for Predicting the Hydraulic Conductivity of Unsaturated Soils1. *Soil Sci. Soc. Am. J.* **1980**, *44*, 892. [CrossRef]
54. Spinoni, J.; Vogt, J.; Naumann, G.; Carrao, H.; Barbosa, P. Towards identifying areas at climatological risk of desertification using the Köppen-Geiger classification and FAO aridity index. *Int. J. Climatol.* **2015**, *35*, 2210–2222. [CrossRef]

55. Billesbach, D.P.; Arkebauer, T.J. First long-term, direct measurements of evapotranspiration and surface water balance in the Nebraska SandHills. *Agric. For. Meteorol.* **2012**, *156*, 104–110. [CrossRef]

56. Crosbie, R.S.; Scanlon, B.R.; Mpelasoka, F.S.; Reedy, R.C.; Gates, J.B.; Zhang, L. Potential climate change effects on groundwater recharge in the High Plains Aquifer, USA. *Water Resour. Res.* **2013**, *49*, 3936–3951. [CrossRef]

57. Szilagyi, J.; Zlotnik, V.A.; Gates, J.B.; Jozsa, J. Mapping mean annual groundwater recharge in the Nebraska Sand Hills, USA. *Hydrogeol. J.* **2011**, *19*, 1503–1513. [CrossRef]

58. Lauffenburger, Z.H.; Gurdak, J.J.; Hobza, C.; Woodward, D.; Wolf, C. Irrigated agriculture and future climate change effects on groundwater recharge, northern High Plains aquifer, USA. *Agric. Water Manag.* **2018**, *204*, 69–80. [CrossRef]

59. Greve, P.; Seneviratne, S.I. Assessment of future changes in water availability and aridity. *Geophys. Res. Lett.* **2015**, *42*, 5493–5499. [CrossRef]

60. Thomas, B.F.; Behrangi, A.; Famiglietti, J.S. Precipitation intensity effects on groundwater recharge in the southwestern United States. *Water* **2016**, *8*, 90. [CrossRef]

61. Tabari, H.; Aghajanloo, M.B. Temporal pattern of aridity index in Iran with considering precipitation and evapotranspiration trends. *Int. J. Climatol.* **2013**, *33*, 396–409. [CrossRef]

62. Oglesby, R.; Bathke, D.; Wilhite, D.; Rowe, C. Understanding and assessing projected future climate change for Nebraska and the great plains. *Gt. Plains Res.* **2015**, *25*, 97–107. [CrossRef]

63. Dawes, W.; Zhang, L.; Dyce, P. WAVES V3.5 user manual. Available online: https://research.csiro.au/software/wp-content/uploads/sites/6/2016/07/preface.pdf (accessed on 4 September 2018).

64. Liang, X.; Lettenmaier, D.P.; Wood, E.F.; Burges, S.J. A simple hydrologically based model of land surface water and energy fluxes for general circulation models. *J. Geophys. Res.* **2004**, *99*, 14415–14428. [CrossRef]

65. Nasta, P.; Adane, Z.; Lock, N.; Houston, A.; Gates, J.B. Links between episodic groundwater recharge rates and rainfall events classified according to stratiform-convective storm scoring: A plot-scale study in eastern Nebraska. *Agric. For. Meteorol.* **2018**, *259*, 154–161. [CrossRef]

Article

Speeding up the Computation of the Transient Richards' Equation with AMGCL

Robert Pinzinger [1],* and René Blankenburg [2]

[1] Institut für Grundwasserwirtschaft, Technische Universität Dresden, Bergstraße 66, 01069 Dresden, Germany
[2] Ingenieurbüro für Grundwasser GmbH, Nonnenstraße 9, 04229 Leipzig, Germany;
    r.blankenburg@ibgw-leipzig.de
* Correspondence: robert.pinzinger@tu-dresden.de

Received: 4 December 2019; Accepted: 15 January 2020; Published: 18 January 2020

Abstract: The Richards'-equation is widely used for modeling complex soil water dynamics in the vadose zone. Usually, the Richards'-equation is simulated with the Finite Element Method, the Finite Difference Method, or the Finite Volume Method. In all three cases, huge systems of equations are to be solved, which is computationally expensive. By employing the free software library AMGCL, a reduction of the computational running time of up to 79% was achieved without losing accuracy. Seven models with different soils and geometries were tested, and the analysis of these tests showed, that AMGCL causes a speedup in all models with 20,000 or more nodes. However, the numerical overhead of AMGCL causes a slowdown in all models with 20,000 or fewer nodes.

Keywords: Richards'-equation; simulation; algebraic multigrid; preconditioner

---

1. Introduction

The vadose zone extends from the surface of the Earth down to the groundwater-table. This region is a habitat for bacteria and other microorganisms, some of which clean water while it travels through this zone. However, human activities in agriculture pollute the vadose zone and the groundwater with chlorine, nitrate [1], and hormones [2]. Industrial activities pollute the vadose zone and the groundwater with radioactive [3], petrochemical [4], and pharmaceutical [5] compounds. Some of these acts of pollution have the potential to change the properties of the vadose zone and the aquifer irreversibly [6] which in turn threatens the quality of groundwater in some regions permanently.

Land use can influence the amount of water in the vadose zone and the groundwater level [7]. The extraction of groundwater for irrigation can lead to land subsidence, especially in arid regions [8]. However, too much water in a slope may cause slope failure [9].

All this underlines the importance of being able to simulate the vadose zone. Accurate simulations of the vadose zone can aid in the decision on how to react to an incident of pollution. Also, changes in land use and their impact on the recharge of the aquifer can be simulated. Furthermore, the decision to evacuate areas under a slope can be supported with accurate simulations of the vadose zone.

The Richards'-equation is widely used for modeling complex soil water dynamics in the vadose zone [10]:

$$\frac{\partial \theta}{\partial t} = \sum_i \frac{\partial}{\partial i}\left[K\left(\sum_j K_{i,j}^A \frac{\partial h}{\partial j} + K_{i,z}^A\right)\right] - S, \tag{1}$$

Here, θ represents the water content, $i, j \in \{x, z\}$ for a 2-dimensional formulation and $i, j \in \{x, y, z\}$ for a 3-dimensional consideration where z is the vertical dimension, K is the hydraulic conductivity, $K_{i,j}^A$ is the anisotropy tensor, h is the pressure head, and S is a sink term that models the in- and outflow over the system boundaries like evapotranspiration and precipitation. Thereby, Equation (1) formulates, that the change in water content is caused by the seepage flow that is caused by pressure and gravity and by the in- and outflow of water over the system boundaries.

The aim of this study is to show that the numerical simulation of Equation (1) can be sped up by using the free software library AMGCL [11]. In Section 2, an overview on how equation systems that arise from numerical discretizations of Equation (1) are solved is given. In Section 3, AMGCL and PCSiWaPro [12] are introduced. The metrics that are used to compare the performance of the solver in PCSiWaPro and AMGCL are introduced in Section 4. In Section 5, models are presented that are used to make the comparison of the solver in PCSiWaPro and AMGCL. The results obtained from the simulations of these models are presented in Section 6. Section 7 finishes this study with a discussion and conclusion.

2. Solving Equation Systems that Arise from the Numerical Discretization of the Richards' Equation

There are several software packages available for the simulation of Equation (1). The most prominent are FEFLOW [13], FEMWATER [14], HydroGeoSphere [15], HYDRUS [16], PCSiWaPro, STOMP [17], SWMS [18], TOUGH3 [19], VS2D [20], and VSAFT2 [21].

Equation (1) is usually simulated with the Finite Element Method [22], the Finite Difference Method [22] or the Finite Volume Method [23].

Equation (1) is modelled with Finite Elements in FEFLOW, HydroGeoSphere, HYDRUS, PCSiWaPro, SWMS, and VSAFT2. The Finite Difference Method is used in STOMP, TOUGH3 and VS2D. The Finite Volume Method is considered in theoretical works [24,25], and also, there are some implementations of Equation (1) in the free numerical simulation toolbox OpenFOAM that discretize Equation (1) with the Finite Volume Method [26,27].

Whatever scheme is chosen to simulate Equation (1), the result is always an equation system of the form

$$Ax = b, \tag{2}$$

where A is a matrix and x and b are vectors. In general, there are two ways to solve Equation (2) for x: The direct solution with the Gauss algorithm or "relatives" of it [22], or iterative methods [28].

The Gauss algorithm is computationally expensive which is why the Gauss algorithm is mainly used for small systems of equations. A "cousin" of the Gauss algorithm is the LU-decomposition [22], where one employs the Gauss algorithm to compute two matrices L and U where L is a lower triangular matrix and U is an upper triangular matrix, so that $LU = A$. A "cousin" of LU is the Cholesky factorization [22], which demands A to be symmetric, where a lower triangular matrix L is computed so that $LL^T = A$. The LU-decomposition or the Cholesky-decomposition are popular in situations, where one wants to solve an equation system as in Equation (2) with constant matrix A and changing right hand sides b, because once the decompositions are computed, $Ax = b$ can be solved, in case of the LU-decomposition, as $LUx = b$, which in turn can be solved as $Ly = b$ with $y = Ux$. Both of these equation systems are triangular, and therefore these systems of equations can be solved very fast.

In the case of iterative procedures, one computes a sequence of approximate solutions x_i, so that the residual $r_i = b - Ax_i$ vanishes with growing i. There are several different schemes to compute this sequence. The most famous schemes are Gauss–Seidel [28], Krylow Subspace [28] algorithms, the Multigrid [28] method, and Stone's [29] method.

The basic idea of Gauss–Seidel is to decompose the matrix $A = D - E - F$ where D is a diagonal matrix, E is a lower triangular matrix and F is an upper triangular matrix. From this, a recursion can be built: $x_{i+1} = (D - E)^{-1} Fx_i + (D - E)^{-1} b$. Instead of computing $y = (D - E)^{-1} v$ for a given vector $v = Fx_i$ or $v = b$, one solves $(D - E)y = v$ for y which is computationally cheap since $(D - E)$ is a lower triangular matrix.

Krylow Subspace schemes employ the idea to compute the error $r_0 = b - Ax_0$, and to create a subspace of dimension i from this error by defining $K_i = span\{r_0, Ar_0, \cdots, A^{i-1}r_0\}$. Then, one chooses $x_i \in x_0 + K_i$ so that $b - Ax_i$ is perpendicular on L_i, where L_i is another subspace of dimension i. The simplest choice is $L_i = K_i$ which is the starting point for the Conjugate Gradient algorithm [28]. In the Conjugate Gradient algorithm, the update of approximation x_i to x_{i+1} is made by employing a search direction p_i, that is computed with the vectors $r_i = b - Ax_i$ and p_{i-1}. The search directions $p_0, p_1, \ldots$ hold $(Ap_i)^T p_j = 0$ for $i \neq j$, which explains the term "conjugate" in Conjugate Gradient. ORTHOMIN [28] is a Krylow-type algorithm, where the search direction p_i is computed with the vectors r_i and $p_{i-m}, p_{i-m+1}, \ldots, p_{i-1}$, where the memory m begins in the first iteration with $m = 1$ and grows with each iteration i until a maximum k is reached, where m is restarted with $m = 1$. All Krylow Subspace algorithms aim at minimizing the norm of r_i, and the use of multiple past search directions as well as the reset of the number of past search directions every k iterations help ORTHOMIN to not get stuck in local minima. Another Krylow-type algorithm is BICGSTAB [28] where the update of x_i is made with a search direction p_i and a vector s_i that is a scaled version of the resulting residual which in turn is caused by the update of x_i with p_i. Vector s_i is scaled, so that the update of x_i with p_i and s_i minimizes the norm of the residual $r_{i+1} = b - Ax_{i+1}$. This choice of s_i stabilizes the convergence of BICGSTAB. Aside from the Conjugate Gradient algorithm, ORTHOMIN and BICGSTAB, there are many other Krylow Subspace schemes. A good overview over these schemes can be found in [28].

Krylow schemes can be sped up by preconditioning. In each iteration, the approximation error can be computed: $r_i = b - Ax_i$, then $x_i + A^{-1}r_i = x_i + A^{-1}Ax - A^{-1}Ax_i = x_i + x - x_i = x$. But since A^{-1} is usually unknown and computationally expensive to compute, one searches for a matrix M so that $M \approx A^{-1}$. A suitable matrix M is called a preconditioner. The most famous preconditioner is ILU (Incomplete LU [28]), where one attempts to approximate A by $LU \approx A$, where L is a lower triangular matrix and U is an upper triangular matrix. L and U are not computed so that $LU = A$ because this would be too expensive computationally. A vector v is preconditioned with ILU by solving $LUy = v$ for y, which is computationally cheap since L and U are both triangular matrices. However, this scheme cannot be parallelized, because in order to solve a triangular equation system, one has to solve the rows consecutively. Another popular preconditioner is SPAI (sparse approximate inverse [30]). Here, one attempts to compute a matrix M, so that the sum over the squares of the entries of $I - MA$ is minimal, where I is the identity matrix. The advantage of SPAI is, that SPAI is constructed in a way that M can be computed massively parallel. There also exists polynomial preconditioning [28]: consider $\widetilde{A} = \omega A = I - B$, where ω is chosen, so that the largest absolute eigenvalue of B is smaller than 1, then the following relationship holds: $\widetilde{A}^{-1} = \sum_k B^k$. This preconditioner is computationally very expensive since it demands many matrix-matrix-multiplications, but it can be massively parallelized. Also, this preconditioner demands enormous amounts of storage. Usually, the matrices involved with Finite Element, Finite Difference, or Finite Volume models are sparse, which means, that each row of the matrices has only a few entries which need to be kept in storage. If one multiplies matrices with each other, this sparsity is lost in general, which results in huge demands for storage.

Another approach to solve large equation systems is the Multigrid for which there are two variations: The geometric Multigrid and the algebraic Multigrid. In the case of the geometric Multigrid [28], the discretization of the problem underlying Equation (2) is mapped to a sequence of consecutive coarser discretizations, and for each level of discretization, an equation system is created. Thereby, the equation system gets smaller with each level of discretization. On a given level, the current approximation error is reduced by a simple iterative scheme. This reduction of the approximation error is called relaxation. Simple algorithms are preferred for the relaxation, because one does not want to compute the exact solution, but rather a good approximation which can be computed fast. Algorithms like Gauss–Seidel or SPAI0 [31] are popular algorithms for relaxation. SPAI0 defines a smoother $x_{i+1} = x_i - M(Ax_i - b)$, where the matrix M is a diagonal matrix with $M_{k,k} = \dfrac{A_{k,k}}{\sum_j A_{k,j}^2}$. Due to its simple structure, M can be computed in parallel. Also, the smoother SPAI0 can be highly parallelized, since it consists only of matrix-vector products and vector subtractions. On a given level, after the relaxation, the current approximation is either mapped to a coarser level or to a finer level. The mapping of the current approximation from a coarse- to a fine level is called prolongation, the mapping from a fine level to a coarse level is called restriction. There are different schemes like the V-cycle [28], where one starts on the finest level, relaxes and restricts until the coarsest level is reached, and then relaxes and prolongs until the finest level is reached, where the sequence of restrictions and prolongations forms a "V". In the W-cycle [28], one arranges the restrictions and prolongations, so that they form a "W". Since the dimensionality of the equation systems sinks with the coarseness of a level, the operations on the coarse levels are computationally very cheap. To sum up the idea of the geometric Multigrid, one maps a discretized geometry to a sequence of discretized geometries with increasing coarseness. For each of these grids, an equation system is formulated and the solution on each grid is approximated and mapped to the next grid. The problem with the geometric Multigrid is, however, that it shows difficulties with anisotropic coefficients of the underlying partial differential equation [28]. These problems arise from the fact, that if the coefficients of the underlying partial differential equation are anisotropic, then errors are automatically introduced into the solution with the step from the fine grid to the coarse grid and vice versa. In the case of Equation (1), the coefficients are anisotropic, because each point in space has a pressure head which defines the hydraulic conductivity at that point. Therefore, if the pressure heads are anisotropic, then the hydraulic conductivity in Equation (1) is anisotropic. To illustrate the problem, consider the simple model of two nodes, where both nodes have a unique pressure head. If one would coarsen this simple model, one would restrict both nodes to one node. The node on the coarser level, however, can only reflect the hydraulic conductivity of at most one node from the finer level. Therefore, geometric Multigrid solvers perform poorly on discretizations of Equation (1).

In the case of the algebraic Multigrid, one does not bother with the geometry of the discretization, but rather one creates from the original equation system a sequence of equation systems with decreasing sizes [28]. There are several schemes to compute the smaller equation system from a given equation system, the most famous schemes are the ones by Ruge and Stüben [32] and Smoothed Aggregation [33]. The basic idea of Ruge and Stüben is, to interpret a matrix A as a graph, where the entry $A_{k,j}$ tells, how strongly the row k is connected to the row j. Then, one uses this strength of connection to decide, which rows need to be represented in the next coarser level, the C rows, and which rows need not to be represented, the F rows. In the Ruge and Stüben scheme, k is strongly negative coupled to j if $-A_{k,j} \geq \varepsilon \max_{A_{k,l}<0} |A_{k,l}|$ for some $0 < \varepsilon < 1$. This strongly negative coupling is then used to define the set $S_k = \{j : A_{k,j} \neq 0, k \text{ strongly negative coupled to } j\}$. The transpose of this set is $S_k^T = \{j : k \in S_j\}$. Therefore, S_k^T contains all j that are strongly negative coupled to k. Once, the sets S_k^T are computed for each row, one iteratively selects one k, puts this row into the C set, and the rows in S_k^T into the F set. Since this selection has to be performed iteratively, it cannot be parallelized. However, the information from the rows in the F set must not vanish, therefore the information from these rows is interpolated into the rows from the C set. Thereby, the problem with anisotropic coefficients in the geometric Multigrid is circumvented. In the Smoothed Aggregation algorithm, for each row k, the neighborhood is defined by $N_k(\varepsilon) = \{j : |A_{k,j}| \geq \varepsilon \sqrt{A_{k,k}A_{j,j}}\}$. For a given k, k is put into the C set, and the j in $N_k(\varepsilon)$ are put into the F set. As with the Ruge and Stüben algorithm, the information from the rows in the F set is interpolated into the rows from the C set. In both algorithms, the solution is computed as with the geometric Multigrid once the equation systems are created.

The last approach to solve large equation systems that shall be considered in this paper is Stone's Method. The core of Stone's Method is to decompose $A = S - T$. Then, one can formulate the iterative scheme $Sx_{i+1} = Tx_i + b$. Now, one has to solve this equation system for x_{i+1} which is why one chooses S, so that the decomposition $S = LU$, where L is a lower triangular matrix and U is an upper triangular matrix, is easy to compute.

FEFLOW offers direct Gaussian-type solvers, Krylow Subspace schemes and geometric and algebraic Multigrid. The Krylow schemes can be preconditioned with ILU. FEMWATER has Gauss–Seidel and preconditioned Conjugate Gradient solvers. The preconditioners are ILU and polynomial. HydroGeoSphere uses ORTHOMIN with ILU as preconditioner. HYDRUS, PCSiWaPro, STOMP and SWMS employ the Gauss-algorithm and preconditioned Conjugate Gradient with ILU as preconditioner. TOUGH3 offers the Gauss-algorithm and ILU-preconditioned Krylow Subspace schemes. VS2D uses Stone's method as solver, and VSAFT2 has the ILU-preconditioned Conjugate Gradient as solver.

3. AMGCL and PCSiWaPro

As one can see, most software packages that solve the Richards'-equation with Krylow-type algorithms employ the ILU preconditioner. As shown above, ILU suffers from two aspects: On the one hand, ILU itself can in general only be a poor approximation to the real inverse, and on the other hand, ILU cannot be parallelized which makes preconditioning with ILU slow. Hence, there is a need for a highly parallelizable and yet accurate preconditioner for Krylow-type solvers of equation systems arising from discretizations of Equation (1). Because if one had a highly parallelizable and yet accurate preconditioner, then this preconditioner would speed up the solution of equation systems arising from discretizations of Equation (1). This speedup can then be translated into larger models that simulate the system in question in more detail, or one could use this additional time for simulating multiple scenarios.

Demidov published AMGCL, a C++-library with several Krylow-type solvers for which there are several algebraic Multigrid preconditioners available. AMGCL supports parallelization via OpenMP [34], CUDA [35] and MPI [36]. In the standard settings, AMGCL performs preconditioning with an algebraic Multigrid and employs the V-cycle, and the equation system is coarsened until the coarsest level has at most 3000 nodes. On the coarsest level, a LU-solver is employed. According to the standard settings of AMGCL, on each level two iterations of relaxation are performed. AMGCL offers algebraic Multigrid preconditioning according to the Ruge and Stüben scheme and Smoothed Aggregation. For the relaxation, there are Gauss–Seidel-, ILU- and SPAI-smoothers.

The development of PCSiWaPro began in the early 2000s when a group of scientists at the Technical University of Dresden took the numerical simulation code of SWMS_2D and extended this code by coupling the numerical kernel of SWMS_2D with a GUI that allows the user to create, modify and save 2-dimensional Finite Element models of the Richards' equation. Also, PCSiWaPro extends SWMS_2D by offering a database that contains the properties of various soils according to DIN 4220 [37] which is a standard for the designation, classification and deduction of soil parameters in Germany. The parameters of the soils can also be set manually. In order to extend the modelling capabilities of PCSiWaPro, the numerical kernel of SWMS_3D [38] was added in 2019. Furthermore, the extension "Weather Generator" allows to create atmospheric boundary conditions from measurement data. PCSiWaPro uses the van Genuchten (which shall be abbreviated by VG) model [39,40] to describe the retention curve and the unsaturated hydraulic conductivity function of the soils in the model. As stated above, PCSiWaPro uses a Gaussian-type solver (for equation systems with less than 500 dimensions) and the ILU-preconditioned Conjugate Gradient (for equation systems with 500 or more dimensions) to solve equation systems that arise from the numerical approximation of Equation (1). A relative and an absolute tolerance of 1×10^{-6} and a maximum of 1000 iterations are used as criteria for convergence and divergence. PCSiWaPro supports varying time step sizes. If the computation of a time step costs less than 4 calls of the solver, then the time step size is increased by a user defined factor. If the computation of a time step costs more than 6 calls of the solver, then the time step size is reduced by a user defined factor.

In order to investigate whether AMGCL can speed up the simulation of Equation (1), AMGCL was integrated into PCSiWaPro. The AMGCL solver was set up to employ BICGSTAB with an algebraic Multigrid-preconditioner which is based on Smoothed Aggregation where SPAI0 is used for relaxation.

BICGSTAB was chosen, due to its stabilized convergence behavior which makes the solver more robust. Initial tests revealed, that BICGSTAB preconditioned with algebraic Multigrid based on Smoothed Aggregation computes faster than BICGSTAB preconditioned with algebraic Multigrid based on the algorithm by Ruge and Stüben, therefore algebraic Multgrid with Smoothed Aggregation was chosen as preconditioner. SPAI0 was chosen for the relaxation, because SPAI0 was designed with the intention of massive parallelization, whereas, for example, the Gauss–Seidel scheme or ILU cannot be parallelized really well since triangular equation systems have to be solved.

Since the performance of AMGCL is to be compared with the ILU-preconditioned Conjugate Gradient solver, the absolute and relative tolerance of 1×10^{-6} and a maximum of 1000 iterations were coded into AMGCL to have a fair comparison.

PCSiWaPro with the AMGCL-solver with BICGSTAB as solver and algebraic Multigrid preconditioning based on the Smoothed Aggregation scheme with SPAI0 relaxation will be from now on referenced as PCSiWaPro AMGCL, whereas PCSiWaPro with the ILU-preconditioned Conjugate Gradient will be referenced as PCSiWaPro Original.

4. Method

PCSiWaPro AMGCL and PCSiWaPro Original were compared on 7 synthetic models which will be described in the next section. These models were tuned with computations run in PCSiWaPro Original. The number of nodes in the models ranges from 7000 nodes to 2,400,000 nodes. Two models consider simple rectangular shapes, one model considers a rectangular cuboid, whereas the other four models consider more complex geometries. Four models consist of only one type of soil, the other three models each have at least two different kinds of soil.

All computations for the 6 2-dimensional models were run on a desktop PC running Windows 8 with an Intel Core i7-6700 CPU and 8 GB of RAM. During the computations, all 8 cores that are available on the i7-6700 CPU were used since PCSiWaPro and AMGCL both support parallelization with OpenMP. The 3-dimensional model was simulated on a computer running Windows Server 2012 r2 with an Intel Core i7-6800K CPU and 32 GB of RAM, and all 6 cores were used during the computations.

For the comparison of PCSiWaPro AMGCL with PCSiWaPro Original, three metrics were considered: the computational running time of the simulations which was measured in seconds, the R^2-value between the pressure heads computed with PCSiWaPro Original and PCSiWaPro AMGCL, and the relative cumulative mass balance error (which shall abbreviated with RCMBE) computed with these two programs.

The R^2-value between the pressure heads computed with PCSiWaPro Original and PCSiWaPro AMGCL was chosen as a metric, because PCSiWaPro solves Equation (1) for the pressured heads.

$$R^2(x_i, y_i) = 1 - \frac{\sum_j \left(x_i^j - y_i^j\right)^2}{\sum_j \left(x_i^j - \mu_i\right)^2},\tag{3}$$

The R^2-value was computed with Equation (3), where x_i is the vector of the pressure heads computed with PCSiWaPro Original at time point t_i, and x_i^j is the j-th entry of this vector. μ_i is the mean of the values in x_i. y_i is the vector of the pressure heads computed with PCSiWaPro AMGCL at time point t_i, and y_i^j is the j-th entry of this vector. The nominator in Equation (3) is the squared error metric that computes the squared distance between the vectors x_i and y_i. This distance is normalized by the denominator. The normalization allows the R^2-value of different time steps or models to be compared. The best possible R^2-value is 1 when x_i and y_i are identical. The smaller the R^2-value, the greater the distance between the vectors x_i and y_i. The R^2-value was computed for every time step, and thereby, for each model, a time series of R^2-values was computed. The R^2-values were rounded to 4 digits after the decimal point.

$$RCMBE(t_i) = \frac{\left|V_{t_i} - V_{t_0} + \int_{t_0}^{t_i} RWU(t)dt - \int_{t_0}^{t_i} \sum_{j\in\Gamma} Q_j(t)dt\right|}{\max\left(\sum_e \left|V_{t_i}^e - V_{t_0}^e\right|, \int_{t_0}^{t_i}\left(RWU(t) + \sum_{j\in\Gamma}|Q_j(t)|\right)dt\right)}\tag{4}$$

The RCMBE was computed with Equation (4). Here, V_{t_0} is the amount of water in the model at time point t_0. V_{t_i} is the amount of water in the model at time point t_i. $RWU(t)$ is the root water uptake at time point t. Γ is the set of boundary nodes of the model. $Q_j(t)$ is the flow over the boundary in boundary node j at time point t. $V_{t_0}^e$ is the amount of water in element e at time point t_0, $V_{t_i}^e$ is the amount of water in element e at time point t_i. Therefore, the nominator in Equation (4) computes the absolute cumulative mass balance error at time point t_i by considering the difference in water mass between the first time point t_0 and the current time point t_i, the root water uptake between the first time point t_0 and the current time point t_i, and the in- and outflow of water over the system boundaries between the first time point t_0 and the current time point t_i. The denominator compares two terms and chooses the bigger one. The first term is the sum of the absolute changes in water mass in each element between the first time point t_0 and the current time point t_i. The second term is the integral over the root water uptake and the sum of the absolute values of the mass transfer over system boundaries in the boundary nodes. The RCMBE is a proxy to how well the numerical approximation works, because if there were no numerical approximation and no rounding errors, the nominator in Equation (4) would be 0, because the change in water mass between the first and the current time point would be explained by the amount of water removed by the plants and the amount of water that left and entered the system over the boundaries. If the nominator in Equation (4) is not equal to 0, then the numerical approximation and rounding errors created or destroyed water artificially. The code of PCSiWaPro Original was used during the tuning of the models that will be presented in the next section. One aim of the tuning was, to keep the RCMBE below 1% for all time points in the simulation period.

5. Description of Synthetic Models

Figure 1 gives a first overview over the 2-dimensional models used in the comparison. The properties of their soils can be read in Table 1. Figure 2 gives an overview over Model 7.

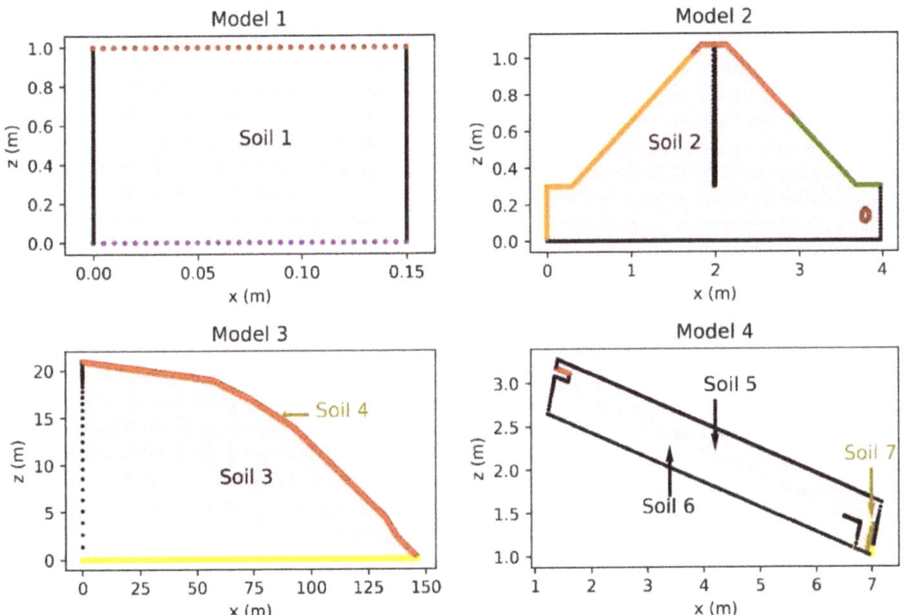

Figure 1. *Cont.*

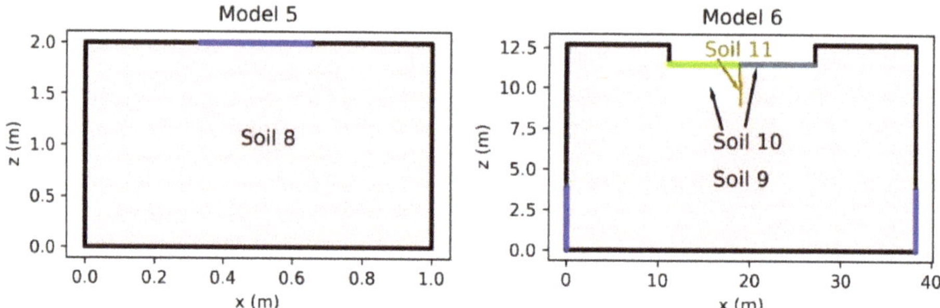

Figure 1. An overview over the shape of models 1 to 6, their sizes and the layout of the soil layers. The colors mistyrose, lightcyan, and darkgoldenrod are reserved for soils. Black depicts a no-flow boundary condition, blue depicts a constant pressure head boundary condition. Red, green, and grey depict a time dependent infiltration rate, darkviolet and brown depict a system dependent boundary condition: As long as the nodes are not saturated, darkviolet and brown nodes depict a no-flow boundary condition, and once the nodes are saturated, the boundary condition switches to a boundary condition with a constant pressure head of 0 m. Orange depicts a time dependent pressure head, olive depicts a seepage face, yellow depicts a constant pressure head and only outflow is allowed in yellow nodes.

Table 1. Overview over the parameters of the soils in Figures 1 and 2.

Soil	Porosity	Permeability	Residual Air Content	Residual Water Content	VG α	VG n	VG m	VG λ
-	%	m/s	%	%	1/m	-	-	-
1	39	5×10^{-5}	0	5	3	2.2	0.545	0.5
2	40	1.67×10^{-5}	0	2	4	1.9	0.47	0.5
3	30	1×10^{-3}	0	1	55	2	0.5	0.5
4	40	1×10^{-5}	0	5	1	1.6	0.33	0.5
5	40	1.06×10^{-4}	0	4	4.5	4	0.75	0.5
6	30	1.83×10^{-2}	0	1	100	3	0.67	0.5
7	30	1.06×10^{-4}	0	4	4.5	4	0.75	0.5
8	43	2.89×10^{-6}	0	7.8	3.6	1.56	0.36	0.5
9	41	2×10^{-5}	0	5.7	12.4	2.28	0.561	0.5
10	36	1.65×10^{-3}	0	1	35	3	0.666	0.5
11	43	1.16×10^{-8}	0	7.8	3.6	1.56	0.641	0.5
12	43	2.89×10^{-6}	0	7.8	3.6	1.56	0.36	0.5

5.1. Model 1

Model 1 is a column of 1 m in height and 0.15 m in width. It is based on a setup of an infiltration experiment with polluted water. The upper boundary condition is described by a time-dependent infiltration rate (depicted in red in Model 1 in Figure 1) whereas the lower boundary condition is system dependent (depicted in darkviolet in Model 1 in Figure 1): while the lower boundary is unsaturated, it is handled as a no-flow boundary condition and the pressure head is calculated. As soon as the lower boundary is saturated, the boundary condition switches to a boundary condition of the first type with a prescribed pressure head of zero, and thus, the outflow is calculated. The model is discretized with 7178 nodes. The parameters of Soil 1 from Table 1 were used to describe the hydraulic properties of the soil (depicted in mistyrose in Model 1 in Figure 1). The black nodes in Model 1 in Figure 1 depict a no-flow boundary condition.

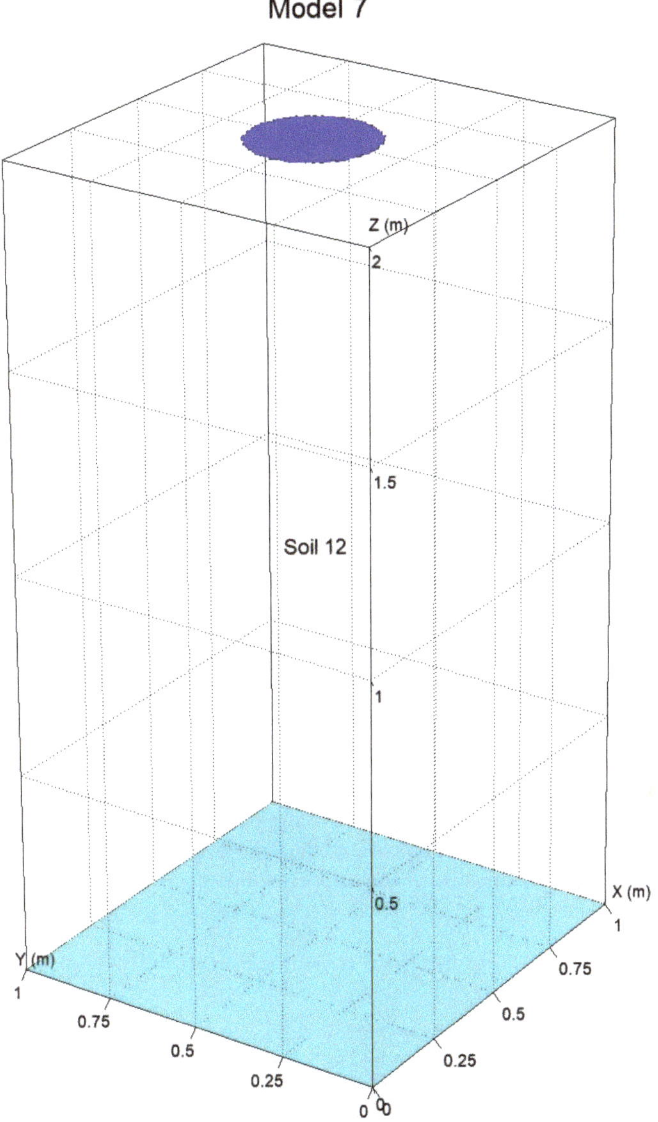

Figure 2. Overview over Model 7. Blue depicts a constant pressure head boundary condition, cyan depicts a free drainage boundary condition. The rest of the surface of the cuboid has a no-flow boundary condition which is not depicted.

5.2. Model 2

Model 2 is based on a physical experiment setup [41] and describes a small-scale levee of a homogenous soil. The foot-width of the model-scale levee is 3.38 m, the slope angle of 1:2 according to a height of 0.77 m. At the levee-crest, a rubber wall is installed having an anchoring depth of 100% of the dam height. Below the levee, a berm with a height of 0.3 m was added at both, the water and the valley side of the dam, to allow for exchange of the inflowing water with the groundwater. On the downstream face, a drainage tube was implemented to effectively discharge excess water (depicted in brown in Model 2 in Figure 1). The model is discretized into 11,896 nodes. The parameters from Soil 2 in Table 1 were used for the hydraulic properties of the homogenous soil layer, the soil is depicted in mistyrose in Model 2 in Figure 1. The black nodes in Model 2 in Figure 1 depict a no-flow boundary condition. A flood is simulated by a time dependent pressure head depicted in orange nodes in Model 2 in Figure 1. Precipitation is modeled by a time dependent infiltration rate depicted in red nodes in Model 2 in Figure 1. The olive nodes in Model 2 in Figure 1 model a seepage face where the flow is 0 and the pressure head is computed.

5.3. Model 3

Model 3 is a landfill site of 20 m height and 150 m width which is constructed with a surface sealing system using the principle of a capillary barrier. It consists of a coarse layer as capillary block (filling material, Soil 3 in Table 1, depicted in mistyrose in Model 3 in Figure 1) and a fine layer as cover (Soil 4 in Table 1, depicted in brown in Model 3 in Figure 1). The aim of the model is to prove the functionality of the sealing system under different precipitation scenarios and thicknesses of the cover layer. Therefore, the upper boundary condition (depicted in red in Model 3 in Figure 1) infiltrates water with varying intensity directly into the cover layer whose thickness ranges from 0.1 m to 0.3 m. It is modelled with 53,795 nodes. The yellow nodes in Model 3 in Figure 1 depict a constant pressure boundary condition where only outflow is allowed, whereas the black nodes in Model 3 in Figure 1 model a no-flow boundary condition.

5.4. Model 4

Model 4 is based on a physical experiment of a tipping trough. Its purpose is to simulate the proper functionality of a capillary barrier under different inflow scenarios, slope angles, and hydraulic properties of the built-in soil materials. The model used for the tests in this paper has a height of 0.6 m, a length of 6 m and a slope angle of 0.277 rad. Three different soil types are used: a cover layer (capillary layer, Soil 5 in Table 1, depicted in mistyrose in Model 4 in Figure 1), a bottom layer (capillary block, Soil 6 in Table 1, depicted in lightcyan in Model 4 in Figure 1) and a layer of filter sand (Soil 7 in Table 1, depicted in brown in Model 4 in Figure 1). On the upper left side, a water inlet with a time-dependent influx (depicted in red nodes in Model 4 in Figure 1) is simulated while balancing the outflow in the capillary block layer and in the filter sand (depicted in yellow nodes in Model 4 in Figure 1). The black nodes in Model 4 in Figure 1 portray a no-flow boundary condition. To account for the large gradient expected during the simulation run, the model is relatively fine discretized into 21,221 nodes.

5.5. Model 5

Model 5 is a column of 2 m height and 1 m in width. The properties of the soil are described as Soil 8 in Table 1, depicted in mistyrose in Model 5 in Figure 1. The model is infiltrated by a point source on the top of the column, depicted in blue in Model 5 in Figure 1. Aside from this, there is a no-flow boundary condition along the boundary of the model, depicted in black nodes in Model 5 in Figure 1. It is a synthetic, medium-scale model whose purpose is mainly to check the scalability of the numerical approach. It consists of 501,501 nodes.

5.6. Model 6

Model 6 simulates a mini sewage plant with two neighboring infiltration basins. One basin is for the infiltration of treated sewage water, the other is for the infiltration of rain. The dimensions of the model are 38 m in width and 12 m in height. The infiltration basins both have a width of 6.5 m. Under the infiltration basin for the rain water, there is a layer of gravel of 1.5 m thickness which is modelled with Soil 10 in Table 1, depicted in lightcyan in Model 6 in Figure 1. Under the infiltration basin for the treated sewage water, there is a layer of gravel with a thickness of 0.5 m which is modelled with Soil 10 from Table 1, depicted in lightcyan in Model 6 in Figure 1. Between these two basins, a barrier

(Soil 11 from Table 1, depicted in brown in Model 6 in Figure 1) is installed to prevent immediate mixing of the infiltrated waters. The rest of the model is modelled with the sandy Soil 9 from Table 1, depicted in mistyrose in Model 6 in Figure 1. The model is infiltrated through both infiltration basins with individual infiltration rates and following individual time patterns, depicted in green (for the infiltration basin for the rain) and grey (for the infiltration basin for the treated sewage water) in Model 6 in Figure 1. On the vertical boundaries of the model, there are two constant pressure head boundary conditions to simulate steady-state groundwater, depicted in blue in Model 6 in Figure 1. Aside from these two boundary conditions and the infiltration basins, there is a no-flow boundary condition along the boundary of the model, depicted in black in Model 6 in Figure 1. The model consists of 54,763 nodes.

5.7. Model 7

Model 7 is a rectangular cuboid of 2 m height, 1 m width, and 1 m depth. The properties of the soil are described as Soil 12 in Table 1. The model is infiltrated by a circular source with a diameter of 0.33 m on the top of the cuboid (portrayed in blue in Figure 2). On the bottom, there is a free drainage boundary condition (depicted in cyan in Figure 2), and aside from this boundary condition and the circular source, there is a no-flow boundary condition along the boundary of the model. It is a synthetic, large-scale model whose purpose is mainly to check the scalability of the numerical approach. It consists of 2,377,026 nodes.

6. Results and Discussion

6.1. Comparison of the Computational Running Time

In the comparison in Table 2, the median speedup of PCSiWaPro AMGCL over PCSiWaPro Original is 16.8%. However, there is a huge variance in the speedup. Figure 3 shows the relationship between the speedup achieved with PCSiWaPro AMGCL over PCSiWaPro Original in relation to the number of nodes in the model. One clearly sees, that PCSiWaPro AMGCL outperforms PCSiWaPro Original in models with 20,000 or more nodes in terms of speed. This can be explained in the following way: The creation of the algebraic Multigrid costs computational time. Also, operations on the grids cost computational time, but these costs are negligible for very coarse grids if the original grid is very fine. In AMGCL, the coarsest grid has at most 3000 nodes, therefore, if the original equation system has 10,000 nodes, operations on the coarsest grid cost one third of the computational cost of the same operation on the finest grid. If the original equation system has 300,000 nodes, the operations on the coarsest grid will cost one percent of the computational cost of the same operation on the finest grid. Therefore, with growing size of the problem, the cost for operating on the coarser grids becomes negligible. And this explains why the speedup of PCSiWaPro AMGCL over PCSiWaPro Original grows with the number of nodes in the model. The last two columns in Table 2 support this reasoning for models 1 to 3 and models 5 and 6: in the simulation runs of these models, PCSiWaPro Original and PCSiWaPro AMGCL called the solver approximately equally often, however, there is the huge difference in computational running time documented in the first two columns of Table 2. Since the code of PCSiWaPro AMGCL is identical with the code of PCSiWaPro Original except for the solver, this difference in computational running time can only be explained by the computational running time that is spent with the respective solvers. Since BICGSTAB and the Conjugate Gradient algorithm are similar in their computational complexity, the difference in computational running time can only be attributed to the computational complexity of the preconditioner and the quality with which the preconditioner approximates the inverse matrix.

Table 2. Overview over the computational running time and the number of calls of the solver of the simulations of models 1 to 7 with PCSiWaPro Original and PCSiWaPro AMGCL.

Model	Computational Running Time of PCSiWaPro Original	Computational Running Time of PCSiWaPro AMGCL	Speedup of PCSiWaPro AMGCL in Relation to PCSiWaPro Original	Number of Calls of the Solver in PCSiWaPro Original	Number of Calls of the Solver in PCSiWaPro AMGCL
-	s	s	%	-	-
1	168	258	−53.6	26,103	26,082
2	4	8	−100	370	370
3	1751	784	55.2	10,824	9459
4	5698	3696	35.1	239,754	126,298
5	1856	382	79.4	464	490
6	9318	7755	16.8	118,526	113,476
7	5730	4860	15.2	733	836

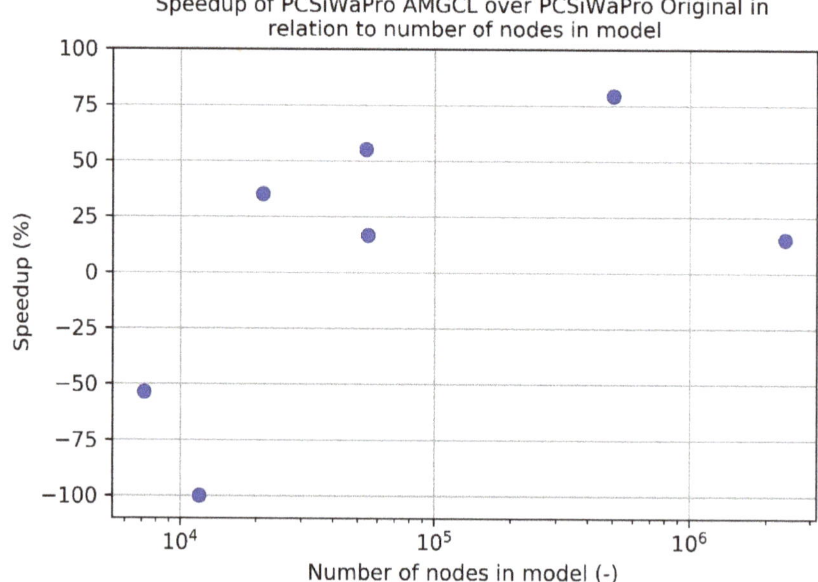

Figure 3. Comparison of the speedup achieved with PCSiWaPro AMGCL over PCSiWaPro Original in relation to the number of nodes in the model. One sees that PCSiWaPro AMGCL outperforms PCSiWaPro Original for models with 20,000 or more nodes.

During the simulation of Model 4 however, PCSiWaPro Original called the solver almost twice as often as PCSiWaPro AMGCL. One learns from Figure 4, that PCSiWaPro Original calls the solver more frequently in the second half of the simulation period than PCSiWaPro AMGCL. Thus, PCSiWaPro AMGCL achieves convergence easier than PCSiWaPro Original in the second half of the simulation period in Model 4.

The columns considering the number of solver calls in Table 2 tell, that PCSiWaPro AMGCL has problems with convergence in comparison with PCSiWaPro Original in Model 7 since PCSiWaPro AMGCL needs 14% more solver calls than PCSiWaPro Original. This also shows, that the average time spent with the solution of a given equation system arising from Model 7 is shorter in PCSiWaPro AMGCL than in PCSiWaPro Original, since PCSiWaPro AMGCL computes faster in general in Model 7 while making more calls to the solver than PCSiWaPro Original.

Therefore, models 4 and 7 show, that there is no single best solver: whereas in Model 4 PCSiWaPro AMGCL could achieve a speedup in comparison with PCSiWaPro Original due to its solver, in Model 7 the AMGCL solver achieves convergence with more solver calls than PCSiWaPro Original and thereby limits the speedup.

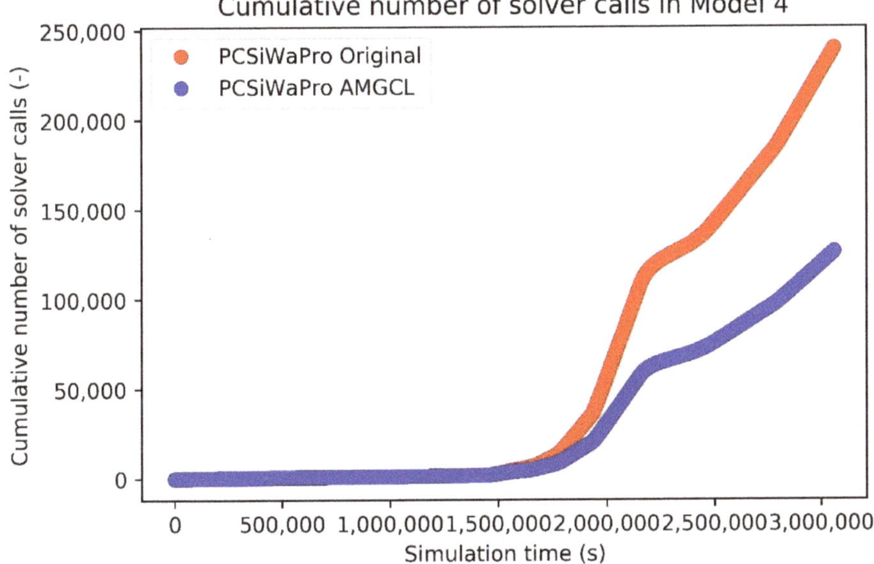

Figure 4. Comparison of the cumulative number of solver calls in Model 4. In the second half of the simulation period, PCSiWaPro AMGCL (blue) calls the solver less frequent than PCSiWaPro Original (red). This tells, that PCSiWaPro Original has greater problems with convergence than PCSiWaPro AMGCL.

6.2. Comparison of the R^2-Values between the Pressure Heads Computed with PCSiWaPro Original and PCSiWaPro AMGCL

Table 3 shows, that PCSiWaPro Original and PCSiWaPro AMGCL basically compute the same pressure heads. This means, that the speedup documented in Table 2 was not bought with a worsening of the quality of the computational results.

Table 3. Overview over the R^2-statistics between the pressure heads computed with PCSiWaPro Original and PCSiWaPro AMGCL.

Model	Minimal R^2	Maximal R^2	Mean R^2	Median R^2
-	-	-	-	-
1	1.0	1.0	1.0	1.0
2	1.0	1.0	1.0	1.0
3	1.0	1.0	1.0	1.0
4	1.0	1.0	1.0	1.0
5	1.0	1.0	1.0	1.0
6	0.9998	1.0	1.0	1.0
7	1.0	1.0	1.0	1.0

6.3. Comparison of the RCMBE Computed with PCSiWaPro Original and PCSiWaPro AMGCL

In Table 4, a comparison of the RCMBE computed with PCSiWaPro Original and PCSiWaPro AMGCL is given. As one can see from Table 4, the models were tuned quite well, which can be read from the columns that represent the RCMBE statistics for PCSiWaPro Original, because this code was used during the tuning of the models.

Table 4. Comparison of the RCMBE computed with PCSiWaPro Original and PCSiWaPro AMGCL.

Model	Relative Cumulative Mass Balance Error							
	%							
-	PCSiWaPro Original				PCSiWaPro AMGCL			
	Minimal	Maximal	Mean	Median	Minimal	Maximal	Mean	Median
1	0.0006	0.1621	0.0322	0.031	0.0004	0.7444	0.0164	0.013
2	0.0001	0.0043	0.0015	0.0013	0.0	0.0123	0.0027	0.0022
3	0.3390	0.70	0.4189	0.3983	0.4113	0.8826	0.6628	0.6623
4	0.0	0.0451	0.0037	0.0028	0.0004	0.2962	0.0192	0.0131
5	0.0116	0.9663	0.5696	0.6197	0.0055	1.8195	1.0265	1.1373
6	0.0	0.0373	0.0226	0.0201	0.0	0.2589	0.2097	0.223
7	0.0008	0.1787	0.0674	0.0602	0.0003	0.3137	0.1332	0.155

According to Table 4, only Model 5 needs to be investigated, since the RCMBE computed with PCSiWaPro AMGCL surpasses the threshold of 1%. In all the other models, this threshold is not violated by the computational results computed with PCSiWaPro AMGCL.

Figure 5 shows the RCMBE computed with PCSiWaPro Original (red) and with PCSiWaPro AMGCL (blue) in Model 5. In the first quarter of the simulation period, PCSiWaPro AMGCL violates the rule of keeping the RCMBE below 1%.

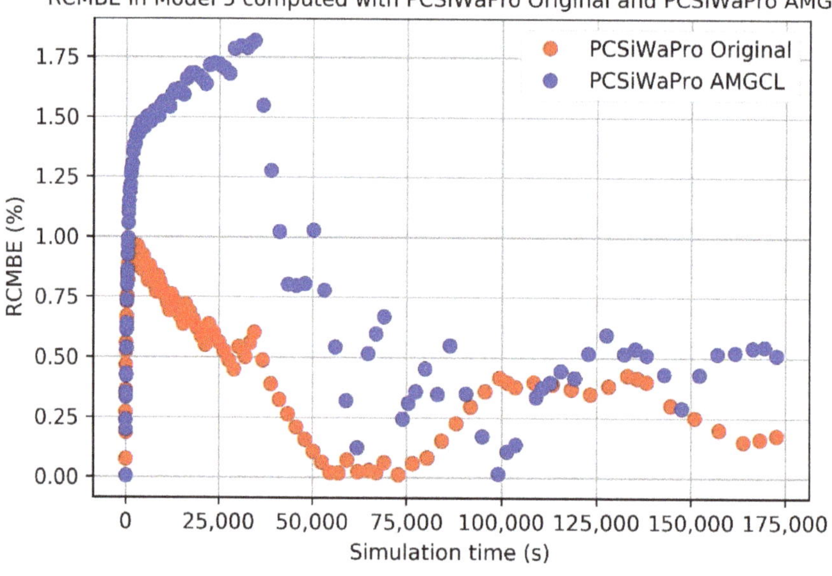

Figure 5. The RCMBE in Model 5 computed with PCSiWaPro Original (red) and PCSiWaPro AMGCL (blue). The results computed with PCSiWaPro Original do not violate the aim of a RCMBE of at most 1%. However, the results computed with PCSiWaPro AMGCL violate this rule in the first quarter of the simulation period.

Figure 6 shows the relationship between the computed influx and the computed RCMBE in both programs PCSiWaPro Original (red) and PCSiWaPro AMGCL (blue) for the first 10% of the simulation period in Model 5. PCSiWaPro solves Equation (1) for the pressure heads, therefore, since the boundary condition that creates the influx is a constant pressure head boundary condition, the computed pressure heads near this boundary condition must be very similar in both programs, otherwise the computed influx would not be identical in both programs. Therefore, the difference in the RCMBE computed with PCSiWaPro Original and PCSiWaPro AMGCL can only be explained by different computational results for the pressure heads of nodes that are not near the constant pressure head boundary condition. The existence of such nodes can be seen in the R^2-statistics in Table 3.

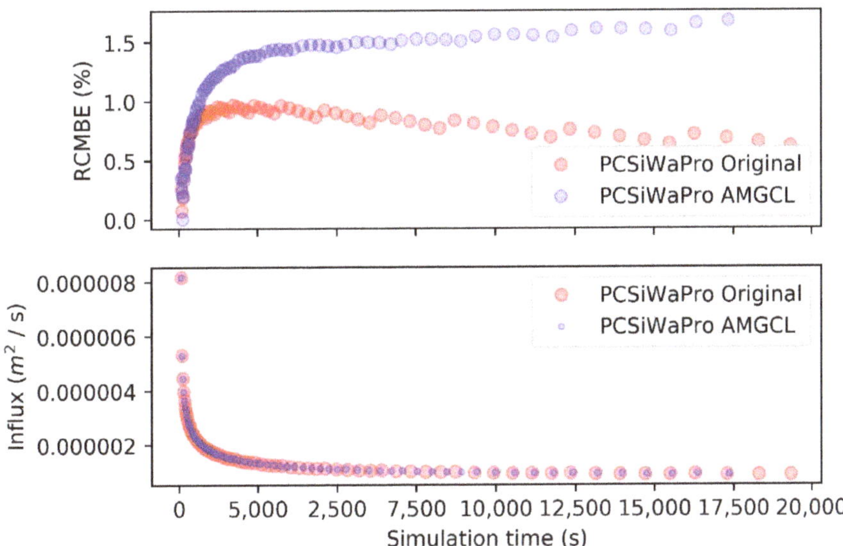

Figure 6. The relationship between the RCMBE computed with PCSiWaPro Original (red) and PCSiWaPro AMGCL (blue) and the computed influx for the first 10% of the simulation period in Model 5. Since the computed influx is identical in both programs, the difference in the mass balance must be explained by differently computed pressure heads.

This comparison of RCMBE shows that PCSiWaPro AMGCL performs slightly worse than PCSiWaPro Original.

7. Conclusions

The comparisons above show, that PCSiWaPro AMGCL computes results that are of similar quality as the results computed with PCSiWaPro Original while being faster for larger models (20,000+ nodes). In all cases considered in this paper, the R^2-values between the pressure heads computed with PCSiWaPro Original and PCSiWaPro AMGCL were in most models 1, and always greater than 0.999. Yet in one case, PCSiWaPro AMGCL computed results that violate against the rule to keep the RCMBE under 1%.

Also, for smaller models (<20,000 nodes), PCSiWaPro AMGCL computes slower than PCSiWaPro Original. Therefore, for larger models, PCSiWaPro Original may be replaced with PCSiWaPro AMGCL in order to speed up the computation.

The comparison of the number of calls of the solver showed, that the slower computational speed of PCSiWaPro AMGCL in comparison with PCSiWaPro Original in smaller models can only be explained by the numerical complexity of computing the algebraic Multigrid preconditioner. Therefore, for smaller models, the ILU-preconditioned Conjugate Gradient algorithm is a more efficient choice than BICGSTAB with algebraic Multigrid preconditioning.

This study only considered seepage flow that is modelled by Equation (1), but since BICCGSTAB can also deal with non-symmetric matrices [28], BICGSTAB preconditioned with an algebraic Multigrid with Smoothed Aggregation and SPAI0 relaxation may be tested on transport problems in the vadose zone.

Another direction for future research is to evaluate whether further tuning of the models can increase the quality of the computational results computed with PCSiWaPro AMGCL while still being faster than PCSiWaPro Original.

Since PCSiWaPro uses the numerical discretization of the Richards' equation that is also used in SWMS and HYDRUS, the findings of this study should also hold for SWMS and HYDRUS.

However, one finding of this study is the poor speedup caused by AMGCL with the 3-dimensional Model 7. The speedup of 15% is too little to be of practical relevance. Therefore, the performance of PCSiWaPro AMGCL has to be further evaluated for larger models. Unless more promising results are produced, we can only recommend to apply AMGCL in the simulation of small and medium scale problems (20,000 to 500,000 nodes).

Author Contributions: Conceptualization, R.P.; methodology, R.P.; software R.P. and R.B.; validation, R.P. and R.B.; formal analysis, R.P. and R.B.; investigation, R.P. and R.B., resources, R.P. and R.B.; data curation, R.P. and R.B.; write-original draft preparation, R.P. and R.B.; writing-review and editing, R.P. and R.B.; visualization, R.P. and R.B. All authors have read and agreed to the published version of the manuscript.

Funding: This work was funded by the Saxon Development Bank (grant number 100301390), the European Social Funds of the European Union, and the German Research Foundation (grant number LI 727/24-1). Open Access Funding by the Publication Fund of the TU Dresden

Acknowledgments: We thank Denis Demidov for writing, maintaining and publishing of AMGCL, the solver library that was used in this study. We also want to express our gratitude to him due to his eagerness to answer our questions. Also, we would like to thank Falk Händel for providing Model 6. Furthermore, we thank Peter-Wolfgang Gräber and Rudolf Liedl for fruitful discussions about numerical mathematics.

Conflicts of Interest: The authors declare no conflict of interest. The funders had no role in the design of the study; in the collection, analyses, or interpretation of data; in the writing of the manuscript, or in the decision to publish the results.

References

1. Saffigna, P.G.; Keeney, D.R. Nitrate and Chloride in Ground Water under Irrigated Agriculture in Central Wisconsin. *Groundwater* **1977**, *15*, 170–177. [CrossRef]

2. Arnon, S.; Dahan, O.; Elhanany, S.; Cohen, K.; Pankratov, I.; Gross, A.; Ronen, Z.; Baram, S.; Shore, L.S. Transport of Testosterone and Estrogen from Dairy-Farm Waste Lagoons to Groundwater. *Environ. Sci. Technol.* **2008**, *42*, 5521–5526. [CrossRef] [PubMed]

3. McKinley, J.P.; Zachara, J.M.; Wan, J.; McCready, D.E.; Heald, S.M. Geochemical Controls on Contaminant Uranium in Vadose Hanford Formation Sediments at the 200 Area and 300 Area, Hanford Site, Washington. *Vadose Zone J.* **2007**, *6*, 1004–1017. [CrossRef]

4. Konečný, F.; Boháček, Z.; Müller, P.; Kovářová, M.; Sedláčková, I. Contamination of soils and groundwater by petroleum hydrocarbons and volatile organic compounds—Case study: ELSLAV BRNO. *Bull. Geosci.* **2003**, *78*, 225–239.

5. Richardson, M.L.; Bowron, J.M. The fate of pharmaceutical chemicals in the aquatic environment. *J. Pharm. Pharmacol.* **1985**, *37*, 1–12. [CrossRef]

6. Yaron, B.; Dror, I.; Berkowitz, B. Contaminant-induced irreversible changes in properties of the soil-vadose-aquifer zone: An overview. *Chemosphere* **2008**, *71*, 1409–1421. [CrossRef]

7. Zhang, Y.K.; Schilling, K.E. Effects of land cover on water table, soil moisture, evapotranspiration, and groundwater recharge: A Field observation and analysis. *J. Hydrol.* **2006**, *319*, 328–338. [CrossRef]

8. Schumann, H.H.; Poland, J.F. Land subsidence, earth fissures, and ground-water withdrawal in south-central Arizona. In Proceedings of the 1st international symposium on land subsidence, Tokyo, Japan, 17–22 September 1969; International Association of Hydrological Sciences: Wallingford, UK, 1970; pp. 295–302.

9. Iverson, R.M.; Major, J.J. Groundwater Seepage Vectors and the Potential for Hillslope Failure and Debris Flow Mobilization. *Water Resour. Res.* **1986**, *22*, 1543–1548. [CrossRef]

10. Richards, L.A. Capillary Conduction of Liquids through Porous Mediums. *Physics* **1931**, *1*, 318–333. [CrossRef]

11. Demidov, D. AMGCL: An Efficient, Flexible, and Extensible Algebraic Multigrid Implementation. *Lobachevskii J. Math.* **2019**, *40*, 535–546. [CrossRef]

12. Gräber, P.W.; Blankenburg, R.; Kemmesies, O.; Krug, S. SiWaPro DSS-Beratungssystem zur Simulation von Prozessen der unterirdischen Zonen. In *Simulationen in Umwelt- und Geowissenschaften*; Wittmann, J., Müller, M., Eds.; Shaker: Leipzig, Germany, 2006; pp. 225–234.

13. Diersch, H.J.G. FEFLOW. Finite Element Modeling of Flow, Mass and Heat Transport in Porous and Fractured Media. Springer: Berlin/Heidelberg, Germany; New York, NY, USA; Dordrecht, The Netherlands; London, UK, 2014; pp. 1–1018.

14. Lin, H.C.J.; Richards, D.R.; Talbot, C.A.; Yeh, G.T.; Cheng, J.R.; Cheng, H.P.; Jones, N.L. *FEMWATER: A Three-Dimensional Finite Element Computer Model for Simulating Density-Dependent Flow and Transport in Variably Saturated Media*; Technical Report CHL-97-12; U.S. Army Corps of Engineers, Waterways Experiment Station: Vicksburg, MS, USA, 1997; pp. 1–143.

15. Therrien, R.; McLaren, R.G.; Sudicky, E.A.; Panday, S.M. *HydroGeoSphere. A Three-Dimensional Numerical Model Describing Fully-Integrated Subsurface and Surface Flow and Solute Transport*; Groundwater Simulation Group: Leve, QC Canada, 2010; pp. 1–430.

16. Šimůnek, J.; Šejna, M. *HYDRUS. Technical Manual. Version 2*; PC-Progress: Prague, Czech Republic, 2011; pp. 1–230.

17. White, M.D.; Oostrom, M.; Lenhard, R.J. Modeling fluid flow and transport in variably saturated porous media with the STOMP simulator. 1. Nonvolatile three-phase model description. *Adv. Water Resour.* **1995**, *18*, 353–364. [CrossRef]

18. Šimůnek, J.; Vogel, T.; van Genuchten, M.T. *The SWMS_2D Code for Simulating Water Flow and Solute Transport in Two-Dimensional Variably Saturated Media. Version 1.21*; U.S. Department of Agriculture, Agricultural Research Service, U.S. Salinity Laboratory: Riverside, CA, USA, 1994; pp. 1–197.

19. Jung, Y.; Pau, G.S.H.; Finsterle, S.; Doughty, C. *TOUGH3 User's Guide. Version 1.0*; Energy Geosciences Division, Lawrence Berkeley National Laboratory, University of Berkeley: Berkeley, CA, USA, 2018; pp. 1–163.

20. Lappala, E.G.; Healy, R.W.; Weeks, E.P. *Documentation of Computer Program Vs2d to Solve the Equations of Fluid Flow in Variably Saturated Porous Media*; Water-Resources Investigations Report 83-4099; U.S. Geological Survey: Denver, CO, USA, 1987; pp. 1–184.

21. Yeh, T.C.J.; Srivastava, R.; Guzman, A.; Harter, T. A Numerical Model for Water Flow and Chemical Transport in Variably Saturated Porous Media. *Ground Water* **1993**, *31*, 634–644. [CrossRef]

22. Burden, R.L.; Faires, J.D. *Numerical Analysis*, 9th ed.; Brooks/Cole: Boston, MA, USA, 2011; pp. 358–430, 684–696, 746–760.

23. Moukalled, F.; Mangani, L.; Darwish, M. *The Finite Volume Method in Computational Fluid Dynamics. An Advanced Introduction with OpenFOAM® and Matlab®*; Springer International Publishing: Cham/Heidelberg, Germany; New York, NY, USA; Dordrecht, The Netherlands; London, UK; Cham Switzerland, 2016; pp. 103–136.

24. Eymard, R.; Gutnic, M.; Hilhorst, D. The finite volume method for Richards equation. *Comput. Geosci.* **1999**, *3*, 259–294. [CrossRef]

25. Casulli, V.; Zanolli, P. A Nested Newton-Type Algorithm for Finite Volume Methods Solving Richards' Equation in Mixed Form. *SIAM J. Sci. Comput.* **2010**, *32*, 2255–2273. [CrossRef]

26. Liu, X. Parallel Modeling of Three-dimensional Variably Saturated Ground Water Flows with Unstructured Mesh using Open Source Finite Volume Platform Openfoam. *Eng. Appl. Comput. Fluid Mech.* **2012**, *7*, 223–238.

27. Orgogozo, L.; Renon, N.; Soulaine, C.; Hénon, F.; Tomer, S.K.; Labat, D.; Pokrovsky, O.S.; Sekhar, M.; Ababou, R.; Quintard, M. An open source massively parallel solver for Richards equation: Mechanistic modelling of water fluxes at the watershed scale. *Comput. Phys. Commun.* **2014**, *185*, 3358–3371. [CrossRef]

28. Saad, Y. *Iterative Methods for Sparse Linear Systems*, 2nd ed.; Society for Industrial and Applied Mathematics: Philadelphia, PA, USA, 2003; pp. 1–547.

29. Stone, H.L. Iterative Solution of Implicit Approximations of Multidimensional Partial Differential Equations. *SIAM J. Numer. Anal.* **1968**, *5*, 530–558. [CrossRef]

30. Grote, M.J.; Huckle, T. Parallel Preconditioning with Sparse Approximate Inverses. *SIAM J. Sci. Comput.* **1997**, *18*, 838–853. [CrossRef]

31. Bröker, O.; Grote, M.J. Sparse approximate inverse smoothers for geometric and algebraic multigrid. *Appl. Numer. Math.* **2002**, *41*, 61–80. [CrossRef]
32. Stüben, K. *Algebraic Multigrid (AMG): An Introduction with Applications*; GMD-Report 70; German National Research Center for Information Technology (GMD), Institute for Algorithms and Scientific Computing (SCAI): St. Augustin, Germany, November 1999; pp. 1–127.
33. Vaněk, P.; Mandel, J.; Brezina, M. Algebraic multigrid by smoothed aggregation for second and fourth order elliptic problems. *Computing* **1996**, *56*, 179–196. [CrossRef]
34. Dagum, L.; Menon, R. OpenMP: An Industry-Standard API for Shared-Memory Programming. *IEEE Comput. Sci. Eng.* **1998**, *5*, 46–55. [CrossRef]
35. Nickolls, J.; Buck, I.; Garland, M.; Skadron, K. Scalable Parallel Programming with CUDA. *Queue* **2008**, *6*, 40–53. [CrossRef]
36. Message P Forum. *MPI: A Message-Passing Interface Standard*; Technical Report; University of Tennessee: Knoxville, TN, USA, 1994; pp. 1–228.
37. DIN, e.V. *DIN 4220:2008-11. Bodenkundliche Standortbeurteilung—Kennzeichnung, Klassifizierung und Ableitung von Bodenkennwerten (Normative und Nominale Skalierungen)*; Beuth: Berlin, Germany, 2008; pp. 1–50.
38. Šimůnek, J.; Huang, K.; van Genuchten, M.T. *The SWMS_3D Code for Simulating Water Flow and Solute Transport in Three-Dimensional Variably Saturated Media. Version 1.0*; U.S. Department of Agriculture, Agricultural Research Service, U.S. Salinity Laboratory: Riverside, CA, USA, 1995; pp. 1–155.
39. Van Genuchten, M.T. A Closed-form Equation for Predicting the Hydraulic Conductivity of Unsaturated Soils. *Soil Sci. Soc. Am. J.* **1980**, *44*, 892–898. [CrossRef]
40. Luckner, L.; van Genuchten, M.T.; Nielsen, N.R. A Consistent Set of Parametric Models for the Two-Phase Flow of Immiscible Fluids in the Subsurface. *Water Resour. Res.* **1989**, *25*, 2187–2193. [CrossRef]
41. Aigner, D. *Auswertung von Untersuchungen Über den Einsatz Einer Gummispundwand Sowie Einer Sickerleitung an Einem Durchströmten Modelldeich*; Technische Universität Dresden, Institut für Wasserbau und Technische Hydromechanik: Dresden, Germany, 2004.

Technical Note

Exploring Compatibility of Sherwood-Gilland NAPL Dissolution Models with Micro-Scale Physics Using an Alternative Volume Averaging Approach

Scott K. Hansen

Zuckerberg Institute for Water Research, Ben-Gurion University of the Negev, Beer-Sheva 84105, Israel; skh@bgu.ac.il

Received: 21 May 2019; Accepted: 16 July 2019; Published: 23 July 2019

Abstract: The dynamics of NAPL dissolution into saturated porous media are typically modeled by the inclusion of a reaction term in the advection-dispersion-reaction equation (ADRE) with the reaction rate defined by a Sherwood-Gilland empirical model. This stipulates, among other things, that the dissolution rate is proportional to a power of the NAPL volume fraction, and also to the difference between the local average aqueous concentration of the NAPL species and its thermodynamic saturation concentration. Solute source models of these sorts are ad hoc and empirically calibrated but have come to see widespread use in contaminant hydrogeology. In parallel, a number of authors have employed the method of volume averaging to derive upscaled transport equations describing the same dissolution and transport phenomena. However, these solutions typically yield forms of equations that are seemingly incompatible with Sherwood-Gilland source models. In this paper, we revisit the compatibility of the two approaches using a radically simplified alternative volume averaging analysis. We begin from a classic micro-scale formulation of the NAPL dissolution problem but develop some new simplification approaches (including a physics-preserving transformation of the domain and a new geometric lemma) which allow us to avoid solving traditional closure boundary value problems. We arrive at a general, volume-averaged governing equation that does not reduce to the ADRE with a Sherwood-Gilland source but find that the two approaches do align under straightforward advection-dominated conditions.

Keywords: NAPL; volume averaging; upscaling; mass transfer

---

1. Introduction

We consider the modeling of dissolution of residual non-aqueous phase liquids (NAPL)—that is, liquids that are essentially immiscible in water and which are trapped by capillary forces in porous media—into pore water and the subsequent transport of solute originating in the NAPL. Transport can be modeled without difficulty by means of the advection-dispersion-reaction equation (ADRE):

$$\frac{\partial c}{\partial t} + v \cdot \nabla c - \nabla \cdot D_{\text{eff}} \nabla c = Q, \tag{1}$$

where v is the pore water velocity, D_{eff} is an effective Fickian dispersion coefficient capturing the effects of local-scale hydrodynamic dispersion and diffusion, and Q is a source term. (See the nomenclature in Appendix A for units and definitions of all symbols.) However, it is necessary to have a model of Q that captures the physics of NAPL dissolution. To this end, a number of similar empirical models of Q have been proposed [1–6] which are of the form $Q = K\left(c_{\text{sat}} - c\right)$, where c_{sat} is the thermodynamic saturation aqueous concentration of the NAPL chemical species and K is a mass transfer coefficient. These models, known as Sherwood-Gilland models [7], most generally express K as

$$K = \alpha \frac{\mathscr{D}}{d_m^2} \eta_n^\beta \text{Re}^\gamma, \tag{2}$$

where α, β and γ are fitting parameters, $\mathscr{D}$ is the Fickian diffusivity, d_m is a characteristic length of the porous media, η_n is the bulk volume fraction of NAPL, and Re is the Reynolds number. Certain empirical studies combine terms and do not calibrate all of α, β and γ, but with the exception of [6], all contain terms η_n^β, or equivalently S_n^β (with NAPL saturation expressed as a fraction of pore volume rather than bulk volume), with calibrated values of 0.6 to 1.24 for β ([8], see Table 2) The theory underlying the form of such models is out of scope for this paper but is discussed in more detail in References [1,8]. The importance of the Sherwood-Gilland approach lies in the relation of Q to local concentration in the water phase and a small number of measurable system parameters.

By contrast with the Sherwood-Gilland empirical approach, the upscaled dynamics of residual NAPL dissolution have also been considered using volume averaging theory [9]. While there is not room for a full account here, volume averaging operates by defining a superficial volume average operator

$$\langle \cdot \rangle = \frac{1}{\mathscr{V}} \int_{V(x)} \cdot \, dV, \tag{3}$$

where V is a simply-connected region of volume $\mathscr{V}$ centered at x. From the superficial average, phase-intrinsic volume average operators are defined for each phase i according to $\langle \cdot \rangle = \eta_i \langle \cdot \rangle^i$, where i stands in for any of the three phases in the system: water (w), NAPL (n), and solid (s) and $\eta_i(x)$ is the volume fraction of $V(x)$ occupied by the i phase. Inside $V(x)$, the surface between phases i and j is denoted by Γ_{ij} and an infinitesimal element of that surface is denoted by dA_{wn}. The unit vector normal to Γ_{ij}, pointing into phase j, is denoted by n_{ij}. Although all averaged quantities are functions of location, x, and time, t, these parameters are generally suppressed. Employing these concepts, we may state a volume averaging theorem [9–11] for c (and analogously for other scalar quantities),

$$\langle \nabla c \rangle = \nabla \langle c \rangle + \sum_{j \neq w} \frac{1}{\mathscr{V}} \int_{\Gamma_{wj}} c n_{wj} dA_{wj}, \tag{4}$$

and for v (and analogously for other vector-valued quantities),

$$\langle \nabla \cdot v \rangle = \nabla \cdot \langle v \rangle + \sum_{j \neq w} \frac{1}{\mathscr{V}} \int_{\Gamma_{wj}} v \cdot n_{wj} dA_{wj}. \tag{5}$$

The essence of the volume averaging technique is that spatially-nonuniform dependent variables are partitioned into volume averages and small-scale fluctuations and the governing PDE for the un-averaged dynamics is rewritten in terms of both the volume average quantities and spatial integrals of the fluctuations using (4) and (5). Finally, an equation in terms of the volume averaged terms alone is developed by application of physical and mathematical knowledge to rewrite the fluctuation integrals in terms of the volume averaged quantities [12]. Commonly, the closure is accomplished by specifying boundary value problems (BVPs) for the fluctuations in which the volume average quantities participate as forcing functions, as in Reference [9]. If one is only interested in the form of the volume averaged PDEs, these IBVPs do not need to be solved explicitly. If quantitative results are required, then analytical [13] or numerical [14,15] solution of the IBVP in a characteristic cell may be indicated, using for example, Lattice-Boltzmann methods [16].

The volume averaging approach has been applied to the dissolution of residual NAPL, most notably in a pair of papers by Quintard and Whitaker [13,17], each employing a different micro-scale formulation but both arriving at the same upscaled (i.e., volume averaged) equation, which is apparently incompatible with (1):

$$\eta_w \frac{\partial \langle c \rangle^w}{\partial t} + \left[\eta_w \langle v \rangle^w - u \right] \cdot \nabla \langle c \rangle^w - \nabla \cdot \left(D_w^* \cdot \nabla \langle c \rangle^w \right) + \nabla \eta_w \cdot \mathscr{D} \nabla \langle c \rangle^w$$
$$= \nabla \cdot \left[d_w \left(\langle c \rangle^w - c_{\text{sat}} \right) \right] - \kappa \left(\langle c \rangle^w - c_{\text{sat}} \right). \quad (6)$$

Equation (6) contains two quantities, D_w^* and d_w, that are functions of the formal solutions to two distinct closure problems, along with u and κ which are forcing quantities in those respective closure problems and which are themselves formally defined in terms of micro-scale surface integrals that must be expressed in terms of larger-scale quantities to fully close the problem. See Reference [13] for full details on the underlying assumptions and the volume averaging procedure. An identical treatment of the water-NAPL mass transfer equations appears in the four-phase (also including biofilm) analysis of Reference [18], with a minor simplification appearing in Reference [19], which embeds an apparent assumption that d_w is spatially-uniform. An alternative analysis which makes different closure assumptions [20] also arrives at an equation with two sources that seemingly cannot be re-expressed in Sherwood-Gilland form.

In Reference [17], the authors show that an ADRE with Sherwood-Gilland source term can be derived by imposing four scale restrictions that they dub "dominant convective transport." However, some of these scale restrictions involve the quantities D_w^* and d_w, which do not have a straightforward interpretation, and so it is difficult to know when or if they are satisfied. In a similar study on biofilm growth (where the "NAPL" grows according to Monod kinetics, rather than shrinks by dissolution) [21], the authors obtained a closed, classic ADRE and in a follow-up study [22], the authors showed that either ADRE or non-classical volume average behavior occurred, depending on the coupling assumptions employed.

Informed by these works, in this note we reconsider the compatibility of volume averaged micro-scale physics and the ADRE with a Sherwood-Gilland source. We begin from the microscopic formulation employed by Reference [13]. However, we perform different simplifying analysis, obviating the need for formal solution of closure boundary value problems relating fluctuations to volume averaged quantities, yielding a slightly different volume averaged governing equation in terms of independent variables η_n and $\langle c \rangle^w$, still ostensibly with multiple sources, which is incompatible with (1). We then show how using any Sherwood-Gilland model as a second governing equation relating η_n and $\langle c \rangle^w$ causes one of the source terms to be negligible under advection dominated conditions much more straightforward than those in [17], allowing emergence of the expected form of upscaled ADRE:

$$\frac{\partial \langle c \rangle^w}{\partial t} + \langle v \rangle^w \cdot \nabla \langle c \rangle^w - \nabla \cdot D_{\text{eff}} \nabla \langle c \rangle^w \propto \eta_n^\beta \left(c_{\text{sat}} - \langle c \rangle^w \right). \quad (7)$$

2. Formulation

2.1. Governing Equations

We are attempting to model a three-phase system, a characteristic portion of which is illustrated in Figure 1, each phase of which occupies a respective volume fraction of averaging domain $V(x)$ $\eta_w(\mathbf{x}, t)$, $\eta_n(\mathbf{x}, t)$, and η_s, which satisfy

$$\eta_w + \eta_n + \eta_s = 1. \quad (8)$$

We assume that the Darcy flux, q, for the system as a whole is fixed and satisfies $\nabla \cdot q = 0$. Then, it follows from the chain rule and the fact that $\langle v \rangle^w = q / \eta_w$, that

$$\nabla \cdot \langle v \rangle^w = -\frac{q}{\eta_w^2} \cdot \nabla \eta_w. \quad (9)$$

The ADE applies in the water phase:

$$\frac{\partial c}{\partial t} + \nabla \cdot (cv) = \nabla \cdot (\mathcal{D}\nabla c).$$ (10)

The solid and NAPL phases are homogeneous and do not need internal mass balance equations. However, the NAPL phase shrinks over time, yielding space to the water phase. This mass balance can be expressed by:

$$\rho_n \frac{\partial \eta_n}{\partial t} = \frac{1}{\mathcal{V}} \int_{\Gamma_{wn}} \mathcal{D}\nabla c \cdot n_{wn} dA_{wn}.$$ (11)

We work extensively with this term and so make the definition

$$Q \equiv -\rho_n \frac{\partial \eta_n}{\partial t},$$ (12)

where Q represents the upscaled source per unit time from the NAPL phase into the water phase.

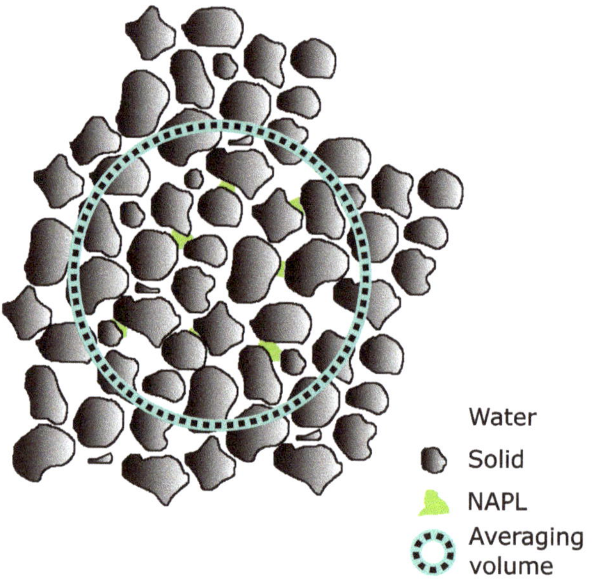

Figure 1. Schematic diagram of blobs of residual NAPL in pores defined by solid media otherwise saturated with water, with averaging volume superimposed.

2.2. Coupling Conditions

We assume no-slip boundary conditions, local equilibrium at the surface of the NAPL and no flux into the solid. Mathematically, these boundary conditions are expressed:

$$v = 0 \ x \epsilon \Gamma_{wn},$$ (13)

$$v = 0 \ x \epsilon \Gamma_{ws},$$ (14)

$$c = c_{\text{sat}} \ x \epsilon \Gamma_{wn},$$ (15)

$$\nabla c \cdot n_{wn} = 0 \ x \epsilon \Gamma_{ws}.$$ (16)

The solid volume fraction does not change with time, so the volume fractions can be coupled according to the following relationship:

$$\frac{\partial \eta_w}{\partial t} = -\frac{\partial \eta_n}{\partial t}. \tag{17}$$

3. Simplified Volume Averaging Analysis

We begin by applying the superficial averaging operator $\langle \cdot \rangle$ to (10), yielding

$$\frac{\partial \langle c \rangle}{\partial t} + \langle \nabla \cdot (cv) \rangle = \langle \nabla \cdot (\mathscr{D}\nabla c) \rangle. \tag{18}$$

We manipulate each of the three terms separately to write them all in terms of intrinsic volume averages, simplifying each. We then assemble them and simplify further.

3.1. Accumulation Term

There is next to no analysis required here. The accumulation term can be expressed in terms of intrinsic volume averages directly

$$\frac{\partial \langle c \rangle}{\partial t} = \eta_w \frac{\partial \langle c \rangle^w}{\partial t} + \frac{\partial \eta_w}{\partial t} \langle c \rangle^w. \tag{19}$$

3.2. Advection Term

From volume averaging theorem (5):

$$\langle \nabla \cdot (cv) \rangle = \nabla \cdot \langle (cv) \rangle + \frac{1}{\mathscr{V}} \int_{\Gamma_{wn}} cv \cdot n_{wn} dA_{wn} + \frac{1}{\mathscr{V}} \int_{\Gamma_{ws}} cv \cdot n_{ws} dA_{ws}, \tag{20}$$

from which it follows from no-slip conditions (13) and (14), that $\langle \nabla \cdot (cv) \rangle = \nabla \cdot \langle cv \rangle = \nabla \cdot \eta_w \langle cv \rangle^w$. Further simplification follows from writing c and v in terms of their volume average and a perturbation: $c = \langle c \rangle^w + \tilde{c}$, and $v = \langle v \rangle^w + \tilde{v}$. Then, we may write $\langle cv \rangle^w = \langle c \rangle^w \langle v \rangle^w + \langle \tilde{c}\tilde{v} \rangle^w$. We have

$$\begin{aligned}
\langle \nabla \cdot (cv) \rangle &= \nabla \cdot \left(\eta_w \langle c \rangle^w \langle v \rangle^w \right) + \nabla \cdot \left(\eta_w \langle \tilde{c}\tilde{v} \rangle^w \right) \\
&= \nabla \eta_w \cdot \langle c \rangle^w \langle v \rangle^w + \eta_w \nabla \langle c \rangle^w \cdot \langle v \rangle^w + \eta_w \langle c \rangle^w \nabla \cdot \langle v \rangle^w + \eta_w \nabla \cdot \langle \tilde{c}\tilde{v} \rangle^w + \nabla \eta_w \cdot \langle \tilde{c}\tilde{v} \rangle^w
\end{aligned} \tag{21}$$

Note that we can manipulate this by applying (9), so that

$$\eta_w \langle c \rangle^w \nabla \cdot \langle v \rangle^w = \eta_w \langle c \rangle^w \left(-\frac{q}{\eta_w^2} \cdot \nabla \eta_w \right) = -\nabla \eta_w \langle c \rangle^w \cdot \langle v \rangle^w, \tag{22}$$

which can be back-substituted into (21), canceling and leading to the conclusion

$$\langle \nabla \cdot (cv) \rangle = \eta_w \nabla \langle c \rangle^w \cdot \langle v \rangle^w + \eta_w \nabla \cdot \langle \tilde{c}\tilde{v} \rangle^w + \nabla \eta_w \cdot \langle \tilde{c}\tilde{v} \rangle^w. \tag{23}$$

For our purposes, we are primarily interested in the form of the source terms. It may thus be reasonable to employ a mean field approximation and discard the small double fluctuation term $\langle \tilde{c}\tilde{v} \rangle$, however we retain it presently, later to be incorporated into an effective dispersion.

3.3. Diffusion Term

Again applying volume averaging theorem (5), we see that

$$\langle \nabla \cdot (\mathscr{D}\nabla c)\rangle = \nabla \cdot \langle \mathscr{D}\nabla c\rangle + \frac{\mathscr{D}}{\mathcal{V}} \int_{\Gamma_{wn}} \nabla c \cdot \boldsymbol{n}_{wn} dA_{wn} + \frac{\mathscr{D}}{\mathcal{V}} \int_{\Gamma_{ws}} \nabla c \cdot \boldsymbol{n}_{ws} dA_{ws}, \tag{24}$$

or, by application of the no-flux boundary conditions, that

$$\langle \nabla \cdot (\mathscr{D}\nabla c)\rangle = \nabla \cdot \mathscr{D} \langle \nabla c\rangle + \frac{\mathscr{D}}{\mathcal{V}} \int_{\Gamma_{wn}} \nabla c \cdot \boldsymbol{n}_{wn} dA_{wn}, \tag{25}$$

where we observe by direct inspection that the second term is a volumetric NAPL sink, representing diffusion of solute from the water phase *into* the NAPL. We thus replace it with the symbol $-Q$, as defined in (12), emphasizing this role. By another application of an averaging theorem (4), it follows that

$$\nabla \cdot \mathscr{D} \langle \nabla c\rangle = \mathscr{D}\nabla \cdot \nabla \langle c\rangle + \mathscr{D}\nabla \cdot \left[\frac{1}{\mathcal{V}} \int_{\Gamma_{ws}} c\boldsymbol{n}_{ws} dA_{ws}\right] + \mathscr{D}\nabla \cdot \left[\frac{1}{\mathcal{V}} \int_{\Gamma_{wn}} c\boldsymbol{n}_{wn} dA_{wn}\right], \tag{26}$$

We now strongly diverge from the approach in Reference [13]. We eliminate the middle term on the RHS by noting that the average of $\boldsymbol{n}_{ws}$ at the elements of Γ_{ws} intersecting any plane is $\boldsymbol{0}$, and that V can be partitioned into planar slices, each orthogonal to $\nabla \langle c\rangle$, on which the expected value of c is constant. Assuming ergodicity, we conclude that the portion in square brackets is $\boldsymbol{0}$, independent of location, meaning that its divergence is also zero. To simplify the third term, we apply a result developed in Appendix B ($\frac{1}{\mathcal{V}} \int_{\Gamma_{wn}} \boldsymbol{n}_{wn} dA_{wn} = -\nabla \eta_w$). Observing that $c = c_{sat}$ everywhere on Γ_{wn}, we see that:

$$\nabla \cdot \mathscr{D} \langle \nabla c\rangle = \eta_w \mathscr{D}\nabla^2 \langle c\rangle^w + 2\mathscr{D}\nabla \eta_w \cdot \nabla \langle c\rangle^w + \mathscr{D}\nabla^2 \eta_w \left[\langle c\rangle^w - c_{sat}\right]. \tag{27}$$

Thus,

$$\langle \nabla \cdot (\mathscr{D}\nabla c)\rangle = \eta_w \mathscr{D}\nabla^2 \langle c\rangle^w + 2\mathscr{D}\nabla \eta_w \cdot \nabla \langle c\rangle^w + \mathscr{D}\nabla^2 \eta_w \left[\langle c\rangle^w - c_{sat}\right] - Q. \tag{28}$$

3.4. Assembly and Simplification of Governing Equation

Inserting (19), (23), and (28) into (18), and placing the variables in each of the terms in a consistent order yields:

$$\eta_w \frac{\partial \langle c\rangle^w}{\partial t} + \frac{\partial \eta_w}{\partial t} \langle c\rangle^w + \eta_w \langle \boldsymbol{v}\rangle^w \cdot \nabla \langle c\rangle^w + \eta_w \nabla \cdot \langle \tilde{c}\tilde{\boldsymbol{v}}\rangle^w + \nabla \eta_w \cdot \langle \tilde{c}\tilde{\boldsymbol{v}}\rangle^w = -\mathscr{D}\nabla^2 \eta_w c_{sat} - Q. \tag{29}$$

We here diverge again from the customary volume averaging approach by immediately making the closure approximation $\langle \tilde{c}\tilde{\boldsymbol{v}}\rangle^w = -k\nabla \langle c\rangle^w$. Customarily, this relationship would emerge from solution of a closure problem, as in Reference [9]. However, this particular dispersive flux relationship is "common practice" [23] (p. 174) textbook material (see also Reference [24], p. 244) and does not affect the form of the NAPL source term, which is the focus of the analysis in this note. Thus, we take the liberty of skipping its derivation.

$$\eta_w \frac{\partial \langle c\rangle^w}{\partial t} + \frac{\partial \eta_w}{\partial t} \langle c\rangle^w + \eta_w \langle \boldsymbol{v}\rangle^w \cdot \nabla \langle c\rangle^w - k\eta_w \nabla^2 \langle c\rangle^w - k\nabla \eta_w \cdot \nabla \langle c\rangle^w$$
$$= \eta_w \mathscr{D}\nabla^2 \langle c\rangle^w + 2\mathscr{D}\nabla \eta_w \cdot \nabla \langle c\rangle^w + \mathscr{D}\nabla^2 \eta_w \left[\langle c\rangle^w - c_{sat}\right] - Q. \tag{30}$$

This can be rearranged with its LHS containing only standard transport terms, and its RHS "source" terms relating to the dissolution of the NAPL:

$$\eta_w \frac{\partial \langle c \rangle^w}{\partial t} + \eta_w \langle v \rangle^w \cdot \nabla \langle c \rangle^w - \eta_w \left(\mathscr{D} + k \right) \nabla^2 \langle c \rangle^w$$
$$= \left(2\mathscr{D} + k \right) \nabla \eta_w \cdot \nabla \langle c \rangle^w + \mathscr{D} \nabla^2 \eta_w \left[\langle c \rangle^w - c_{\text{sat}} \right] - Q - \frac{\partial \eta_w}{\partial t} \langle c \rangle^w . \tag{31}$$

We simplify this equation by noting that $Q = \rho_n \frac{\partial \eta_w}{\partial t}$, and since the water phase solute concentration is dilute, $\langle c \rangle^w \ll \rho_n$. It follows immediately that $\left| \frac{\partial \eta_w}{\partial t} \langle c \rangle^w \right| \ll |Q|$:

$$\eta_w \frac{\partial \langle c \rangle^w}{\partial t} + \eta_w \langle v \rangle^w \cdot \nabla \langle c \rangle^w - \eta_w \left(\mathscr{D} + k \right) \nabla^2 \langle c \rangle^w$$
$$= \left(2\mathscr{D} + k \right) \nabla \eta_w \cdot \nabla \langle c \rangle^w + \mathscr{D} \nabla^2 \eta_w \left[\langle c \rangle^w - c_{\text{sat}} \right] - Q . \tag{32}$$

The LHS of this equation resembles the desired advection-dispersion equation (multiplied through by η_w). A necessary condition for this to reduce to an ADRE with Sherwood-Gilland source (which depends on only the concentration difference $\left[\langle c \rangle^w - c_{\text{sat}} \right]$) to apply, the equation must simplify so that the mixed gradient term (proportional to $\nabla \eta_w \cdot \nabla \langle c \rangle^w$) on the RHS is eliminated. Doing so requires special assumptions. We consider one such case below.

3.5. Special Case: Advection Domination

In light of the failure to find an ADRE with Sherwood-Gilland source under general conditions, we follow Reference [17] in considering strongly advective conditions, which we here define as conditions in which $\left\| \langle v \rangle^w \right\| \gg \left(2\mathscr{D} + k \right) / L$. As η_w is less than unity, has a system minimum value above zero, and is a volume averaged quantity, $\left\| \nabla \eta_w \right\| < L^{-1}$. Thus, under sufficiently strongly advective conditions, $\left\| \eta_w \langle v \rangle^w \right\| \gg \left\| \left(2\mathscr{D} + k \right) \nabla \eta_w \right\|$, allowing neglect of the $\nabla \langle c \rangle^w$ term on the RHS of (32), as its coefficient is dominated by the one on the LHS. In such a case, one may simplify:

$$\frac{\partial \langle c \rangle^w}{\partial t} + \langle v \rangle^w \cdot \nabla \langle c \rangle^w - \left(\mathscr{D} + k \right) \nabla^2 \langle c \rangle^w = \frac{1}{\eta_w} \left[\mathscr{D} \nabla^2 \eta_n \left[c_{\text{sat}} - \langle c \rangle^w \right] - Q \right] . \tag{33}$$

In order to close the system, we need a second equation to relate $\langle c \rangle^w$ and η_n. Note also that we are aiming to upscale the dissolution dynamics, and we only have a microscopic equation for Q, which is expressed in terms of the local concentration c, rather than the needed, upscaled $\langle c \rangle^w$. To solve both problems, we require an ansatz relating Q to $\langle c \rangle^w$ and η_w, for which we consider the Sherwood-Gilland relation. We stress, lest there be any confusion that we "get out what we put in," we are not *deriving* the Sherwood-Gilland formulation from first principles in this section, only considering whether it is consistent with the microscopic formulation presented in Section 2. In the context of our analysis, this means that substitution into (33) ought to transform it to be similar to (7). For our analysis, we combine certain quantities defining the mass-transfer coefficient K in (2) into a single parameter, ω, and write

$$Q = -\omega \eta_n^\beta \mathscr{D} \left[c_{\text{sat}} - \langle c \rangle^w \right] . \tag{34}$$

From dimensional considerations, it is clear that ω has units of L^{-2} and we observe that it plays the role in the above equation of the squared reciprocal of a characteristic length ("boundary layer") for Fickian diffusion. Substituting (12), the definition for Q, into (34) yields:

$$\frac{\partial \eta_n}{\partial t} = -\eta_n^\beta \frac{\omega \mathscr{D}}{\rho_n} \left[c_{\text{sat}} - \langle c \rangle^w \right] . \tag{35}$$

This is a second, independent equation in terms of our dependent variables $\langle c \rangle^w$ and η_w. To arrive at (7), it then suffices to establish that $\eta_n^\beta \omega \gg \nabla^2 \eta_n$. For $\beta = 1$, a characteristic value in the middle of the range of proposed values of β (recall that these ranged from 0.6 to 1.24), (35) can be solved explicitly using the usual strategy for first-order non-homogeneous equations to generate

$$\eta_n(\mathbf{x}, t) = \eta_n(\mathbf{x}, 0) \exp \left\{ - \int\limits_0^t \frac{\omega \mathscr{D}}{\rho_n} \left[\langle c \rangle^w (\mathbf{x}, \tau) - c_{\text{sat}} \right] d\tau \right\}. \tag{36}$$

By differentiating (36), we observe that

$$\nabla^2 \eta_n = \eta_n \omega \left[\omega \left(\frac{\mathscr{D}}{\rho_n} \right)^2 \left(\int_0^t \nabla \langle c \rangle^w (\mathbf{x}, \tau) d\tau \cdot \int_0^t \nabla \langle c \rangle^w (\mathbf{x}, \tau) d\tau \right) + \left(\frac{\mathscr{D}}{\rho_n} \right) \int_0^t \nabla^2 \langle c \rangle^w (\mathbf{x}, \tau) d\tau \right]. \tag{37}$$

Recalling that L is the length scale of the averaging volume, we see that

$$\nabla^2 \eta_n \; < \; \eta_n \omega \left(\frac{1}{L^2} \right) \left[\omega \left(\mathscr{D}t \right)^2 \left(\frac{c_{\text{sat}}}{\rho_n} \right)^2 + \left(\mathscr{D}t \right) \left(\frac{c_{\text{sat}}}{\rho_n} \right) \right] \tag{38}$$

$$< \; \eta_n \omega \left[(\omega \mathscr{D}t)^2 \left(\frac{c_{\text{sat}}}{\rho_n} \right)^2 + (\omega \mathscr{D}t) \left(\frac{c_{\text{sat}}}{\rho_n} \right) \right], \tag{39}$$

where the last inequality follows because the length scale of pore scale diffusion is small relative to the length scale of volume averaging. Because NAPL are sparingly soluble, it follows that $\frac{c_{\text{sat}}}{\rho_n} \ll 1$. We may understand $\omega \mathscr{D}t$ as an expected number of boundary layer length scales traversed by a solute particle due to diffusion since source emplacement. As long as $(\omega \mathscr{D}t) \left(\frac{c_{\text{sat}}}{\rho_n} \right) \ll 1$, it follows that $\eta_n \omega \gg \nabla^2 \eta_n$ and thus that

$$\frac{\partial \langle c \rangle^w}{\partial t} + \langle v \rangle^w \cdot \nabla \langle c \rangle^w - (\mathscr{D} + k) \nabla^2 \langle c \rangle^w = \frac{\omega \eta_n \mathscr{D}}{\eta_w} \left(c_{\text{sat}} - \langle c \rangle^w \right). \tag{40}$$

Under these specific conditions, to the extent that η_w can be treated as approximately constant (for instance, if everywhere $\eta_w \gg \eta_n$), we do recover the ADRE with Sherwood-Gilland source term.

4. Conclusions

In this note, we consider the dissolution of residual NAPL in saturated porous media by means of a nonstandard volume averaging analysis. Our approach applies a variety of strategies (including direct application of boundary conditions, a physics-preserving domain simplification, elimination of coefficient-dominated terms, a geometric lemma, and some other physically-grounded approximations) to simplify the phase boundary jump discontinuity terms that arise under superficial volume averaging and perform analysis using only the volume averaged equation. This is in contrast with the traditional volume averaging approach which entails developing a separate equation for concentration fluctuations which contains volume averaged terms such as "sources," and formally closing the system by solving one or more boundary value problems to determine the functional form of the dependence of the fluctuations on the volume averaged quantities and then substituting these back into the volume averaged equation. The analysis here is radically simplified.

The particular problem we consider is whether Sherwood-Gilland empirical models are consistent with the micro-scale physics. Existing volume averaging analyses have led to equations that look much unlike the ADRE with a Sherwood-Gilland-type source, so this was a topic worth considering explicitly. The only attempt to recover a Sherwood-Gilland formula that we are aware of in the previous literature is due to Reference [17]: the form is recovered by applying four separate scale constraints that are defined in terms of the formal solutions to fluctuation boundary value problems formulated as part of the volume averaging process. Unfortunately, it is difficult to predict a priori whether these constraints are satisfied. Above, we derive the volume-averaged transport equation using the simplified volume averaging approach and find it is not an ADRE with a Sherwood-Gilland source term. We subsequently consider a physical restriction of advection-dominated behavior, but unlike Reference [17], it only requires a single restriction on the strength of advective transport relative to dispersive transport. We explicitly corroborate usage of the Sherwood-Gilland source model with the ADRE under such conditions when $\beta = 1$.

Funding: This research received no external funding; SKH holds the Helen Unger Career Development Chair in Desert Hydrogeology.

Conflicts of Interest: The author declares no conflict of interest.

Appendix A. Nomenclature

Here, we summarize the algebraic symbols and special operators introduced in this note. Phase subscripts or superscripts i and j are placeholders for any of the three phases in the system: n for NAPL, s for solid, and w for water. Bold is used to denote vector and matrix quantities (with the indicated dimensions applying to each component), with ordinary-weight Roman text denoting scalars. Dimensions are encoded in accordance with the SI standard: M for mass, L for length, and T for time, with a one indicating a dimensionless quantity.

Symbol	Dimensions	Description
$\langle \cdot \rangle$		Superficial volume average operator
$\langle \cdot \rangle^i$		Intrinsic volume average operator for i phase
Γ_{ij}		Locus of interface between phases i and j within V
$V(\boldsymbol{x})$		Spatial region with centroid at $\boldsymbol{x}$ over which volume averaging occurs
α	[1]	Fitted constant in empirical model of mass transfer coefficient K
β	[1]	Fitted exponent in empirical model of mass transfer coefficient K
γ	[1]	Fitted exponent in empirical model of mass transfer coefficient K
η_i	[1]	Fraction of V occupied by i phase
κ	$[\mathrm{T}^{-1}]$	Mass transfer coefficient in (6), as derived in [13]
ρ_n	$[\mathrm{ML}^{-3}]$	Density of NAPL
ω	$[\mathrm{L}^{-2}]$	Mass transfer coefficient
c	$[\mathrm{ML}^{-3}]$	Chemical concentration in the water phase
c_{sat}	$[\mathrm{ML}^{-3}]$	Thermodynamic saturation chemical concentration in the water phase
d_m	[L]	Characteristic length of porous media
$\boldsymbol{d}_w$	$[\mathrm{LT}^{-1}]$	Velocity-like closure variable in (6), as derived in [13]
$\boldsymbol{D}_{\mathrm{eff}}$	$[\mathrm{L}^2\mathrm{T}^{-1}]$	Effective Fickian dispersion coefficient
$\boldsymbol{D}_w^*$	$[\mathrm{L}^2\mathrm{T}^{-1}]$	Dispersion-like closure variable in (6), as derived in [13]
dA_{ij}	$[\mathrm{L}^2]$	Infinitesimal surface element on Γ_{ij}
dV	$[\mathrm{L}^3]$	Infinitesimal volume element
$\mathscr{D}$	$[\mathrm{L}^2\mathrm{T}^{-1}]$	Fickian diffusion constant
$I_i(\boldsymbol{x})$	[1]	Indicator function for location $\boldsymbol{x}$ belonging to the i phase
$\boldsymbol{k}$	$[\mathrm{L}^2\mathrm{T}^{-1}]$	Closure variable representing dispersive flux
K	$[\mathrm{T}^{-1}]$	Mass transfer coefficient from NAPL to water phase
L	[L]	Length scale of averaging volume; $\mathscr{V} = O(L^3)$
$\boldsymbol{n}_{ij}$	[1]	Normal vector on interface Γ_{ij} directed from i phase into j phase
$\boldsymbol{q}$	$[\mathrm{LT}^{-1}]$	Darcy velocity near NAPL source zone

Symbol	Dimensions	Description
Q	$[\mathrm{ML^{-3}T^{-1}}]$	Strength of NAPL-to-water mass source
Re	$[1]$	Reynolds number
S_n	$[1]$	NAPL saturation ($= \eta_n/(\eta_n + \eta_w)$)
t	$[\mathrm{T}]$	Time
u	$[\mathrm{LT^{-1}}]$	Velocity-like term in (6), as derived in [13]
v	$[\mathrm{LT^{-1}}]$	Pore water velocity
$\mathscr{V}$	$[\mathrm{L^3}]$	Physical volume of V
x	$[\mathrm{L}]$	Spatial position vector

Appendix B. Derivation of a Geometric Lemma

This is a simplification of a "geometric lemma" presented by Whitaker on p. 17 of [9]. The derivation follows from use of a conceptual simplification of the domain—eliminating all NAPL-solid interfaces by imagining an infinitesimally thin water layer between the NAPL and the solid—that we show does not alter the flow or transport in the system or the volume averaging mathematics. Physically, because of the no-slip boundary conditions, there is no flow in the new notional, infinitesimally-thin water layers. Because any solute in these layers is at c_{sat}, no net NAPL-water mass transfer takes place there. And because the "new" water has infinitesimal volume, it represents a negligible source of mass into the existing pore water phase at the former NAPL-water-solid three-way interfaces. Mathematically, this simplification eliminates the interface Γ_{ns}, therefore eliminating terms of the form $\frac{1}{\mathscr{V}} \int_{\Gamma_{ws}} f n_{ws} dA_{ws}$, for arbitrary f, when the appropriate volume averaging theorem is applied. Instead, new (collocated) interfaces Γ^*_{nw} and Γ^*_{ws} come into being. However, $\frac{1}{\mathscr{V}} \int_{\Gamma^*_{ns}} f n_{nw} dA_{nw} + \frac{1}{\mathscr{V}} \int_{\Gamma^*_{ws}} f n_{ws} dA_{ws} = 0$ for any f, because Γ^*_{nw} and Γ^*_{ws} are the same surface, and the two integrals have opposite-directed normal vectors. Thus, the mathematics of volume averaging are unchanged.

Having established that we are always justified in assuming no NAPL-solid interface, the geometric lemma follows from application of volume averaging theorem (4) to an indicator function, I_w, representing the presence of the water phase:

$$\langle \nabla I_w \rangle = \nabla \langle I_w \rangle + \frac{1}{\mathscr{V}} \int_{\Gamma_{wn}} n_{wn} dA_{wn} + \frac{1}{\mathscr{V}} \int_{\Gamma_{ws}} n_{ws} dA_{ws}, \tag{A1}$$

noticing that everywhere (except for jump discontinuities) I_w is a constant, so $\langle \nabla I_w \rangle = 0$ and $\langle I_w \rangle = \eta_w$. Thus,

$$\nabla \eta_w = -\frac{1}{\mathscr{V}} \int_{\Gamma_{wn}} n_{wn} dA_{wn} - \frac{1}{\mathscr{V}} \int_{\Gamma_{ws}} n_{ws} dA_{ws}. \tag{A2}$$

We may apply identical analysis to the solid phase:

$$\nabla \eta_s = -\frac{1}{\mathscr{V}} \int_{\Gamma_{sn}} n_{sn} dA_{sn} - \frac{1}{\mathscr{V}} \int_{\Gamma_{sw}} n_{sw} dA_{sw}, \tag{A3}$$

but notice that because porosity is constant, $\nabla \eta_s = 0$. Furthermore, because we have notionally eliminated the NAPL-solid interface by imagining an infinitesimal layer of saturated, immobile water phase between them,

$$\frac{1}{\mathscr{V}} \int_{\Gamma_{sw}} n_{sw} dA_{sw} = 0. \tag{A4}$$

Noting that $\frac{1}{\mathcal{V}} \int_{\Gamma_{ws}} \boldsymbol{n}_{ws} dA_{ws} = -\frac{1}{\mathcal{V}} \int_{\Gamma_{sw}} \boldsymbol{n}_{sw} dA_{sw} = 0$, it follows immediately from (A2) that

$$\frac{1}{\mathcal{V}} \int_{\Gamma_{wn}} \boldsymbol{n}_{wn} dA_{wn} = -\nabla \eta_w. \tag{A5}$$

References

1. Miller, C.; Poirer-McNeill, M.; Mayer, A.S. Dissolution of Trapped Nonaqueous Phase Liquids: Mass Transfer Characteristics. *Water Resour. Res.* **1990**, *26*, 2783–2796. [CrossRef]
2. Imhoff, P.T.; Jaffe, P.R.; Pinder, G.F. An experimental study of complete dissolution of a nonaqueous phase liquid in saturated porous media. *Water Resour. Res.* **1993**, *30*, 307–320. [CrossRef]
3. Powers, S.E.; Abriola, L.M.; Weber, W.J. An experimental investigation of nonaqueous phase liquid dissolution in saturated subsurface systems: Transient mass transfer rates. *Water Resour. Res.* **1994**, *30*, 321–332. [CrossRef]
4. Saba, T.; Illangasekare, T.H. Effect of groundwater flow dimensionality on mass transfer from entrapped nonaqueous phase liquid contaminants. *Water Resour. Res.* **2000**, *36*, 971–979. [CrossRef]
5. Nambi, I.M.; Powers, S.E. Mass transfer correlations for nonaqueous phase liquid dissolution from regions with high initial saturations. *Water Resour. Res.* **2003**, *39*. [CrossRef]
6. Hossain, S.Z.; Mumford, K.G.; Rutter, A. Laboratory study of mass transfer from diluted bitumen trapped in gravel. *Environ. Sci. Process. Impacts* **2017**, *19*, 1583–1593. [CrossRef]
7. Sherwood, T.K.; Gilliland, E.R. Diffusion of Vapors through Gas Films. *Ind. Eng. Chem.* **1934**, *26*, 1093–1096. [CrossRef]
8. Kokkinaki, A.; O'Carroll, D.M.; Werth, C.J.; Sleep, B.E. An evaluation of Sherwood-Gilland models for NAPL dissolution and their relationship to soil properties. *J. Contam. Hydrol.* **2013**, *155*, 87–98. [CrossRef]
9. Whitaker, S. *The Method of Volume Averaging*; Springer: Berlin/Heidelberg, Germany, 1999; p. 221.
10. Whitaker, S. Diffusion and Dispersion in Porous Media. *AIChE J.* **1967**, *13*, 420–427. [CrossRef]
11. Slattery, J.C. Flow of Viscoelastic Fluids Through Porous Media. *AIChE J.* **1967**, *13*, 1066–1071. [CrossRef]
12. Wood, B.D. The role of scaling laws in upscaling. *Adv. Water Resour.* **2009**, *32*, 723–736. [CrossRef]
13. Quintard, M.; Whitaker, S. Convection, dispersion, and interfacial transport of contaminants: Homogeneous porous media. *Adv. Water Resour.* **1994**, *17*, 221–239. [CrossRef]
14. Porta, G.M.; Riva, M.; Guadagnini, A. Upscaling solute transport in porous media in the presence of an irreversible bimolecular reaction. *Adv. Water Resour.* **2012**, *35*, 151–162. [CrossRef]
15. Porta, G.M.; Ceriotti, G.; Thovert, J.F. Comparative assessment of continuum-scale models of bimolecular reactive transport in porous media under pre-asymptotic conditions. *J. Contam. Hydrol.* **2016**, *185–186*, 1–13. [CrossRef]
16. Ma, Y.; Mohebbi, R.; Rashidi, M.M.; Manca, O.; Yang, Z. Numerical investigation of MHD effects on nanofluid heat transfer in a baffled U-shaped enclosure using lattice Boltzmann method. *J. Therm. Anal. Calorim.* **2019**, *135*, 3197–3213. [CrossRef]
17. Quintard, M.; Whitaker, S. Dissolution of an Immobile Phase during Flow in Porous Media. *Ind. Eng. Chem. Res.* **1999**, *38*, 833–844. [CrossRef]
18. Bahar, T.; Golfier, F.; Oltéan, C.; Benioug, M. An Upscaled Model for Bio-Enhanced NAPL Dissolution in Porous Media. *Transp. Porous Media* **2016**, *113*, 653–693. [CrossRef]
19. Bahar, T.; Golfier, F.; Oltéan, C.; Lefevre, E.; Lorgeoux, C. Comparison of theory and experiment for NAPL dissolution in porous media. *J. Contam. Hydrol.* **2018**, *211*, 49–64. [CrossRef]
20. Kechagia, P.E.; Tsimpanogiannis, I.N.; Yortsos, Y.C.; Lichtner, P.C. On the upscaling of reaction-transport processes in porous media with fast or finite kinetics. *Chem. Eng. Sci.* **2002**, *57*, 2565–2577. [CrossRef]
21. Golfier, F.; Wood, B.D.; Orgogozo, L.; Quintard, M.; Buès, M. Biofilms in porous media: Development of macroscopic transport equations via volume averaging with closure for local mass equilibrium conditions. *Adv. Water Resour.* **2009**, *32*, 463–485. [CrossRef]

22. Orgogozo, L.; Golfier, F.; Buès, M.; Quintard, M. Upscaling of transport processes in porous media with biofilms in non-equilibrium conditions. *Adv. Water Resour.* **2010**, *33*, 585–600. [CrossRef]
23. Rubin, Y. *Applied Stochastic Hydrogeology*; Oxford University Press: New York, NY, USA, 2003; p. 391.
24. Gelhar, L.W. *Stochastic Subsurface Hydrology*; Prentice-Hall: Englewood Cliffs, NJ, USA, 1993; p. 390.

MDPI

St. Alban-Anlage 66

4052 Basel

Switzerland

Tel. +41 61 683 77 34

Fax +41 61 302 89 18

www.mdpi.com

Water Editorial Office

E-mail: water@mdpi.com

www.mdpi.com/journal/water